PRINCIPLES OF

Quantum Mechanics

PRINCIPLES OF

Quantum Mechanics

Hans C. Ohanian

RENSSELAER POLYTECHNIC INSTITUTE

PRENTICE HALL / *Englewood Cliffs, New Jersey 07632*

Library of Congress Cataloging-in-Publication Data

Ohanian, Hans C.
    Principles of quantum mechanics / Hans C. Ohanian.
       p. cm.
    Includes bibliographical references.
    ISBN 0-13-712795-2:
    1. Quantum theory.   I. Title.
QC174.12.O33   1990
530.1'2—dc20   89–23149
               CIP

Editorial/production supervision: N. C. Romanelli
Cover designer: Ben Santora
Manufacturing buyer: Paula Massenaro

Cover photo: A segment from Fig. 8.5, page 229. Courtesy of A. F. Burr and A. Fisher, New Mexico State University

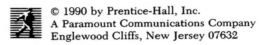

© 1990 by Prentice-Hall, Inc.
A Paramount Communications Company
Englewood Cliffs, New Jersey 07632

Printed in the United States of America

10 9 8 7 6

ISBN 0-13-712795-2

Prentice-Hall International (UK) Limited, *London*
Prentice-Hall of Australia Pty. Limited, *Sydney*
Prentice-Hall Canada Inc., *Toronto*
Prentice-Hall Hispanoamericana, S.A., *Mexico*
Prentice-Hall of India Private Limited, *New Delhi*
Prentice-Hall of Japan, Inc., *Tokyo*
Simon & Schuster Asia Pte. Ltd., *Singapore*
Editora Prentice-Hall do Brasil, Ltda., *Rio de Janeiro*

To C. B. A.

Contents

Appendices

Preface

THIS BOOK is an introduction to the theory of quantum mechanics for advanced undergraduate students. I assume that the student has had some contact with quantum physics and with elementary wave mechanics. Since the first three chapters review the experimental foundations of quantum physics and deal with simple applications of the Schrödinger wave equation, previous contact is not absolutely necessary, but it is certainly helpful.

In their first contact with the quantum–mechanical wavefunction and the Schrödinger wave equation, students often become infected with the misconception that these are the essence of quantum mechanics—if you understand the wavefunction and the Schrödinger wave equation and its solution by separation of variables, you understand everything. To suppress this misconception, I have deliberately placed the wavefunction and the Schrödinger wave equation in a subordinate role, as merely one special representation of the kinematics and the dynamics of a quantum–mechanical system. In this, I adhere to the tradition laid down by Dirac in his great book, *The Principles of Quantum Mechanics*, which is now more than fifty years old, but nevertheless retains its freshness, and still repays close study. Following Dirac, I emphasize the abstract formulation of quantum mechanics in terms of state vectors and operators, and I employ operator techniques for the solution of eigenvalue problems. In most undergraduate textbooks, operator techniques are reserved for the harmonic oscillator and for the treatment of angular momentum, which can be conveniently handled by raising and lowering operators. But, as was shown by Schrödinger in the 1940s, suitable raising and lowering operators can be constructed for all the familiar eigenvalue problems usually solved by separation of variables with the wave equation. I have adopted Schrödinger's factorization method, expanding his discussions to make them accessible to undergraduates. To some extent, my presentation of the factorization method imitates that by H. S. Green.*

My book developed out of successive editions of lecture notes I produced for my classes at Rensselaer Polytechnic Institute over several years. I am indebted to the students for their enthusiasm and patience. And I owe a long-standing debt to Professor Frank S. Crawford, whose lectures many years ago, when I was an undergraduate at Berkeley, first gave me an appreciation for the power of operator methods.

H. C. O.

---

*H. S. Green, *Matrix Methods in Quantum Mechanics* (Barnes & Noble, New York, 1965).

PRINCIPLES OF

Quantum Mechanics

1

The Origins
of Quantum Mechanics

Quantum mechanics lies at the root of all of the physics of today. It attained this fundamental place early this century, when it supplanted classical mechanics. Quantum mechanics has proved extremely successful in describing the behavior of the world of matter, and it is likely to retain its fundamental place in physics for a long time to come. However, quantum mechanics has some serious deficiencies in that it leads to unacceptable infinite results in some calculations of interactions of particles. Although physicists have devised divers tricks for bypassing or hiding these infinities, it is possible that some radical revision of quantum mechanics will ultimately become inevitable. One speculative suggestion for such a revision is the superstring theory now being investigated by physicists.

In this book we will deal only with nonrelativistic quantum mechanics, and we will concentrate on systems consisting of a single particle interacting with a potential. The relativistic quantum theory of systems of several interacting particles is quite complicated, because the interactions can create and destroy particles. Hence the number of particles in the system is not fixed, and it becomes necessary to consider concurrently a large ensemble of systems with different numbers of particles. The treatment of systems with a variable number of particles requires special mathematical techniques ("second" quantization).

This first chapter is a historical sketch of the origins of quantum mechanics. Full explanations of the concepts and formulas mentioned here are reserved for later chapters.

1.1 *Early Quantum Theory*

Quantum theory began in 1900 with Planck's postulate of the quantization of energy in blackbody radiation, or cavity radiation (the genealogical chart inside the front cover gives a summary of the historical development of quantum physics). Planck's theory of the blackbody spectrum was the culmination of many years of intensive efforts by experimental and theoretical physicists. It had been known for some time that the spectrum of thermal radiation contained in a cavity in thermal equilibrium must be a universal function of the temperature, completely independent of the material of the walls of the cavity. By 1900, detailed measurements by Rubens, Kurlbaum, Lummer, and Pringsheim had determined the shape of the spectrum, and Planck was able to make a clever guess at an empirical formula representing this spectrum. However, he was unable to provide a theoretical derivation of this formula within the context of classical physics. For low frequencies, the spectrum actually did agree with the prediction of classical physics; but for high frequencies, classical physics predicted a monotonic increase of the spectral energy density, whereas the measured energy density decreased toward zero (see Fig. 1.1).

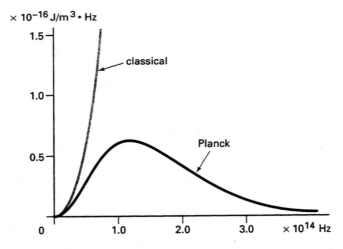

Fig. 1.1 Spectral distribution of the energy density as a function of frequency for the thermal radiation in a cavity at 2000 K. The spectral distribution plotted here is the energy per unit volume and per unit frequency interval. The classical prediction, shown in gray, is based on the equipartition theorem, according to which each mode of electromagnetic oscillation in the cavity should have an average energy $\frac{1}{2}kT$.

Planck finally found that he could derive his empirical formula from a postulate of quantization of energy. He adopted a simple model for the walls of the cavity: the walls consist of a large number of electrically charged harmonic oscillators of all possible frequencies (since the final result is independent of the material of the walls, he was free to adopt any convenient model). The oscillators exchange energy with the radiation in the cavity, and hence, at thermal equilibrium, the energy distribution of the radiation matches the energy distribution of the oscillators. Planck postulated that the energy of an oscillator of frequency v could only assume one or another of the values

$$E = nhv \qquad n = 0, 1, 2, 3, \ldots . \tag{1}$$

where h is a universal constant of proportionality, later called Planck's constant. From this quantization condition, Planck derived the thermal energy distribution of the oscillators and of the radiation. Qualitatively, we can understand how the quantization condition leads to a decrease of the spectral energy at high frequencies ($hv \gg kT$); the energy quantum hv is then so large that the typical thermal energy kT is insufficient to provide the minimum energy hv required to excite the oscillator from the ground state $E = 0$ to the first excited state $E = hv$; thus, the oscillator is likely to remain quiescent, and it then does not emit radiation of this frequency into the cavity.

Although Planck quantized the oscillators in the walls of the cavity, he treated the electromagnetic radiation in the cavity as completely smooth and continuous, according to classical electromagnetic theory, that is, according to Maxwell's equations. A few years later, Einstein took quantization a step further by proposing that the electromagnetic radiation exists in the form of packets of energy hv, which came to be called photons. This meant he could view the radiation as a gas of photons. With this picture of the radiation, Einstein could apply statistical mechanics to investigate the behavior of the photon gas at thermal equilibrium, and he was able to supply an alternative derivation of Planck's formula.

Einstein further exploited the concept of photons to explain the puzzling features of the photoelectric effect. Measurements by Hertz, Lenard, and others had shown that when light strikes the surface of a metal, it ejects electrons, or photoelectrons, with an energy that depends on the frequency of the light, but not on its intensity. This contradicts classical electromagnetic theory, according to which the energy available in the light is proportional to the intensity, and does not depend at all on the frequency. But if light consists of a stream of photons of energy $h\nu$, then the maximum energy that an electron in the metal can absorb in a collision with a photon is $h\nu$. Of this maximum energy, the electron must give up some fixed amount $e\phi$ to escape from the metal, where ϕ is a characteristic amount of energy per unit charge, called the work function of the metal. The final maximum kinetic energy of an ejected photoelectron is then

$$K_{\max} = h\nu - e\phi \tag{2}$$

This equation for the energy of the photoelectrons was later verified in detail by Millikan in a series of meticulous experiments.

The photoelectric effect confirmed that the energy in light is packaged in discrete amounts $h\nu$. The Compton effect, discovered by Compton in 1922, established that the momentum in light is also packaged in discrete amounts. In an experimental investigation of the scattering of a beam of X rays by a carbon target, Compton noticed that the X rays deflected by large angles always emerged with increased wavelengths. He was able to explain this wavelength shift by assuming that the X-ray photons have not only an energy $h\nu$, but also a momentum $h\nu/c$. When such a photon collides with an electron, it suffers a loss of energy (and an increase of wavelength) that depends on the angle of deflection.

Einstein also applied quantization to the calculation of the specific heat of solids. Each atom or molecule in a crystalline solid can be regarded as held in its place by springs, and hence each atom or molecule is a three-dimensional harmonic oscillator. According to the equipartition theorem of classical statistical mechanics, at thermal equilibrium each such oscillator should have an average energy of $3kT$. But this prediction is contradicted by the measured values of the specific heat, which are found to be much less than $3kT$ at low temperatures. Einstein explained the dependence of the specific heat on temperature by appealing to the quantization condition (1). Qualitatively, if the temperature is low ($kT \ll h\nu$), the typical thermal disturbances are insufficient to excite oscillations, and the atoms fail to acquire thermal energy.

In 1913, Bohr extended the quantization of energy to the hydrogen atom, with spectacular success. From the experiments on the scattering of alpha particles by atoms, Rutherford had deduced that the atom must consist of a very small, very massive nucleus around which orbit the electrons. But this nuclear model of the atom was in conflict with classical electrodynamics, since an orbiting, accelerated electron should radiate and quickly lose all its orbital energy. Bohr had no explanation for why orbiting electrons do not radiate; but he blithely postulated that they did not. He postulated that the electrons are locked in certain preferred stationary states, with circular orbits, and that they emit radiation only when they make transitions from one stationary state to another. The possible stationary states are characterized by quantized values of the orbital angular momentum:

$$L = \frac{nh}{2\pi} \equiv n\hbar \tag{3}$$

This quantization of angular momentum leads to the quantization of the orbital energy of the hydrogen atom:

$$E_n = -\frac{1}{(4\pi\varepsilon_0)^2} \frac{m_e e^4}{2\hbar^2 n^2} \tag{4}$$

In a transition from one stationary state to another, the atom emits a single photon of an energy equal to the difference between the orbital energies. For instance, in the transition from the first excited state (of energy E_2) to the ground state (of energy E_1), the energy of the emitted photon is $h\nu = E_2 - E_1$, and the frequency of the emitted spectral line is

$$\nu = \frac{E_2 - E_1}{h} \tag{5}$$

This result for the frequency is in accord with the combination principle, discovered empirically by Rydberg and Ritz, which states that the frequencies of the spectral lines of the atoms can be expressed as differences between terms, taken two at a time.

An experiment performed by Franck and Hertz provided direct evidence for energy quantization in atoms. In this experiment, the atoms in mercury vapor were subjected to collisions with low-energy electrons. Franck and Hertz found that if the electron energy was below 4.9 eV, all the collisions were elastic; the incident electrons did not have enough energy to excite the atoms from the ground state to the first excited state, and they therefore merely bounced off the atoms without any loss of energy. But if the electron energy was above 4.9 eV, some of the incident electrons would give up 4.9 eV to excite the atoms, and the excited atoms would subsequently reradiate this energy in the form of ultraviolet light.

In his simple theory of the hydrogen atom, Bohr had considered only circular orbits. Sommerfeld and, independently, Wilson generalized Bohr's quantization rule for angular momentum to elliptical orbits and to any kind of periodic motion. The general Sommerfeld–Wilson quantization rule states that

$$\int p \, dq = n\hbar \tag{6}$$

where q and p are the canonical coordinate and momentum for the motion, respectively.

1.2 *Wave Mechanics*

The early, or "old," quantum theory relied heavily on classical mechanics, but sought to supplement Newton's law with extra quantization conditions for the selection of the preferred stationary states. Roughly, we can say that the old quantum theory accepted Newtonian kinematics, but sought to modify Newtonian dynamics with supplementary conditions. In the 1920s, physicists finally recognized that this attempt to graft a quantum structure on the Newtonian roots was unworkable, and they recognized that both Newtonian kinematics and dynamics had to be discarded.

The first step toward the new quantum mechanics was de Broglie's conjecture that electrons and other "particles" have wave properties. De Broglie was led to this conjecture by a formal analogy between geometrical optics and mechanics. He noticed that the equations determining the rays of geometrical optics are analogous to the equations determining the trajectories of particles in classical mechanics. Since geometrical optics is the limiting case of wave optics, he conjectured that classical mechanics is the limiting case of some wave motion. De Broglie postulated that the frequency of the wave associated with a particle is related to the energy of the particle by the same equation as for the light wave associated with a photon:

$$\nu = \frac{E}{h} \tag{7}$$

He then exploited the relativistic connection between energy and momentum and frequency and wavelength to deduce that the wavelength of the wave must be related to the momentum of the particle:

$$\lambda = \frac{h}{p} \tag{8}$$

This is the de Broglie relation. Thus, for the wave associated with a particle moving in the x direction, de Broglie proposed the harmonic wavefunction[1]

$$\psi = \sin 2\pi \left(\nu t - \frac{x}{\lambda} \right) = \sin \frac{2\pi}{h} (Et - px) \qquad (9)$$

He suggested that the wave properties of particles could be confirmed experimentally by the observation of diffraction effects when electrons are incident on crystals. In fact, some data on scattering of electrons by crystals were already available, and Elsasser and Franck recognized diffraction peaks in these data. However, their interpretation was not widely accepted until Davisson and Germer did further detailed experiments with crystals. At about the same time, Thomson succeeded in demonstrating electron diffraction in scattering experiments with thin films of metals. Figure 1.2 shows such a diffraction pattern produced by electrons that have passed through a thin film of aluminum consisting of very many crystallites oriented at random. Additional evidence for the

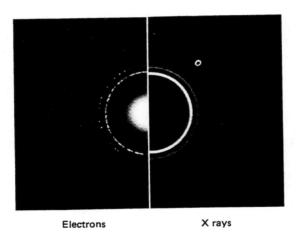

Electrons   X rays

Fig. 1.2 This composite photograph strikingly demonstrates the similarity of the diffraction patterns produced by X rays and by electrons upon passage through a thin film of aluminum. The X rays were registered on a photographic plate placed beyond the film of aluminum; the electrons were registered by a fluorescent screen, similar to a television screen. (*Project Physics*, 1975; Holt, Rinehart, Winston, and Education Development Center, Inc., Newton, MA)

---

[1] The phase velocity of this wave is $\lambda\nu = E/p$; it does not coincide with the velocity of the particle. The group velocity is dE/dp; this coincides with the velocity of the particle (phase and group velocities will be discussed in Chapter 2).

wave properties of electrons was provided by the experiments of Rupp, who found that electrons obliquely incident on an optical grating give rise to the familiar multiple-slit interference pattern.

The diffraction and interference patterns observed in these experiments not only established that electrons have wave properties, but also that the waves obey the superposition principle: when waves from two or more sources—such as two or more slits—arrive at some point simultaneously, the net wave amplitude is the sum of the individual wave amplitudes.

Meanwhile, Schrödinger formulated the wave equation obeyed by the de Broglie waves, and he demonstrated that the quantization conditions emerge from the solution of the eigenvalue problem for this wave equation. Schrödinger obtained his wave equation by taking as starting point the standard classical wave equation

$$\frac{\partial^2}{\partial x^2} \psi(x, t) - \frac{1}{u^2} \frac{\partial^2}{\partial t^2} \psi(x, t) = 0 \tag{10}$$

Here, u is the phase velocity, $u = \lambda\nu = h\nu/p$. For a particle of energy E moving in a potential $V(x)$, the momentum is $p = \sqrt{2m[E - V(x)]}$, and the phase velocity is $u = h\nu/\sqrt{2m[E - V(x)]}$, so

$$\frac{\partial^2}{\partial x^2} \psi(x, t) - \frac{2m[E - V(x)]}{h^2} \frac{1}{\nu^2} \frac{\partial^2}{\partial t^2} \psi(x, t) = 0 \tag{11}$$

Since the frequency of the wave is ν, the second time derivative of $\psi(x, t)$ is $-(2\pi\nu)^2 \psi(x, t)$, and Eq. (11) reduces to

$$\frac{\partial^2}{\partial x^2} \psi(x, t) + \frac{2m[E - V(x)]}{h^2} \psi(x, t) = 0 \tag{12}$$

This is called the (time-independent) Schrödinger wave equation.

Note that Eqs. (9) and (10) are actually wrong—we will see in the next chapter that the wavefunction ψ is a *complex* function, not a real function; and the time-dependent Schrödinger equation involves the *first* derivative with respect to time, not the second derivative. Nevertheless, Schrödinger's approximate argument led him to the correct equation (12). He applied this equation or, rather, its three-dimensional version to the hydrogen atom, and he found that both the quantization of angular momentum and the quantization of energy emerge from the equation. In essence, the Schrödinger equation yields a discrete set of possible energies (and a discrete set of frequencies $\nu = E/h$) for the same reason that the one-dimensional classical wave equation for a string fixed at both ends yields a discrete set of possible standing waves and possible frequencies. For the string, the standing wave must fit the length of the string; for the hydrogen atom, the standing Schrödinger wave must fit into the potential.

Schrödinger at first thought that the wavefunction of an electron represents an actual distribution of electric charge throughout space, with a charge density $\rho = -e|\psi|^2$. However, this interpretation proved untenable, since the wavefunction of a free electron gradually spreads out over a large volume, and such an electron could then be cut in two by some obstacle. An acceptable interpretation of the wavefunction was soon given by Born, who proposed that the absolute magnitude $|\psi|^2$ of the wavefunction represents the probability distribution for the position of the electron. Born was led to this probabilistic interpretation by an analogy with light waves: the intensity of a light wave at some position indicates the density of light quanta at this position, and it therefore seemed self-evident that the intensity $|\psi|^2$ of the quantum-mechanical wave should indicate the probability density for electrons or other particles.

The wave properties of particles and the description of particles by probability waves implied a profound revision of the foundations of physics. Instead of specifying the state of a particle by position as a function of time, we now have to describe the state by a wavefunction, and we can never predict exactly where the particle will move as a function of time—we can predict only the probabilities for motion from one position to another.

1.3 *Matrix Mechanics*

The wave mechanics based on Schrödinger's equation was to become the most popular formulation of the new quantum mechanics. But even before Schrödinger published his equation, an alternative formulation of quantum mechanics had been conceived by Heisenberg. This alternative was called matrix mechanics, because it relied on the mathematics of matrices instead of the solution of differential equations.

Heisenberg's point of departure was his conviction that in the atomic realm classical quantities—such as x and p—are meaningless, since they cannot be measured. He therefore decided to replace these classical quantities by some new quantities directly related to the quantum-mechanical stationary states. Since the observable frequencies emitted in transitions always involve the differences between the energies of two states, Heisenberg guessed that the relevant new quantities ought to depend on *two* states. Thus, instead of the classical quantity x, he introduced a set of new quantities $x_{11}, x_{12}, x_{13}, x_{21}, x_{22}, x_{23}, \ldots$, where the subscripts indicate the two states. He then investigated how these sets of quantities ought to be multiplied. Exploiting divers results from theoretical calculations of intensities of spectral lines in emission and absorption, he deduced the multiplication rule for his sets of new quantities. Heisenberg used a rather clumsy notation, and his multiplication rules looked very complicated. However, Born and Jordan recognized that Heisenberg's multiplication was nothing but matrix multiplication. They wrote the quantities $x_{11}, x_{12}, x_{13}, \ldots$ as a matrix:

$$x = \begin{pmatrix} x_{11} & x_{12} & x_{13} & \cdot \\ x_{21} & x_{22} & x_{23} & \cdot \\ x_{31} & x_{32} & x_{33} & \cdot \\ \vdots & \vdots & \vdots & \vdots \end{pmatrix} \tag{13}$$

and a similar matrix for p, and they established that Heisenberg's result for the product of x and p could be expressed in the concise form

$$px - xp = \frac{\hbar}{i} I \tag{14}$$

Thus, in matrix mechanics, products of quantities are not commutative.

In a collaborative effort, Born, Heisenberg, and Jordan demonstrated that Eq. (14) leads to the quantization conditions for energy and angular momentum, and they demonstrated that the energies of the stationary states can be obtained by a calculation of the eigenvalues of a matrix representing the energy. These results were also obtained independently by Dirac.

The wave mechanics of Schrödinger and the matrix mechanics of Heisenberg at first seemed quite irreconcilable. According to Schrödinger, the atom is described by a continuous wavefunction whose normal modes of oscillation correspond to the stationary states. We can visualize the three-dimensional pattern of wave crests and wave troughs of this wave much as we can visualize the wave crests and wave troughs of a sound wave. In contrast, according to Heisenberg, the atom must be described abstractly, by matrices that represent the position of the electron, its momentum, energy, and so on. This abstract description emphasizes the discrete aspects of the atom, and it defies visualization. However, in spite of the drastic differences between wave mechanics and matrix mechanics, these two theories lead to the same results for all the observable quantities of the atom. Schrödinger soon discovered that wave mechanics and matrix mechanics are mathematically equivalent. He proved that any formula of wave mechanics could be translated into a corresponding formula of matrix mechanics, and conversely. He constructed this proof by recognizing that multiplication of the wavefunction by the momentum is equivalent to multiplication by the differential operator $(\hbar/i)(d/dx)$, and that the commutation relation (14) of matrix mechanics is equivalent to the following identity for the differential operator $(\hbar/i)(d/dx)$:

$$\left(\frac{\hbar}{i}\frac{d}{dx}\right) x - x \left(\frac{\hbar}{i}\frac{d}{dx}\right) = \frac{\hbar}{i} \tag{15}$$

By using such operator relations, Schrödinger showed that all of Heisenberg's matrices could be calculated from the wavefunction, and that, conversely, the wavefunction could be calculated from the matrices.

1.4 *Particle vs. Wave; the Uncertainty Relations*

In classical mechanics, a particle is a pointlike mass. At each instant of time, such a classical particle has a well-defined position $\mathbf{r}(t)$. The motion of the particle proceeds along a well-defined trajectory, and the motion is completely described by specifying how the position $\mathbf{r}(t)$ varies with time. In contrast, a classical wave is an extensive disturbance in a medium. The medium may consist of a distribution of particles (for example, air serves as the medium for sound waves) or it may consist of classical fields (for example, the electric and magnetic fields surrounding a charge serve as the medium for electromagnetic waves; the wave may be regarded as a propagating kink in the field lines surrounding the charge). Such a disturbance in a medium is endowed with energy and with momentum, and the motion of the disturbance transports this energy and this momentum through the medium. However, the wave has no well-defined position and no well-defined trajectory. Only under exceptional circumstances, when the wavelength is very short compared with the relevant dimensions of any obstacles or apertures, is it possible to identify an approximate trajectory for the wave (in geometrical optics, such an approximate trajectory is called a ray). But even when the wave has an approximate trajectory, the wave can still be distinguished from a particle by its characteristic interference and diffraction effects—when several waves come together, they combine constructively or destructively according to their phase relationships, and when a wave passes through a small aperture, it deflects and spreads out into the shadow zone.

Nineteenth-century physicists observed interference and diffraction effects in light, and they found that electrons appeared to follow definite trajectories in cathode-ray tubes. They therefore concluded that light is a wave and that electrons are particles. The discovery of particle properties of light and the discovery of interference and diffraction effects of electrons demolished this tidy distinction between particles and waves. Light sometimes behaves like a classical particle, and sometimes like a classical wave. Electrons—as well as protons, neutrons, and other "particles"—sometimes behave like classical particles, and sometimes like classical waves. Whether electrons exhibit particle or wave properties depends on circumstances—it depends on what measurement we perform. For instance, in Thomson's electron-diffraction experiment, the electrons behave like waves while passing through the crystallites in the metallic film; but they behave like particles when they strike the fluorescent screen. The pattern of rings shown in Fig. 1.2 is made up of the impacts of many electrons on the fluorescent screen; each individual electron impact yields a pointlike flash of light, as expected for the impact of a particle. Electrons are said to exhibit *duality*: they have the properties of both classical particles and classical waves. However, as Heisenberg has emphasized, electrons are entities of one kind, and the apparent duality arises from the limitations of our language and our intuition. Our language was invented to describe the processes we observe in our everyday life, and all such processes involve macroscopic bodies with a large number of atoms. The processes that occur at the atomic level are outside the realm of our direct experience. We lack the words to describe these processes, and we lack the intuition to visualize them. If we attempt to describe the behavior of electrons in terms of the familiar concepts of classical particles and classical waves, we find that neither is adequate by itself, and only some particle–wave hybrid will give a crude approximation to the behavior of electrons. In view of this, Eddington proposed that electrons be called *wavicles*; but this quite apt neologism has not gained wide acceptance. In modern physics, electrons (or protons, or neutrons, etc.) are often called quantum-mechanical particles.

Heisenberg recognized that, since a quantum-mechanical particle has wave properties, it cannot have sharply defined position and velocity, or position and momentum. The wave packet describing the state of the quantum-mechanical particle has some width, that is, it spans a spread of positions. Furthermore, since a wave packet is a superposition of a number of harmonic waves, it contains a spread of wavelengths, or a spread of momenta. Heisenberg demonstrated that the spread of positions and the spread of momenta are subject to the inequality

$$\Delta x \, \Delta p_x \geq \tfrac{1}{2}\hbar \tag{16}$$

The spreads of position and momentum within the wave packet represent uncertainties in the possible outcomes of simultaneous measurements of position or momentum. Thus, Eq. (16) is called *Heisenberg's uncertainty relation.* Similar uncertainty relations are, of course, valid for the y and z components of the position and the momentum. Note that according to Eq. (16), if the position is defined very accurately (small Δx), then the momentum is poorly defined (large Δp_x), and conversely.

Heisenberg illustrated the uncertainty relation with several *Gedankenexperimente.* The simplest of these involves the measurement of the position of an electron by means of a slit. Suppose that a beam of electrons approaches a horizontal slit of vertical width a (see Fig. 1.3). If an electron passes through this slit, then

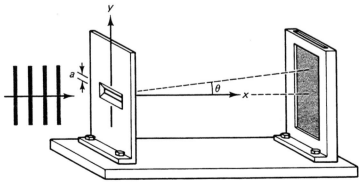

Fig. 1.3 Determination of the vertical position of an electron by means of a horizontal slit.

its vertical position is known to within an uncertainty $\Delta y = a$. However, because of its wave properties, the electron will suffer diffraction at the slit, that is, the electron wave will spread up and down over a range of angles. As a rough measure of the uncertainty of the direction of motion we can take the angular width θ of the central maximum of the diffraction pattern; this angular width is given by an equation familiar from wave optics:

$$a \sin \theta = \lambda \tag{17}$$

The vertical component of the momentum is then uncertain by

$$\Delta p_y = p \sin \theta = \frac{h}{\lambda} \sin \theta \simeq \frac{h}{a} \tag{18}$$

and the product of the simultaneous uncertainties Δy and Δp_y for the measurement of position and momentum in this experimental arrangement is

$$\Delta y \, \Delta p_y = a \, \frac{h}{a} = h \tag{19}$$

This is consistent with the uncertainty relation (16). Note that Eq. (18) shows quite explicitly how the uncertainty in momentum is affected by the choice of a. If we make a small, and thereby achieve a precise measurement of position, then Δp_y will become large.

Another *Gedankenexperiment* by Heisenberg seeks to measure the position of an electron by means of a hypothetical microscope operating with light of extremely short wavelength, or gamma rays. Figure 1.4 shows such a gamma-ray microscope aimed at an electron. The gamma rays are scattered by the electron into the objective lens of the microscope (in practice, no lenses for gamma rays are available, but we will ignore this petty technicality). According to wave optics, the resolving power of the microscope is

$$\Delta x = \frac{\lambda}{\sin \alpha} \tag{20}$$

where α is the angle subtended by the objective lens (see Fig. 1.4). However, the measurement of position is impossible unless at least one photon strikes the electron. When this happens, the electron acquires a recoil momentum and an uncertainty in the momentum, by the Compton effect. The magnitude of the mo-

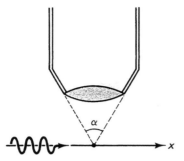

Fig. 1.4 Heisenberg's gamma-ray microscope.

mentum of the scattered photon is h/λ;[2] since the direction of motion of the scattered photon can fall anywhere within the angle α, the horizontal component of the momentum of this photon is uncertain by $\Delta p_x \simeq (h/\lambda) \sin \alpha$, and this must be the uncertainty in the momentum of the electron. The product of the simultaneous uncertainties Δx and Δp_x is then

$$\Delta x \, \Delta p_x = \left(\frac{\lambda}{\sin \alpha}\right)\left(\frac{h}{\lambda} \sin \alpha\right) \simeq h \tag{21}$$

which is, again, consistent with the general uncertainty relation (16).

The uncertainty relation tells us that the position and the momentum of a quantum-mechanical particle are complementary variables; if we perform a measurement that determines one of these with high accuracy, then the other will be poorly determined. Since the momentum is directly related to the wavelength, we can also say that the particle aspect (position) and the wave aspect (wavelength) are complementary. Thus, in any given measurement, either the particle aspect will be displayed or the wave aspect, but not both together. This principle of complementarity was formulated by Bohr.

The following *Gedankenexperiment* neatly illustrates the complementarity of the particle and wave aspects. Consider an opaque plate with two thin, parallel slits and a fluorescent screen placed beyond the plate (see Fig. 1.5). If a beam of monoenergetic electrons is incident on the plate, the electron waves passing through the two slits will form a typical interference pattern on the

---

[2] Here, and also in Eq. (20), λ is the wavelength of the photon *after* it is scattered by the electron.

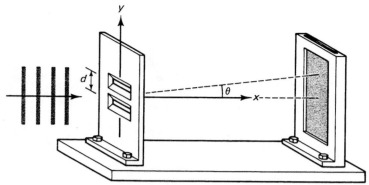

Fig. 1.5   An electron interference experiment involving two slits.

screen. The angular locations of the maxima of this interference pattern are given by a formula familiar from wave optics:

$$d \sin \theta = n\lambda \qquad n = 0, 1, 2, \dots \tag{22}$$

where d is the distance between the slits. Thus, the experimental arrangement of Fig. 1.5 brings out the wave aspect of the electrons. But suppose we now ask the question: Through which of the two slits does each individual electron pass? The experimental arrangement of Fig. 1.5 does not give us any information on this point, and if we want to investigate the passage of the electrons through the slits, we need some extra experimental equipment, for instance, we might aim a Heisenberg gamma-ray microscope at the slits and watch electrons passing through. Since we want to decide through which of the slits an electron passes, the gamma-ray microscope must measure the position of the electron with an uncertainty no larger than d, that is, $\Delta y \simeq d$. But this implies $\Delta p_y \simeq h/d$, and consequently the direction of motion of the electron acquires an uncertainty.

$$\Delta(\sin \theta) = \frac{\Delta p_y}{p} \simeq \frac{h}{d}\frac{\lambda}{h} \simeq \frac{\lambda}{d} \tag{23}$$

Comparing this with Eq. (22), we see that the uncertainty in the angle is as large as the angular separation between one maximum of the interference pattern and the next. Thus, the interference pattern will be completely washed out—the electrons will strike the fluorescent screen more or less at random. This *Gedankenexperiment* illustrates that we can either let the electrons display their wave properties (if we do not check through which slit the electrons pass) or we can let the electrons display particle properties (if we check through which slit the electrons pass). But if we investigate the particle properties, then the wave properties will remain hidden. Note that what equipment we use to check on the passage of electrons through the slits is irrelevant—any kind of equipment will yield the uncertainty relation $\Delta p_y \approx h/d$, and therefore lead to the conclusion (23).

PROBLEMS   1. What is the energy (in eV) of photons of violet light? Red light?

2. Estimate the number of photons emitted per second by a 100-W light bulb.

3. Planck's formula for the spectral distribution of the flux (or energy per unit area and unit time) emitted by a blackbody is

$$S_\nu = \frac{2\pi h}{c^2} \frac{\nu^3}{e^{h\nu/kT} - 1}$$

(a) From this formula deduce that the total flux is proportional to the fourth power of the temperature, that is,

$$\int_0^\infty S_\nu \, d\nu \propto T^4$$

(b) Deduce that the maximum in the spectral distribution occurs at a frequency proportional to the temperature, that is

$$\nu_{\max} \propto T$$

4. The work function of zinc is 4.24 eV. Can visible light eject photoelectrons from zinc?

5. Prove that when a photon collides with a free electron it cannot be (completely) absorbed. (Hint: Examine the conservation of energy and of momentum in a reference frame in which the electron is initially at rest.)

6. In the context of Bohr's theory, derive the quantization of energy [Eq. (4)] from the quantization of angular momentum [Eq. (3)].

7. According to Bohr's theory, what is the frequency of the light emitted by an electron in a transition from the first excited state to the ground state in hydrogen? Compare this with the frequency of the orbital motion of the electron in the first excited state.

8. (a) What is the acceleration of an electron in a circular Bohr orbit of radius $4\pi\varepsilon_0\hbar^2/m_e e^2$ in a hydrogen atom?

(b) According to classical electrodynamics, what is the rate at which the electron loses energy by radiation? (Hint: The classical formula for energy lost by electromagnetic radiation is $dE/dt = -e^2a^2/6\pi\varepsilon_0 c^3$.)

(c) If the electron continues to radiate at this rate, how long does it take to reduce its energy from -13.6 eV to -2×13.6 eV?

9. What is the wavelength of X rays of 20 keV, that is, photons of 20 keV? What is the wavelength of electrons of kinetic energy 20 keV?

10. What is the wavelength of electrons of kinetic energy 0.038 eV, the average classical thermal kinetic energy at room temperature? What is the wavelength of conduction electrons in copper, of energy 7.04 eV?

11. What is the wavelength of thermal neutrons from a nuclear reactor? Assume that the temperature is 500 K.

12. Calculate the de Broglie wavelength for a billiard ball of 0.05 kg moving at 0.1 m/s. Calculate the de Broglie wavelength for the Earth orbiting around the Sun.

13. A diffraction grating used in optics has 20,000 rulings per centimeter. Suppose that we use such a grating for electron diffraction. What would be the angular separation between the (principal) interference maxima if the electron energy is 100 eV?

14. The electron in a hydrogen atom may be regarded as having a position uncertainty of $\Delta x \simeq 10^{-10}$ m. Find the momentum uncertainty Δp_x required by the uncertainty principle. Compare this uncertainty with the magnitude of the momentum of an electron in the ground state of the atom, according to Bohr's theory.

15. At high temperatures and/or low densities, the behavior of a gas can be described by classical statistical mechanics, because, to a good approximation, the atoms move along classical trajectories. However, when the temperature of the gas is low and/or the density is high, the classical approximation fails, and quantum effects become important. The classical approximation breaks down completely when the de Broglie wavelength of a typical atom becomes comparable with the average interatomic distance (defined as $n^{-1/3}$, where n is the density of atoms).

(a) Show that this happens when

$$\frac{h}{\sqrt{mkT}} \simeq n^{-1/3}$$

(b) Consider the gas of conduction electrons in a typical metal. According to this criterion, is the classical approximation valid?

16. At any given instant of time, the position and the momentum of the Earth are somewhat uncertain because of quantum effects. This implies that the year, defined as the time it takes for the Earth to com-

plete one orbit, is somewhat uncertain. Give a *rough* estimate for the latter uncertainty.

17. An electron microscope operates with electrons of an energy of 100 keV and it has a resolution of 20 Å. What is the effective value of the aperture angle α, according to Eq. (20)?

18. The position of a virus of mass 5×10^{-21} kg can be observed with an electron microscope to within 20 Å. What is the minimum uncertainty in the momentum of the virus? What is the corresponding uncertainty in kinetic energy? Assume that the kinetic energy before the measurement is the thermal kinetic energy $\frac{3}{2}kT$ at room temperature.

19. A particle of mass m moving in one dimension in a harmonic-oscillator potential has an energy $\frac{1}{2}p^2/m + \frac{1}{2}kx^2$. The uncertainty principle implies that the energy of the lowest state cannot be zero. Estimate the lowest energy compatible with the uncertainty principle.

20. Suppose that in the Heisenberg gamma-ray microscope (see Fig. 1.4), a photon of wavelength λ approaches the electron horizontally and is scattered into the lens of the microscope. What are the consequent uncertainties in the horizontal and the vertical components of the momentum of the electron? Neglect the wavelength change of the photon during the collision; this is a good approximation if $\lambda \gg h/m_e c$.

21. In Section 1.4 we examined the two-slit electron interference experiment and we found that if we attempt to observe the passage of the electrons through the slits with a Heisenberg gamma-ray microscope, we destroy the interference pattern. To circumvent this limitation, we might attempt to use the apparatus shown in Fig. 1.6. The two parallel slits are cut in two separate plates, with independent suspensions. We can then detect through which plate an electron went by

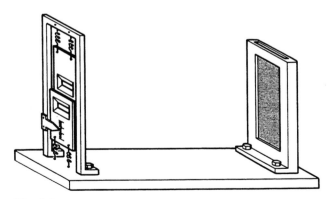

Fig. 1.6

measuring the vertical recoil the plate acquires when the electron passes through and is diffracted. This measurement does not disturb the electron, and therefore might permit us to deduce through which slit the electron went, while we can simultaneously see the interference pattern. Prove that the uncertainty in the momentum of the *plate* prevents us from detecting the recoil, unless the uncertainty in the position of the *plate* is so large that the interference pattern is washed out.

2

The Free Particle
in Wave Mechanics

In the first chapter we reviewed the experimental evidence for the wave behavior of electrons and other quantum-mechanical particles. Now we will establish the mathematical properties of the waves that describe these particles. Our arguments will be intended merely to make the results plausible, with emphasis on physical principles rather than on mathematical rigor. This chapter and the next will provide the motivation for the rigorous postulates and theorems of Chapter 4.

2.1 *The Wave Equation*

We begin by constructing a suitable wave equation for free particles. For the sake of simplicity, we will deal only with motion in one direction, say, along the x axis. This means that all our waves are plane waves, with wave fronts perpendicular to the x axis. We will sometimes call this kind of wave one-dimensional, because the wavefunction depends on only one spatial variable. However, a plane wave propagating in the x direction necessarily has a very large extent in the transverse directions, and when we call such a wave one-dimensional, we must take care to remember that we are not referring to the physical extent of the wave.

According to the proposals of de Broglie, the frequency of the wave that describes a quantum-mechanical particle of energy E and momentum p is

$$\nu = \frac{E}{h} \tag{1}$$

and the wavelength is

$$\lambda = \frac{h}{p} \tag{2}$$

For convenience, we introduce the angular frequency ω and the wave number k, and we rewrite these equations as

$$\omega \equiv 2\pi\nu = \frac{E}{\hbar} \tag{3}$$

and

$$k \equiv \frac{2\pi}{\lambda} = \frac{p}{\hbar} \tag{4}$$

By appealing to the theory of Special Relativity, it can be established that Eqs. (3) and (4) are mathematically equivalent—each implies the other. Energy and momentum are related by Lorentz transformations in precisely the same way as frequency and wave number; because of this, the proportionality of the frequency and the energy implies the proportionality of the wave number and the momentum, and conversely.[1]

Since we are dealing with a free particle with $E = p^2/2m$, we can also write Eq. (3) as

$$\omega = \frac{p^2}{2m\hbar} \tag{5}$$

For the description of a particle of momentum p traveling in the $+x$ direction, we will have to use a wave traveling in that direction. The conceivable harmonic waves are

$$e^{ikx-i\omega t}, \quad e^{-ikx+i\omega t}, \quad \sin(kx - \omega t), \quad \text{and} \quad \cos(kx - \omega t) \tag{6}$$

To settle which of these wavefunctions is the one we want, let us exploit the superposition principle. Suppose that we try to describe a particle traveling in the $+x$ direction by the wave $\sin(kx - \omega t)$. Then a particle traveling in the $-x$ direction will have to be described by $\sin(kx + \omega t)$. By the superposition princi-

---

[1] Relativistically, the energy E includes the rest-mass energy mc^2. This rest-mass energy can be regarded as an additive constant in the energy, and the frequency (3) contains a corresponding additive constant $mc^2/\hbar$. Such an additive constant has no observable consequences, and we will ignore it hereafter.

ple, the net wave resulting from the simultaneous presence of these two waves is simply their sum,

$$\sin(kx - \omega t) + \sin(kx + \omega t) \tag{7}$$

Physically, this superposition represents a state in which the particle has equal probabilities for motion in the $+x$ direction and the $-x$ direction; such a state can be arranged by reflecting the wave at a barrier at infinity. By combining the two sine functions in the sum (7), we obtain

$$2 \sin kx \cos \omega t \tag{8}$$

But obviously, the expression (8) is *not* a satisfactory wavefunction, because at $t = \pi/2\omega$ this wavefunction vanishes identically everywhere, and this means there is *zero* probability for finding the particle anywhere. Since the particle must be somewhere, we cannot accept this wavefunction, and we therefore conclude that $\sin(kx - \omega t)$ is not the kind of wavefunction we want. A similar argument rules out $\cos(kx - \omega t)$.

Next, let us consider $e^{ikx-i\omega t}$. The superposition of this with $e^{-ikx-i\omega t}$ gives

$$e^{ikx-i\omega t} + e^{-ikx-i\omega t} = 2e^{-i\omega t} \cos kx \tag{9}$$

which is never zero everywhere. Therefore, $e^{ikx-i\omega t}$ is an acceptable wavefunction.

What about $e^{-ikx+i\omega t}$? This is an equally acceptable wavefunction. However, here we must make a choice between $e^{ikx-i\omega t}$ and $e^{-ikx+i\omega t}$; if we adopt the former, then we must exclude the latter, and conversely. The reason is that the superposition of the former and the latter wavefunctions gives

$$2 \cos(kx - \omega t)$$

which, as we saw above, is not an acceptable wavefunction. Traditionally, the wavefunction $e^{ikx-i\omega t}$ has been adopted as the standard harmonic wavefunction for a free particle, and the wavefunction $e^{ikx+i\omega t}$ must then be excluded.

If $e^{ikx-i\omega t}$ is the correct wavefunction describing a free particle of given momentum, what is the wave equation satisfied by this wavefunction? In general, to discover the wave equation, we must take derivatives of the wavefunction with respect to x and t, and we must check which of these derivatives are proportional. The constant of proportionality must be independent of k and ω, so the wave equation is valid for waves of any wavelength and frequency. It is easy to check that the *second* derivative of $e^{ikx-i\omega t}$ with respect to x and the *first* derivative with respect to time are proportional,

$$-\frac{\hbar^2}{2m}\frac{\partial^2}{\partial x^2}e^{ikx-i\omega t} = i\hbar\frac{\partial}{\partial t}e^{ikx-i\omega t} \tag{10}$$

In this equation, the factors multiplying the derivatives are independent of k and of ω. The left side of Eq. (10) is

$$\frac{\hbar^2 k^2}{2m}e^{ikx-i\omega t}$$

and the right side is

$$\hbar\omega e^{ikx-i\omega t}$$

In consequence of Eq. (5), $\omega = p^2/2m\hbar = \hbar k^2/2m$, and the left and the right sides of Eq. (10) are, indeed, equal. Equation (10) is the simplest wave equation satisfied by the given wavefunction.[2]

Note that Eq. (10) is a *linear* equation, and therefore the solutions will be compatible with the superposition principle. This tells us that any function $\psi(x, t)$ that can be written as a superposition of a finite or infinite number of waves of the type $e^{ikx-i\omega t}$ (with different wavelengths) will satisfy the differential equation

$$-\frac{\hbar^2}{2m}\frac{\partial^2}{\partial x^2}\psi(x, t) = i\hbar\frac{\partial}{\partial t}\psi(x, t) \tag{11}$$

This is *Schrödinger's wave equation* for the free particle. We will find the most general solution of this equation in a later section.

Exercise 1. What would have been the Schrödinger equation if we had adopted the wavefunction $e^{-ikx+i\omega t}$?

The above argument illustrates the power of the superposition principle. This principle demands that the wavefunction be *complex*, and it determines the differential equation that must be satis-

---

[2] We can construct other, more complicated, wave equations; for instance, the fourth derivative with respect to x is proportional to the second derivative with respect to time, and so on. But all such other wave equations can be regarded as consequences of Eq. (10).

fied. The superposition principle is one of the most important principles of quantum mechanics.

It is instructive to compare the Schrödinger equation with the wave equation for classical waves, say, the equation for sound waves,

$$\frac{\partial^2 \phi}{\partial x^2} = \frac{1}{v^2} \frac{\partial^2 \phi}{\partial t^2} \tag{12}$$

The Schrödinger equation differs from this classical wave equation in that it has only a first time derivative, and it has an imaginary coefficient. Such a first time derivative in conjunction with an imaginary coefficient leads to oscillatory behavior in the solutions, that is, it leads to waves [an equation of the type (11) with a real coefficient would lead to diffusive behavior, as in the diffusion of heat]. But although the classical wave equation and the Schrödinger equation both describe waves, they differ in one essential aspect: the speed v of the classical waves obtained from Eq. (12) is a constant, independent of wavelength, whereas the speed of the quantum-mechanical waves depends on the wavelength; that is, it depends on the momentum (the wave speed will be discussed in detail in Section 2.4).

The interpretation of the wavefunction postulated by Born is that the intensity of the wave gives the probability for finding the particle at some point. For a real-valued wavefunction, the intensity is simply the square of the wavefunction. But for a complex wavefunction, the square ψ^2 is not real and not positive. The intensity of a complex wavefunction must therefore be taken to be $|\psi|^2$, rather than ψ^2. The connection between this intensity and the probability is then

$$\left[\begin{array}{c} \text{probability for finding the particle} \\ \text{between } x \text{ and } x + dx \text{ at time } t \end{array}\right] = |\psi(x,\,t)|^2\, dx \tag{13}$$

Because the net probability for finding the particle somewhere must add up to 1, we require that ψ be normalized, so

$$\int_{-\infty}^{\infty} |\psi|^2\, dx = 1 \tag{14}$$

There is one difficulty with this normalization requirement. The wavefunction $e^{i(kx-\omega t)}$ of a particle of momentum p *cannot* be normalized, since

$$\int_{-\infty}^{\infty} |e^{i(kx-\omega t)}|^2\, dx = \infty$$

The trouble with this wavefunction is that it assigns equal probability to all points in space. This is clearly not realistic since we always know that the particle is somewhere in a room, or at least somewhere on the Earth, or in some finite part of the universe. A particle therefore never has the wavefunction $e^{i(kx-\omega t)}$, and therefore never has an absolutely precise momentum. It is possible to construct a normalizable wavefunction (a wave packet) only if the momentum is left somewhat uncertain. The amount by which the momentum must be uncertain can be estimated from Heisenberg's relation. For example, if it is known that an electron is somewhere in a room ($\Delta x \simeq 10$ m), the momentum uncertainty is[3]

$$\Delta p \simeq \frac{\hbar}{10 \text{ m}} \simeq 10^{-14} \text{ MeV}/c \tag{15}$$

This is an extremely small momentum uncertainty (it corresponds to a velocity $v/c \simeq 10^{-14}$), and therefore it imposes no practical limitation on the precision with which the momentum of a beam of electrons can be determined.

Although $e^{i(kx-\omega t)}$ is, strictly speaking, *not* an acceptable wavefunction, we can do many calculations with it as though it were. For instance, when dealing with this wavefunction or with a superposition of several such wavefunctions, we can calculate relative probabilities, that is, we can calculate the ratio of the probabilities for finding the particle at two different positions. Such relative probabilities are well defined even when the wavefunction is not normalizable and its absolute probability distribution is not well defined. In later sections, we will learn some other tricks for dealing with the wavefunction $e^{i(kx-\omega t)}$ and with the superpositions of such wavefunctions.

2.2   *Fourier Analysis*

Fourier analysis permits us to express an arbitrary wavefunction as a superposition of harmonic waves. This means that the time evolution of the wavefunction can be deduced from the known time evolution of the harmonic waves in the superposition.

The decomposition of a wavefunction into harmonic waves rests on *Fourier's integral theorem:*

---

[3] Convenient units of energy and momentum are MeV and MeV/c, respectively. In these units $\hbar = 6.6 \times 10^{-22}$ MeV·s, or $\hbar = 3 \times 10^8 \times 6.6 \times 10^{-22}$ MeV · m/c.

*Suppose that $\psi(x)$ is a square-integrable function, that is,
$\int |\psi|^2 \, dx < \infty$. Then there exists a function $\phi(k)$ such that*

$$\psi(x) = \frac{1}{\sqrt{2\pi}} \int_{-\infty}^{\infty} \phi(k)e^{ikx} \, dk \tag{16}$$

where

$$\phi(k) = \frac{1}{\sqrt{2\pi}} \int_{-\infty}^{\infty} \psi(x)e^{-ikx} \, dx \tag{17}$$

We will not give a rigorous proof of this theorem, but only a more or less plausible derivation, which traces the Fourier integral theorem to the familiar Fourier series theorem. First consider a periodic function of x, with a repeat distance L, such that

$$\psi(x + L) = \psi(x) \tag{18}$$

Obviously, L is the "wavelength" of the function (Fig. 2.1). Such a function can be written as a Fourier series:

$$\psi(x) = \sum_{n=-\infty}^{n=+\infty} c_n e^{2i\pi nx/L} \tag{19}$$

This sum is a superposition of harmonic waves with all possible wavelengths L/n; $n = 1, 2, 3, \ldots$. To find c_m, multiply both sides of Eq. (19) by $(e^{-2i\pi mx/L})/L$ and integrate:

$$\frac{1}{L} \int_{-L/2}^{L/2} \psi(x)e^{-2i\pi mx/L} \, dx = \sum_n c_n \int_{-L/2}^{L/2} e^{2i\pi(n-m)x/L} \frac{dx}{L}$$

$$= \sum_n c_n \frac{1}{2\pi i(n-m)} [e^{2i\pi(n-m)/2} - e^{-2i\pi(n-m)/2}]$$

$$= \sum_n c_n \frac{\sin \pi(n-m)}{\pi(n-m)}$$

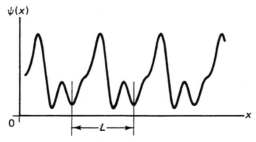

Fig. 2.1 A periodic function of x. The repeat distance, or the wavelength, is L.

The factor $(1/\pi(n - m)) \sin \pi(n - m)$ equals zero if $n \neq m$, and it equals 1 if $n = m$. The infinite sum therefore collapses to the single term c_m, and thus

$$c_m = \frac{1}{L} \int_{-L/2}^{L/2} \psi(x) e^{-2i\pi mx/L} \, dx \tag{20}$$

An arbitrary, nonperiodic function can be regarded as a periodic function of infinite wavelength. We therefore will let $L \rightarrow \infty$ in the above equations. With $k = 2\pi n/L$ and $\Delta k = 2\pi/L$, Eq. (19) becomes

$$\psi(x) = \sum_{k=-\infty}^{k=+\infty} c_n e^{ikx} = \sum_{k=-\infty}^{k=+\infty} \frac{L}{2\pi} c_n e^{ikx} \, \Delta k \tag{21}$$

If we now take the limit at $L \rightarrow \infty$, the sum becomes an integral:

$$\psi(x) = \int_{-\infty}^{\infty} \frac{1}{\sqrt{2\pi}} \frac{Lc_n}{\sqrt{2\pi}} e^{ikx} \, dk \tag{22}$$

where, by Eq. (20),

$$\frac{Lc_n}{\sqrt{2\pi}} = \frac{1}{\sqrt{2\pi}} \int_{-\infty}^{\infty} \psi(x) e^{-ikx} \, dx \tag{23}$$

Note that in this expression, $L \rightarrow \infty$ and $c_n \rightarrow 0$ in such a way that the product Lc_n remains finite. Equation (23) is exactly the statement of Fourier's theorem, provided that we identify $Lc_n/\sqrt{2\pi}$ with $\phi(k)$.

The function $\phi(k)$ is called the *Fourier transform* of $\psi(x)$. Sometimes we will write $\psi(k)$ instead of $\phi(k)$, that is, we will use the argument of the function to distinguish between function (argument x) and Fourier transform (argument k).

By combining Eqs. (16) and (17) we can write

$$\psi(x) = \frac{1}{2\pi} \int_{-\infty}^{\infty} e^{ikx} \left[\int_{-\infty}^{\infty} \psi(x') e^{-ikx'} \, dx' \right] dk \tag{24}$$

Let us change the order of integration in this double integral:

$$\psi(x) = \frac{1}{2\pi} \int_{-\infty}^{\infty} \psi(x') \left[\int_{-\infty}^{\infty} e^{ik(x-x')} \, dk \right] dx' \tag{25}$$

In this equation, the quantity

$$\frac{1}{2\pi} \int_{-\infty}^{\infty} e^{ik(x-x')} \, dk$$

depends on $x - x'$ only. We will use the notation $\delta(x' - x)$ for this quantity:

$$\delta(x' - x) = \frac{1}{2\pi} \int_{-\infty}^{\infty} e^{ik(x-x')} \, dk \tag{26}$$

Then Eq. (25) becomes

$$\psi(x) = \int_{-\infty}^{\infty} \delta(x' - x)\psi(x') \, dx' \tag{27}$$

Unfortunately, there is a difficulty with Eq. (26): the integral $\int_{-\infty}^{\infty} e^{ik(x-x')} \, dx$ does not converge; the integral does not exist. This difficulty can be traced to Eq. (25); our change of the order of integration in this equation is not legitimate. However, although $\delta(x' - x)$ does not exist as a function, we can almost always pretend that it does; that is, we can manipulate it formally almost as though it were an ordinary function.

The object $\delta(x)$ is called the *Dirac delta function*. Equation (26) gives a particular representation of the delta function. The general definition is as follows: $\delta(x)$ is a delta function if, for every square-integrable function $f(x)$,

$$f(0) = \int_{-\infty}^{\infty} \delta(x)f(x) \, dx \tag{28}$$

Intuitively, this says that $\delta(x)$ is zero for all $x \neq 0$, but that at $x = 0$ it must be infinite. If we set $f(x) = 1$, then

$$1 = \int_{-\infty}^{\infty} \delta(x) \, dx \tag{29}$$

Thus, the infinity at $x = 0$ must be large enough so the area under the peak integrates to 1.

The object $\delta(x)$ is not really a function (in rigorous theory it is called a "functional"), but if we want to think of it as a function, we have to pretend that it is zero except for a sharp spike at $x = 0$. There are many ways of generating such a spike. For example, we can define a function $g_b(x)$ by (see Fig. 2.2)

$$g_b(x) = \begin{cases} \dfrac{1}{b} & \text{for } |x| < \dfrac{b}{2} \\[2mm] 0 & \text{for } |x| > \dfrac{b}{2} \end{cases} \tag{30}$$

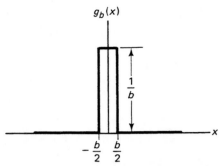

Fig. 2.2 A step function of height $1/b$ and width b.

Obviously, for this function

$$\int g_b(x)\,dx = 1 \tag{31}$$

If we take the limit as $b \to 0$, then $g_b(x)$ becomes a delta function:

$$\delta(x) = \lim_{b \to 0} g_b(x) \tag{32}$$

To check that this agrees with Eq. (28), note that for any arbitrary function $f(x)$,

$$\lim_{b \to 0} \int_{-\infty}^{\infty} g_b(x) f(x)\,dx = \lim_{b \to 0} \int_{-b/2}^{b/2} \frac{1}{b} f(x)\,dx = f(0) \tag{33}$$

In this calculation it is important to take the limit $b \to 0$ *after* performing the integral. This must be made part of the prescription for using this representation of the δ function.

We can verify that the representation (32) can be put into the form given by Eq. (26):

$$\delta(x) = \frac{1}{2\pi} \int_{-\infty}^{\infty} e^{-ikx}\,dk \tag{34}$$

or, as is obvious by a change of variable from k to $-k$,

$$\delta(x) = \frac{1}{2\pi} \int_{-\infty}^{+\infty} e^{ikx}\,dk \tag{35}$$

The Fourier transform of $g_b(x)$ is

$$g_b(k) = \frac{1}{\sqrt{2\pi}} \int_{-\infty}^{\infty} g_b(x) e^{-ikx}\,dx = \frac{1}{\sqrt{2\pi}} \int_{-b/2}^{b/2} \frac{1}{b} e^{-ikx}\,dx$$

$$= \frac{1}{\sqrt{2\pi}} \frac{2}{kb} \sin \frac{kb}{2}$$

In the limit $b \to 0$, this becomes $1/\sqrt{2\pi}$. Hence, in this limit,

$$g_b(x) = \frac{1}{\sqrt{2\pi}} \int_{-\infty}^{\infty} g_b(k) e^{ikx} \, dk = \frac{1}{2\pi} \int_{-\infty}^{\infty} e^{ikx} \, dk$$

which establishes Eq. (35).

There exist many other ways of constructing representations of the delta function (see, for instance, Problem 6).

The delta function has the following general properties:

(i) $\int \delta(x - a) f(x) \, dx = f(a)$ (36)

(ii) $x\delta(x - a) = a\delta(x - a)$ (37)

(iii) $\delta(-x) = \delta(x)$ (38)

(iv) $\delta(ax) = \dfrac{1}{|a|} \delta(x)$ for $a \neq 0$ (39)

(v) $\delta(x^2 - a^2) = \dfrac{1}{2|a|} [\delta(x - a) + \delta(x + a)]$ for $a \neq 0$ (40)

(vi) $\int f(x)\delta'(x) \, dx = -f'(0)$ (41)

Equations (ii)—(v) are true in the sense that if both sides are multiplied by an arbitrary function $f(x)$ and integrated over x, the results are identities. For example, (iii) gives

$$\int_{-\infty}^{\infty} \delta(-x) f(x) \, dx = \int_{-\infty}^{\infty} \delta(x) f(x) \, dx$$

The right side is $f(0)$ because of the definition (28). The left side can be put in the form $\int \delta(x') f(-x') \, dx'$ by a change of variable to $x' = -x$; this is, again, $f(0)$.

Exercise 2. "Prove" (ii), (iv), and (v) by using Eq. (28) and changes of variables.

In (vi) we have introduced the derivative of the delta function. Property (vi) is actually the *definition* of this derivative. But we can verify that (vi) is formally correct. Integrate the left side by parts; this gives

$$f(x)\delta(x) \Big|_{-\infty}^{\infty} - \int_{-\infty}^{\infty} \left[\frac{d}{dx} f(x) \right] \delta(x) \, dx \tag{42}$$

The first term is zero, because $\delta(x)$ is zero at infinity. The second term is $-(df/dx)_{x=0}$, in agreement with (vi).

Exercise 3. Does $\delta(x^2)$ have any meaning? Does $\delta^2(x)$?

In a formal sense, we can use the δ function to establish, or to verify, some theorems in Fourier analysis. For instance, we can use it to establish *Parseval's theorem:*

$$\int_{-\infty}^{\infty} \psi^*(x)\psi(x)\ dx = \int_{-\infty}^{\infty} \phi(k)\phi^*(k)\ dk \tag{43}$$

To "prove" this, we write the left side as[4]

$$\int_{-\infty}^{\infty}\int_{-\infty}^{\infty}\int_{-\infty}^{\infty} \frac{1}{2\pi}\ \phi(k)e^{+ikx}\phi^*(k')e^{-ik'x}\ dk\ dk'\ dx$$

$$= \int_{-\infty}^{\infty}\int_{-\infty}^{\infty} \phi(k)\phi^*(k')\delta(k - k')\ dk\ dk' \tag{44}$$

$$= \int_{-\infty}^{\infty} \phi(k)\phi^*(k)\ dk \tag{45}$$

2.3 *Solution of the Free-Particle Wave Equation*

We will now find the general solution of the Schrödinger wave equation for a free particle. By Fourier's theorem, at one given time t, any normalizable wavefunction can be written as

$$\psi(x,\ t) = \frac{1}{\sqrt{2\pi}}\int_{-\infty}^{\infty} \phi(k,\ t)e^{ikx}\ dk \tag{46}$$

We have indicated a time dependence in the Fourier transform $\phi(k,\ t)$, because although $\psi(x,\ t)$ always can be expressed in the form given in Eq. (46), the Fourier transforms at different times will be different. We will call $\phi(k,\ t)$ the amplitude in momentum space. Next, we have to find $\phi(k,\ t)$. For this, substitute Eq. (46) into the Schrödinger equation (11)

$$-\frac{\hbar^2}{2m}\left[\frac{1}{\sqrt{2\pi}}\int_{-\infty}^{\infty} -k^2\phi(k,\ t)e^{ikx}\ dk\right] = i\hbar\left[\frac{1}{\sqrt{2\pi}}\int_{-\infty}^{\infty} \frac{\partial}{\partial t}\phi(k,\ t)e^{ikx}\ dk\right]$$

---

[4] Note that $\delta(k - k')$ is not the Fourier transform of the delta function. It is the delta function of argument $k - k'$.

If we compare the Fourier transforms, or the coefficients of e^{ikx}, on both sides of this equation, we obtain

$$\frac{\hbar^2 k^2}{2m} \, \phi(k, \, t) = i\hbar \, \frac{\partial}{\partial t} \, \phi(k, \, t) \tag{47}$$

Equation (47) has the obvious solution

$$\phi(k, \, t) = \phi(k, \, 0)e^{-i\hbar k^2 t/2m}$$

or

$$\phi(k, \, t) = \phi(k, \, 0)e^{-iEt/\hbar} \tag{48}$$

where

$$E = \frac{\hbar^2 k^2}{2m} \tag{49}$$

is the energy that corresponds to the momentum $p = \hbar k$. Equation (46) then becomes

$$\psi(x, \, t) = \frac{1}{\sqrt{2\pi}} \int_{-\infty}^{\infty} \phi(k, \, 0)e^{i(kx - \hbar k^2 t/2m)} \, dk \tag{50}$$

This is the general solution of Schrödinger's wave equation for a free particle. The amplitude $\phi(k, \, 0)$ is arbitrary, but must satisfy the normalization condition

$$\int_{-\infty}^{\infty} |\phi(k, \, 0)|^2 \, dk = 1 \tag{51}$$

By Parseval's theorem, this will imply the correct normalization for $\psi(x, \, t)$.

According to Eq. (50), the general solution of the Schrödinger equation is a (continuous) superposition of harmonic waves. This means that the free-particle Schrödinger equation does not produce any solutions except those that can be directly constructed by superposition—the equation gives us nothing new.

If we chose the coefficients $\phi(k, \, 0)$ in Eq. (50) correctly, we can construct a wavefunction ψ that is nonzero only in some small region. Such a wavefunction is called a *wave packet*.

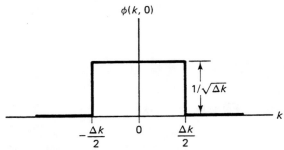

Fig. 2.3   Momentum amplitude for a wave packet.

As an example, take the momentum amplitude (see Fig. 2.3)

$$\phi(k, 0) = \begin{cases} \dfrac{1}{\sqrt{\Delta k}} & \text{if } |k| < \dfrac{\Delta k}{2} \\[3mm] 0 & \text{if } |k| > \dfrac{\Delta k}{2} \end{cases} \qquad (52)$$

Then

$$\psi(x, 0) = \frac{1}{\sqrt{2\pi}} \int_{-\Delta k/2}^{\Delta k/2} \frac{1}{\sqrt{\Delta k}}\, e^{ikx}\, dk$$

$$= \frac{2}{\sqrt{2\pi\,\Delta k}} \frac{1}{x} \sin \frac{x\,\Delta k}{2} \qquad (53)$$

Roughly, we can say that at $t = 0$, $\psi(x, 0)$ is appreciably different from zero only in an interval of the order of magnitude of $4\pi/\Delta k$ (see Fig. 2.4; however, at later times the packet becomes wider). The width $4\pi/\Delta k$ of the distribution in position and the width Δk of the distribution in wave number (see Fig. 2.3) can be regarded,

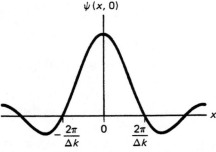

Fig. 2.4   Wave amplitude for the wave packet.

respectively, as the uncertainties in position and in wave number.[5] The product of these uncertainties is

$$\frac{4\pi}{\Delta k} \, \Delta k = 4\pi$$

that is

$$\Delta x \, \Delta k = 4\pi \tag{54}$$

This is an instance of Heisenberg's uncertainty relation. If we want to construct a very narrow packet (small Δx), then Δk will necessarily be large. Conversely, if we want the momentum to be precisely defined (small Δk), then Δx will be large.

The general solution (50) can be rewritten in terms of the momentum $p = \hbar k$,

$$\psi(x, t) = \frac{1}{\sqrt{2\pi\hbar}} \int_{-\infty}^{\infty} \phi(p, 0) e^{i(px/\hbar - p^2 t/2m\hbar)} \, dp \tag{55}$$

with

$$\phi(p, 0) = \frac{1}{\sqrt{2\pi\hbar}} \int_{-\infty}^{\infty} \psi(x, 0) e^{-ipx/\hbar} \, dx \tag{56}$$

Here the function ϕ has been redefined [a factor of $\sqrt{\hbar}$ has been absorbed in ϕ, so as to achieve equal coefficients in front of the integrals in Eqs. (55) and (56)]. The normalization of $\phi(p, 0)$ is

$$\int_{-\infty}^{\infty} |\phi(p, 0)|^2 \, dp = 1 \tag{57}$$

Exercise 4. Derive Eq. (55).

The quantity $|\phi(p, 0)|^2$ can be interpreted as the probability for the particle to have momentum p, that is,

[probability for momentum between p and $p + dp$] $= |\phi(p, 0)|^2 \, dp$

$$\tag{58}$$

This interpretation is plausible, since Eq. (55) shows that $\phi(p, 0)$ is the amplitude with which the plane wave of momentum p appears

---

[5] More precisely, the uncertainties Δx and Δk ought to be defined as root-mean-square deviations from the mean. We will give these precise definitions of the uncertainties in the next section.

in the (continuous) superposition of plane waves that make up the packet, and since Eq. (57) shows that $|\phi(p, 0)|^2$ is correctly normalized for a probability distribution.

2.4   *Wave Packets*

We will now investigate how the properties of particles and wave packets are related. First we ask: How is the velocity of the particle related to the velocity of wave? Consider a harmonic wave

$$e^{ikx-i\omega t} = e^{ikx-i\hbar k^2 t/2m} \tag{59}$$

The *phase velocity* of the wave is[6]

$$v_{ph} = \frac{\omega}{k} = \frac{\hbar k}{2m} = \frac{p}{2m} \tag{60}$$

In classical mechanics, $p/2m$ is one-half of the velocity of the particle. We conclude that the wave property that corresponds to the velocity of the particle cannot be the phase velocity. However, there is another velocity we can consider when dealing with a wave packet: the *group velocity* v_g. This is the velocity with which the packet moves. We will see that this group velocity equals the classical particle velocity.

To understand the meaning of the group velocity v_g, take the wave packet

$$\psi(x, t) = \frac{1}{\sqrt{2\pi\hbar}} \int_{-\infty}^{\infty} \phi(p, 0)e^{ipx/\hbar - ip^2 t/2m\hbar} \, dp \tag{61}$$

where we will assume that $\phi(p, 0)$ is some function with a peak at $p = p_0$ (see Fig. 2.5; actually ϕ could be complex, and Fig. 2.5 shows the *magnitude* of ϕ). Since $\phi(p, 0)$ is different from zero only for values of p near p_0, it is clear that this function $\phi(p, 0)$ will represent a particle with an average momentum p_0 and some spread of momentum about this average. Now examine the argument of the exponential function appearing in (61). We have identically

---

[6] This result depends on the choice of additive constant in the energy. Thus, if we use $E = p^2/2m + mc^2$ (inclusion of the rest-mass energy), then $\omega = p^2/2m\hbar + mc^2/\hbar$ and there is a corresponding change in v_{ph}.

$$\frac{ipx}{\hbar} - \frac{ip^2t}{2m\hbar} \equiv \frac{ix}{\hbar} \left[p_0 + (p - p_0) \right] - \frac{it}{2m\hbar} \left[p_0^2 + 2p_0(p - p_0) + (p - p_0)^2 \right] \qquad (62)$$

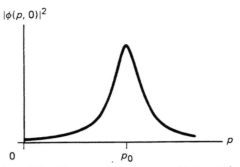

Fig. 2.5 The magnitude squared $|\phi(p, 0)|^2$ of the amplitude in momentum space for a wave packet.

Suppose that we can neglect $(it/2m\hbar)(p - p_0)^2$. This will be possible if

$$\frac{t}{2m\hbar} (\Delta p)^2 \ll 1 \qquad (63)$$

where Δp is the width of the momentum distribution in Fig. 2.5. Equation (62) then reduces to

$$\frac{ipx}{\hbar} - \frac{ip^2t}{2m\hbar} \simeq \frac{ixp_0}{\hbar} - \frac{ip_0^2t}{2m\hbar} + \frac{ix(p - p_0)}{\hbar} - \frac{ip_0(p - p_0)t}{m\hbar} \qquad (64)$$

and Eq. (61) can be written as

$$\psi(x, t) = \frac{1}{\sqrt{2\pi\hbar}} e^{ixp_0/\hbar - ip_0^2t/2m\hbar} \int_{-\infty}^{\infty} \phi(p, 0) e^{ix(p-p_0)/\hbar - ip_0(p-p_0)t/m\hbar} \, dp \qquad (65)$$

It is then easy to check that the absolute value of $\psi(x, t)$ remains unchanged if t increases by Δt and, simultaneously, x by $p_0 \, \Delta t/m$.

Exercise 5. Show that for $\psi(x, t)$ as given by Eq. (65),

$$|\psi(x, t)| = |\psi(x + p_0 \, \Delta t/m, t + \Delta t)| \qquad (66)$$

This says that whenever we increase t by Δt, the only change in $|\psi|$ is that the envelope of the wave packet moves a distance $p_0 \, \Delta t / m$ to the right; that is, the envelope moves with a velocity

$$v_{\mathrm{g}} = \frac{p_0}{m} \qquad (67)$$

This is what we called the group velocity of the wave packet, and we see that it coincides with the classical particle velocity. Note that the result (67) hinges on the energy–frequency relation [Eq. (1)] and on the de Broglie relation [Eq. (2)]. If we were to change one of these equations, then the group velocity of the wave packet would fail to coincide with the velocity of the particle. This interdependence of Eqs. (1) and (2) can be exploited for an alternative derivation of the de Broglie relation.

Exercise 6. Deduce Eq. (2) from Eq. (1) and the requirement that the group velocity of the wave packet should match the classical particle velocity.

The above wave-packet calculation is *not* exact, and it begins to fail at a time

$$t_{\mathrm{c}} \simeq \frac{2m\hbar}{(\Delta p)^2} \qquad (68)$$

after time zero [compare Eq. (63)]. The trouble is that the wave packet not only moves forward at velocity v_{g}, but also gradually changes its shape. The wave packet suffers *dispersion*. After a time interval of the order of magnitude of t_{c}, the shape is completely changed because the phase relations within the wave packet have become quite different. Note that if we use the uncertainty relation $\Delta p \gtrsim \hbar / \Delta x$, we obtain

$$t_{\mathrm{c}} \lesssim \frac{2m}{\hbar} \, (\Delta x)^2 \qquad (69)$$

From this we see that if a packet is initially very small, it will change its shape and spread out in a very short time. For example, an electron with an initial value of $\Delta x \simeq 10^{-10}$ m gives us $t_{\mathrm{c}} \lesssim 10^{-16}$ s. This shows that packets of atomic size change their shape so quickly that after only 10^{-16} s they are unrecognizable. At these small distances a description of an electron as something small that moves with some fairly definite velocity makes no sense. A small packet spreads out into something much bigger almost immediately. For an electron inside an atom the situation is somewhat different because we have to take into account the electric field acting on the electron. The electric field of the nucleus pulls the electron back, and thereby the spreading of the wavefunction is held in check. However, if we consider a wave packet that is initially *small* compared to the atomic dimensions, then the electric field of the nucleus cannot prevent it from spreading out. This means that the Bohr orbits do not exist.

For a macroscopic body, t_c can be quite large and the packet moves with almost no change in shape. For example, a pellet of mass 1 g and an initial uncertainty $\Delta x \simeq 10^{-7}$ m (corresponding to one wavelength of visible light) has $t_c \simeq 10^{10}$ years. Hence the wave packet remains very narrow until the end, or so, of time. The motion of such a wave packet can be adequately described by classical mechanics.

2.5 *The Gaussian Wave Packet*

The Gaussian wave packet has a momentum amplitude given by a Gaussian function,

$$\phi(p, 0) = \frac{1}{\sqrt{b\sqrt{2\pi}}} e^{-(p-p_0)^2/4b^2} \tag{70}$$

where p_0 and b are constants. This wave packet is sufficiently complicated to be interesting and sufficiently simple to be calculable.

Exercise 7. Show that $\phi(p, 0)$ is correctly normalized.

Figure 2.6 is a plot of $|\phi(p, 0)|^2$, the probability for momentum between p and $p + dp$. Obviously, p_0 is the mean momentum of the packet and b characterizes the width of the momentum distribution. The width of the momentum distribution determines the momentum uncertainty. For a precise definition of the uncertainty of any quantity with a probability distribution, we will hereafter adopt the *root-mean square* (rms) *deviation from the mean,* that is, the square root of the average of the square of the deviation from the mean. In the present case, the average of the square of the deviation from the mean is

$$\langle (p - p_0)^2 \rangle = \int_{-\infty}^{\infty} (p - p_0)^2 \, |\phi(p, 0)|^2 \, dp$$

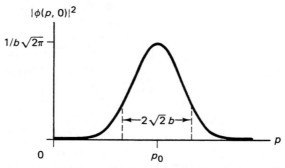

Fig. 2.6   The probability distribution $|\phi(p, 0)|^2$ for momentum in the Gaussian packet. The peak of the momentum distribution is at p_0, and the width of the distribution, measured between the points at which the exponential factor in $|\phi(p, 0)|^2$ equals e^{-1}, is $2\sqrt{2}b$.

Here, the angular brackets are a standard notation that indicates an average. The momentum uncertainty is then

$$\Delta p = \sqrt{\langle (p - p_0)^2 \rangle} = \left[\int_{-\infty}^{\infty} (p - p_0)^2 \, |\phi(p, 0)|^2 \, dp \right]^{1/2} \tag{71}$$

Calculation of the integral gives the result

$$\Delta p = \sqrt{\langle (p - p_0)^2 \rangle} = b \tag{72}$$

Exercise 8. Perform the integration in Eq. (71) and obtain Eq. (72).

Substituting the Gaussian momentum amplitude into Eq. (55), we find the wavefunction

$$\psi(x, t) = \frac{1}{\sqrt{2\pi\hbar}} \int_{-\infty}^{\infty} \frac{1}{\sqrt{\Delta p \sqrt{2\pi}}} \, e^{-(p-p_0)^2/4(\Delta p)^2} e^{i(px/\hbar - p^2 t/2m\hbar)} \, dp \tag{73}$$

The integral can be performed without too much trouble, with the result

$$|\psi(x, t)|^2 = \frac{1}{\sqrt{\pi}} \left[\frac{\hbar^2}{2(\Delta p)^2} + \frac{2(t \, \Delta p)^2}{m^2} \right]^{-1/2} \exp \left[-\frac{(x - p_0 t/m)^2}{\hbar^2/2(\Delta p)^2 + 2(t \, \Delta p)^2/m^2} \right] \tag{74}$$

Equation (74) shows that the Gaussian distribution of the momentum results in a Gaussian distribution of the position.

The mean value of x for the Gaussian wave packet (74) is obviously $p_0 t/m$; thus, the wave packet moves to the right with velocity p_0/m. The uncertainty in x, or the rms deviation of x from its mean value, is

$$\Delta x = \sqrt{\langle (x - p_0 t/m)^2 \rangle} = \left[\left(\frac{\hbar}{2\,\Delta p} \right)^2 + \left(\frac{t\,\Delta p}{m} \right)^2 \right]^{1/2} \qquad (75)$$

From this, we see that at $t = 0$ the product $\Delta x\,\Delta p$ is

$$\Delta x\,\Delta p = \frac{\hbar}{2} \qquad (76)$$

At later times, Δp is still the same, but Δx is larger. We can understand the result (75) for the time dependence of Δx in terms of the presence of several momenta in the packet. The high-momentum part of the packet has an *extra* velocity $\Delta p/m$; hence in a time t it moves ahead by an extra distance $t\,\Delta p/m$ as compared with the rest of the packet. This extra uncertainty in position is combined with *random* sign with the original uncertainty $\Delta x = \hbar/2\Delta p$; such a combination with random signs is equal to the square root of the sum of squares, and the result is (75). Note that for $t < 0$, the packet is also wider, that is, $t = 0$ represents the moment of maximum contraction of the packet. We can construct packets which contract almost to a point, but Eq. (76) then requires that Δp must be very large. Equation (76) actually represents the best we can ever do. In Chapter 4, we will give a general proof that

$$\Delta x\,\Delta p \geq \frac{\hbar}{2} \qquad (77)$$

where Δx, Δp are defined as rms deviations from mean. Thus, at $t = 0$, the Gaussian wave packet satisfies the uncertainty relation with an equal sign (it can be demonstrated that the Gaussian wave packet is the *only* wavefunction that attains the equal sign in the uncertainty relation—it is the wavefunction with the minimum uncertainty in x and p).

Finally, note that the time required for Δx to double from its initial value is given by

$$t = \frac{\sqrt{3}}{2} \frac{m\hbar}{(\Delta p)^2} \qquad (78)$$

This agrees with our earlier estimate [Eq. (68)] for the time it takes a packet to change appreciably.

2.6 *Expectation Values*

We will often be interested in computing the average value of some function of x or p for a given wavefunction. The average value is usually called the *expectation value* because it is the value expected when we perform a large number of repetitive measurements on copies of the given wavefunction.

The expectation value of x for a given wavefunction $\psi(x)$ is simply

$$\langle x \rangle = \int x \, |\psi|^2 \, dx = \int \psi^*(x) x \psi(x) \, dx \tag{79}$$

Here it is understood that the expectation value is calculated at some given time t, but we will not bother to indicate the time dependence of ψ. Likewise, the expectation value of any arbitrary function of x is

$$\langle f(x) \rangle = \int \psi^*(x) f(x) \psi(x) \, dx \tag{80}$$

Next, we want the expectation value of p for a given wavefunction $\psi(x)$. We can find this expectation value easily if we use $\phi(p)$, the Fourier transform of $\psi(x)$. The probability for p to have a value between p and $p + dp$ is $|\phi(p)|^2 \, dp$, and hence

$$\langle p \rangle = \int \phi^*(p) p \phi(p) \, dp \tag{81}$$

Likewise, for any arbitrary function $g(p)$, the expectation value is

$$\langle g(p) \rangle = \int \phi^*(p) g(p) \phi(p) \, dp \tag{82}$$

If we want to use Eq. (81), we first have to calculate $\phi(p)$. But there is a way of expressing $\langle p \rangle$ directly in terms of an integral involving $\psi(x)$. From Eq. (81) we see that

$$\langle p \rangle = \int \left[\frac{1}{\sqrt{2\pi\hbar}} \int \psi(x) e^{-ipx/\hbar} \, dx \right]^* p \left[\frac{1}{\sqrt{2\pi\hbar}} \int \psi(x') e^{-ipx'/\hbar} \, dx' \right] dp$$

$$= \frac{1}{2\pi\hbar} \iiint dx \, dx' \, dp \, \psi^*(x) \psi(x') p e^{ip(x-x')/\hbar}$$

$$= \frac{1}{2\pi\hbar} \iiint dx \, dx' \, dp \, \psi^*(x) \psi(x') \left(-\frac{\hbar}{i} \frac{d}{dx'} \right) e^{ip(x-x')/\hbar} \tag{83}$$

If we integrate by parts on x', we can ignore the contribution from the limit points $x' = \pm\infty$ (where $\psi = 0$), and we obtain

$$\langle p \rangle = \frac{1}{2\pi\hbar} \iiint dx\, dx'\, dp\ \psi^*(x) e^{ip(x-x')/\hbar} \left(\frac{\hbar}{i}\frac{d}{dx'}\right) \psi(x')$$

$$= \iint dx\, dx'\ \psi^*(x)\delta(x - x') \left(\frac{\hbar}{i}\frac{d}{dx'}\right) \psi(x')$$

$$= \int dx\ \psi^*(x) \left(\frac{\hbar}{i}\frac{d}{dx}\right) \psi(x) \tag{84}$$

This formula gives us a convenient way of calculating $\langle p \rangle$ for a given $\psi(x)$. Similar formulas apply for the calculation of the expectation value of any power of p.

Exercise 9. Show that

$$\langle p^2 \rangle = \int dx\ \psi^*(x) \left(\frac{\hbar}{i}\frac{d}{dx}\right)^2 \psi(x) \tag{85}$$

where it is understood that $(\hbar/i\ d/dx)^2 = (\hbar/i)^2\ d^2/dx^2$.

Exercise 10. Show that if $g(p)$ is a polynomial in p, then

$$\langle g(p) \rangle = \int dx\ \psi^*(x) g\left(\frac{\hbar}{i}\frac{d}{dx}\right) \psi(x) \tag{86}$$

It is clear that if we have any function of p that can be written as a power series in p, then we can use Eq. (86) to calculate the average value of the function. If we have a function, such as $\sqrt{p}$, that cannot be expressed as a power series in powers of p, then we cannot use Eq. (86), and we must go back to Eq. (82).

2.7 *Momentum and Position as Operators*

Suppose that we have a wavefunction $\psi(x)$. We want to give some meaning to the operation of "multiplication of $\psi(x)$ by momentum." This is not trivial because $\psi(x)$ will usually be a superposition of harmonic waves with several momenta, and it is not clear by which momentum we should multiply. Our definition of "multiplication by momentum" must somehow take into account the entire distribution of momenta.

In the special case of a wavefunction $e^{ipx/\hbar}$ of definite momentum p, the meaning of "multiplication by momentum" is obvious:

the result is $pe^{ipx/\hbar}$. But in general, the wavefunction $\psi(x)$ at some given time will be a superposition of waves of definite momentum,

$$\psi(x) = \frac{1}{\sqrt{2\pi\hbar}} \int \phi(p)e^{ipx/\hbar} \, dp$$

Then a reasonable definition should lead to

$$\frac{1}{\sqrt{2\pi\hbar}} \int \phi(p)pe^{ipx/\hbar} \, dp$$

that is, each part of ψ of momentum p is simply multiplied by p. This means that "multiplication by momentum" can be defined as multiplication by an *operator* p_{op} such that

$$p_{\mathrm{op}}\psi(x) = \frac{1}{\sqrt{2\pi\hbar}} \int \phi(p)pe^{ipx/\hbar} \, dp$$

But the right side of this equation is the same thing as

$$p_{\mathrm{op}}\psi(x) = \frac{\hbar}{i} \frac{d}{dx} \frac{1}{\sqrt{2\pi\hbar}} \int \phi(p)e^{ipx/\hbar} \, dp$$

$$= \frac{\hbar}{i} \frac{d}{dx} \psi(x) \tag{87}$$

Therefore, the operator that represents multiplication by momentum equals $(\hbar/i) \, d/dx$:

$$p_{\mathrm{op}} = \frac{\hbar}{i} \frac{d}{dx} \tag{88}$$

Thus, momentum can be represented by a *differential operator*. Note that an operator equation, such as Eq. (88), is taken to mean that if $\psi(x)$ is any arbitrary wavefunction, then p_{op} and $(\hbar/i) \, d/dx$ acting on $\psi(x)$ give the same result. Thus, Eq. (88) states no more, and no less, than Eq. (87).

In the special case of a wavefunction $e^{ipx/\hbar}$ of definite momentum p, we recover, of course, the expected result:

$$p_{\mathrm{op}}e^{ipx/\hbar} = \frac{\hbar}{i} \frac{d}{dx} e^{ipx/\hbar} = pe^{ipx/\hbar} \tag{89}$$

This has the form of an *eigenvalue equation:* the momentum operator p_{op} acting on the wavefunction gives p times the wavefunction. We say that $e^{ipx/\hbar}$ is an eigenfunction of the operator p_{op} with the eigenvalue p.

We can now write Eq. (84) as

$$\langle p \rangle = \int \psi^*(x) p_{op} \psi(x) \, dx \qquad (90)$$

More generally, we can write Eq. (86) as

$$\langle g(p) \rangle = \int dx \, \psi^*(x) g(p_{op}) \, \psi(x)$$

This tells us that we can obtain the expectation value of any function involving powers of x and/or p by replacing p by p_{op} and "sandwiching" the function between $\psi^*(x)$ and $\psi(x)$ in an integration.

Since in wave mechanics the momentum is a differential operator, any function of p_{op} will also be an operator. For example, the kinetic energy is the operator

$$K_{op} = \frac{p_{op}^2}{2m} \qquad (91)$$

or equivalently,

$$K_{op} = -\frac{\hbar^2}{2m} \frac{d^2}{dx^2} \qquad (92)$$

Schrödinger's equation can therefore be written as

$$\frac{p_{op}^2}{2m} \psi = i\hbar \frac{\partial}{\partial t} \psi \qquad (93)$$

Note that the wavefunction $e^{ipx/\hbar}$ of definite momentum is an eigenfunction of the kinetic energy operator with the eigenvalue $p^2/2m$, as expected:

$$\frac{p_{op}^2}{2m} e^{ipx/\hbar} = \frac{p^2}{2m} e^{ipx/\hbar}$$

When dealing with operators, we must be careful to keep in mind that multiplication will not always be commutative. For example, xp_{op} and $p_{op}x$ are not the same:

$$xp_{op}\psi(x) = x \frac{\hbar}{i} \frac{d}{dx} \psi(x) \qquad (94)$$

and

$$p_{op}x\psi(x) = \frac{\hbar}{i} \frac{d}{dx} (x\psi(x)) = x \frac{\hbar}{i} \frac{d\psi}{dx} + \frac{\hbar}{i} \psi(x) \qquad (95)$$

Hence we see that

$$(xp_{op} - p_{op}x)\psi(x) = i\hbar\psi(x) \tag{96}$$

Since this is true for any arbitrary function $\psi(x)$, we can write the operator equation,

$$xp_{op} - p_{op}x = i\hbar \tag{97}$$

This is the commutation relation discovered by Heisenberg. In general, let us define the *commutator* of two operators A and B by

$$[A, B] \equiv AB - BA \tag{98}$$

Then

$$[x, p_{op}] = i\hbar \tag{99}$$

By means of this fundamental commutator, we can now calculate the commutator of x and any function of p_{op}. For example:

$$[x, p_{op}^2] = xp_{op}^2 - p_{op}^2x = (xp_{op})p_{op} - p_{op}^2x = (p_{op}x + i\hbar)p_{op} - p_{op}^2x$$

$$= p_{op}(xp_{op}) + i\hbar p_{op} - p_{op}^2x = p_{op}(p_{op}x + i\hbar) + i\hbar p_{op} - p_{op}^2x$$

$$= 2i\hbar p_{op} \tag{100}$$

We can also write this as[7]

$$[x, p_{op}^2] = i\hbar \left(\frac{d}{dp_{op}} p_{op}^2 \right) \tag{101}$$

where the derivative on the right is to be performed according to the standard formula for the differentiation of a power, that is, $d/dp_{op}\ p_{op}^2 \equiv 2p_{op}$.

Exercise 11. Show that

$$[x, p_{op}^n] = i\hbar \frac{d}{dp_{op}} p_{op}^n \tag{102}$$

and hence show that if $f(p_{op})$ is any function of p_{op} that can be written as a power series, then

$$[x, f(p_{op})] = i\hbar \frac{d}{dp_{op}} f(p_{op}) \tag{103}$$

---

[7] The parentheses on the right side of Eq. (101) indicate that the derivative acts only on p, that is, $(d/dp\ p^2) = 2p$, in contrast to $d/dp\ p^2 = 2p + p^2\ d/dp$.

Exercise 12. Show that

$$[p_{\mathrm{op}}, g(x)] = -i\hbar \frac{d}{dx} g(x) \tag{104}$$

To see that $[x, p_{\mathrm{op}}] = i\hbar$ is fundamental in quantum mechanics, let us find out what happens if we describe the state of the particle by the momentum amplitude $\phi(p)$ instead of the wavefunction $\psi(x)$. Since we know that one of these functions determines the other, we can use either one to describe the state of the particle. We have seen that if we use $\psi(x)$, then multiplication by momentum is multiplication by an operator p_{op}. It turns out that if we use $\phi(p)$, it is exactly the other way around: multiplication by momentum is multiplication by an ordinary number, but multiplication by the coordinate is an operator multiplication. The argument is much the same as in the first case. Since

$$\phi(p) = \frac{1}{\sqrt{2\pi\hbar}} \int e^{-ipx/\hbar} \psi(x) \, dx \tag{105}$$

there is no unique value of x associated with $\phi(p)$; that is, $\phi(p)$ is an integral over all values of x. "Multiplication by the coordinate" is therefore meaningless unless we supply a definition of what it should mean. Our definition will be that multiplication by the coordinate is multiplication by an operator x_{op} such that each contribution of definite value of x in the integral (105) is simply multiplied by x:

$$\begin{aligned}
x_{\mathrm{op}}\phi(p) &= \frac{1}{\sqrt{2\pi\hbar}} \int e^{-ipx/\hbar} x\psi(x) \, dx \\
&= -\frac{\hbar}{i} \frac{d}{dp} \frac{1}{\sqrt{2\pi\hbar}} \int e^{-ipx/\hbar} \psi(x) \, dx \\
&= -\frac{\hbar}{i} \frac{d}{dp} \phi(p)
\end{aligned}$$

The position operator therefore has the representation

$$x_{\mathrm{op}} = -\frac{\hbar}{i} \frac{d}{dp} \tag{106}$$

As an example, let us consider a wavefunction that describes a particle at a precisely defined point x_0. For such a wavefunction,

Eq. (106) should give the same as multiplication by x_0. The wavefunction for a precisely localized particle is

$$\psi(x) = \delta(x - x_0) \tag{107}$$

This cannot be normalized ($\int |\psi|^2 \, dx = \infty$), just as the wavefunction describing a particle with precisely defined momentum cannot be normalized; but this difficulty need not concern us for now. The amplitude in momentum space corresponding to the wavefunction (107) is

$$\phi(p) = \frac{1}{\sqrt{2\pi\hbar}} \int e^{-ipx/\hbar} \delta(x - x_0) \, dx$$

$$= \frac{1}{\sqrt{2\pi\hbar}} e^{-ipx_0/\hbar} \tag{108}$$

Multiplying this by the position operator (106), we obtain

$$x_{\text{op}}\phi(p) = \frac{1}{\sqrt{2\pi\hbar}} \left(-\frac{\hbar}{i} \frac{d}{dp} \right) e^{-ipx_0/\hbar}$$

$$= x_0 \frac{1}{\sqrt{2\pi\hbar}} e^{-ipx_0/\hbar} = x_0 \phi(p) \tag{109}$$

We therefore see that this momentum-space wavefunction is an eigenfunction of x_{op} with eigenvalue x_0.

Expectation values of functions of x can be calculated using the wavefunction in momentum space and the operator x_{op}. Thus,

$$\langle x \rangle = \int \phi^*(p) x_{\text{op}} \phi(p) \, dp \tag{110}$$

which is analogous to Eq. (90). The proof is left as an exercise.

Exercise 13. Beginning with Eq. (80), show that

$$\langle x^n \rangle = \int \phi^*(p) \left(-\frac{\hbar}{i} \frac{d}{dp} \right)^n \phi(p) \, dp \tag{111}$$

Let us now check the commutator of x_{op} and p. We have

$$[x_{\text{op}}, p] = -\frac{\hbar}{i} \frac{d}{dp} p + p \frac{\hbar}{i} \frac{d}{dp}$$

$$= -\frac{\hbar}{i} - \frac{\hbar}{i} p \frac{d}{dp} + p \frac{\hbar}{i} \frac{d}{dp} = i\hbar \tag{112}$$

We see that, irrespective of which of the two descriptions we adopt, the *commutator has the same value*. The description of a particle by $\psi(x)$ is usually called the *position representation* and the description by $\phi(p)$, the *momentum representation*.

In the position representation, multiplication by the momentum is an operator multiplication by p_{op}, whereas multiplication by the coordinate is ordinary multiplication by the number x. For formal reasons it is convenient to regard multiplication by the number x as a (trivial) case of operator multiplication. Equation (37),

$$x\delta(x - x_0) = x_0\delta(x - x_0) \tag{113}$$

can then be given the following interpretation: the (trivial) operator x acting on the wavefunction $\delta(x - x_0)$ gives x_0 times the wavefunction, that is, $\delta(x - x_0)$ is an eigenfunction of x with eigenvalue x_0. Note that this interpretation coincides exactly with the conclusion we reached for the corresponding wavefunction in momentum space [see Eq. (109)]. We can therefore make a general statement which holds true both in the position and in the momentum representations: The wavefunction of a particle localized at a point x_0 is an eigenfunction of the position operator with eigenvalue x_0.

In a similar fashion, in the momentum representation it is convenient to regard multiplication by the number p as operator multiplication. In general, we can then say that the wavefunction of a particle of definite momentum p_0 is an eigenfunction of the momentum operator with the eigenvalue p_0. This statement is true in the position representation because of Eq. (89), and it is true in the momentum representation because momentum amplitude for a particle of definite momentum is $\delta(p - p_0)$.

Exercise 14. Verify that the momentum amplitude corresponding to a wavefunction $e^{ip_0x/\hbar}/\sqrt{2\pi\hbar}$ of definite momentum p_0 is $\delta(p - p_0)$.

Table 2.1 summarizes our results in these two representations.

The position and the momentum representations are not the only representations available in quantum mechanics. We will see in the next chapter that there exist other representations, that is, other ways of describing the state of a particle besides the description in terms of $\psi(x)$ or $\phi(p)$.

TABLE 2.1 Summary of Position and Momentum Representations

	Position Representation	Momentum Representation
Particle described by:	$\psi(x)$	$\phi(p)\left[= \dfrac{1}{\sqrt{2\pi\hbar}}\int e^{-ipx/\hbar}\psi(x)\,dx\right]$
Position operator	x (number)	$x_{op} = -\dfrac{\hbar}{i}\dfrac{d}{dp}$
Momentum operator	$p_{op} = \dfrac{\hbar}{i}\dfrac{d}{dx}$	p (number)
Wavefunction for particle localized at x_0	$\delta(x - x_0)$	$\dfrac{e^{-ipx_0/\hbar}}{\sqrt{2\pi\hbar}}$ ⎫
Wavefunction for particle of precise momentum p_0	$\dfrac{e^{ip_0x/\hbar}}{\sqrt{2\pi\hbar}}$	$\delta(p - p_0)$ ⎬ Not normalizable ⎭
Commutator $[x_{op}, p_{op}] =$	$i\hbar$	$i\hbar$

2.8 *Measurement in Quantum Mechanics*

In quantum mechanics, a measurement performed on a wavefunction will usually have a drastic effect of this wavefunction. We can recognize this in the following simple example. Consider a wave packet of some finite width representing a free particle that is initially known to be within some finite region. As time elapses, the packet spreads, and our knowledge about the position of the particle becomes more uncertain. Suppose that we perform a measurement of the particle position after some time has elapsed. The apparatus used for this measurement might be a (small) Geiger counter, or perhaps a fluorescent screen, on which the impact of the particle registers as a scintillation. Just before the measurement, the wavefunction is $\psi(x, t)$, and this wavefunction has an amplitude appreciably different from zero in a fairly wide region. What is the wavefunction just *after* the measurement? Obviously, it cannot be the same wavefunction as before, since immediately after the measurement we know exactly where the particle is located. If the position measurement gave a result x_0 for the coordinate, then immediately after the measurement the new wavefunction must have a very sharp peak at $x = x_0$, and it must be zero everywhere else. Immediately after the measurement, we therefore have a new, very narrow wave packet; that is, the new wavefunction is a delta function. This new wave packet, if not disturbed, will in turn spread out; and if we want to know the position at some later time, we must make a new position measurement.

We therefore arrive at the following picture of the evolution of the wave packet: The particle is described by a wave packet that spreads out as time passes, and whenever we make a position measurement, the wavefunction suffers a "collapse," or a "reduction," into a very sharp peak. This in turn spreads out, collapses upon another measurement, and so on. Each position measurement, and more generally, any measurement of any physical quantity, changes the wavefunction. Measurement destroys the old wavefunction, and creates a new wavefunction.

The changes in the wavefunction are then of two kinds:

(i) Continuous, deterministic changes in the wavefunction of a particle (or other quantum-mechanical system) with time, according to the Schrödinger equation.

(ii) Discontinuous changes brought about by the process of measurement. This kind of change is not deterministic; only probabilities can be given for the outcome.

The process of measurement can be regarded as an interaction of the observed particle with the apparatus. The crucial feature of this interaction is that the result is a macroscopic change of the apparatus, a change large enough for the observer to notice. The "apparatus" used for the measurement will usually consist of a piece of machinery, but it might also consist simply of the sense organs of the observer; for instance, the human eye is a quite sensitive detector of photons. Thus, we can regard the observer as part of the apparatus, and we can regard the measurement as an interaction of the particle with the observer, either directly, or via some auxiliary equipment.

If we were to treat the apparatus and its interaction with the observed particle quantum mechanically, then the change in the wavefunction during measurement would be determined by the Schrödinger equation describing the system consisting of the observed particle plus apparatus. There would then never occur changes of second kind. However, the orthodox interpretation of quantum mechanics, or the *Copenhagen interpretation*, describes the apparatus and the observer in purely classical terms, without any significant quantum-mechanical uncertainties. Thus, the Copenhagen interpretation insists on a sharp dividing line between the observed system and the apparatus. The observed system is described by a wavefunction, and it can be in a superposition of several states, with consequent quantum-mechanical uncertainties; whereas the apparatus is described by classical parameters, and it is always in a well-defined classical state. This dichotomy demands that the observed system suffer changes of the second kind, since during the measurement process the apparatus adopts a well-defined state, and thereby indicates that the observed system has also adopted a corresponding well-defined state.

The changes produced by measurement in quantum mechanics are unlike anything that occurs in classical physics. In the classical case, the measurement of, say, the diameter of a wire by the use of calipers can disturb the measured system—if the calipers are squeezed together too hard, the diameter of the wire might be changed. But classically, we can always make any such disturbances as small as we please by performing a very careful, gentle measurement. In the quantum case, the measurement disturbs the system by an amount over which we have no control. Before the measurement, there were some probabilities for the different possible outcomes. Immediately after measurement, the probabilities are zero for all the possibilities except the one that was actually realized in the measurement. As consequence of the measurement process, we necessarily get a new wavefunction, and just how different this new wavefunction is from the old one *cannot* be predicted. Only probabilities for different possible outcome can be predicted.

Another remarkable property of measurement in quantum mechanics is that a single measurement tells us next to nothing about the wavefunction *before* the measurement. For example, if we measure the position of a particle and find it at a certain point, the only thing we will have learned about the wavefunction *before* the measurement is that $|\psi|^2$ was not zero at the point where we found the particle. We learn nothing about the magnitude of $|\psi|^2$ at that point or elsewhere. Only if we perform many measurements on identically prepared systems do we obtain information about the magnitude $|\psi(x)|^2$ of the wavefunction before the measurement.

However, a single measurement does tell us much about the wavefunction immediately *after* the measurement. If a particle is found at $x = x_0$, then the wavefunction immediately afterward must be proportional to $\delta(x - x_0)$. Another position measurement *immediately* following the first will again find the particle at x_0.

The effect of measurement on a quantum-mechanical system is therefore this: Measurement throws the system into a new state, a state such that *immediate* repetition of the same measurement produces no additional disturbance of the wavefunction.

What state the system is thrown into depends on the kind of measurement we perform. Thus, a measurement of position throws the particle into a state of definite position, a measurement of momentum into a state of definite momentum, a measurement of energy into a state of definite energy. In general, a measurement of any observable quantity throws the system into an eigenstate of that quantity. In a later chapter we will see that any arbitrary state can be expressed as a superposition of eigenstates of any observable quantity, such as position, momentum, energy, angular momentum, and so on. Just before the measurement the particle is in a superposition of some number of eigenstates. If we now perform a measurement which throws the particle into one of the eigenstates in the superposition, then this *one* state is selected and the others are discarded. For example, suppose that the wavefunction of a particle is $\psi(x)$, at some definite time. We can regard this as a superposition of δ functions:

$$\psi(x) = \int \delta(x - x')\psi(x')\, dx' \tag{114}$$

where $\psi(x')$ is the amplitude with which $\delta(x - x')$ occurs in the superposition. Since

$$x\delta(x - x') = x'\delta(x - x') \tag{115}$$

the delta function $\delta(x - x')$ is an eigenfunction of the operator x with eigenvalue x'. Equation (114) is therefore a superposition of eigenfunctions of position. A measurement of position with the result x' amounts to selecting one of these delta functions with this particular value of x', so immediately after measurement the wavefunction is $\delta(x - x')$. This is an instance of the "collapse" of the wave packet mentioned above.

PROBLEMS

1. Derive Eq. (3) from Eq. (4). (Hint: Under Lorentz transformations, $E^2 - c^2p^2$ is invariant, and so is $\omega^2 - c^2k^2$; by evaluating these invariants in the rest frame of the particle, establish that they are proportional.)

2. Consider the wavefunction given by Eq. (9). This is a standing wave. Show that for this standing wave, the probability for finding the particle at some point is independent of time.

3. Find the Fourier transform of the function

$$\psi(x) = \begin{cases} \cos bx & \text{for } |x| < \dfrac{\pi}{2b} \\[2ex] 0 & \text{for } |x| \geq \dfrac{\pi}{2b} \end{cases}$$

4. Find the Fourier transform of the function

$$\psi(x) = \begin{cases} 0 & \text{for } |x \geq a \\[1ex] A & \text{for } 0 < x < a \\[1ex] -A & \text{for } 0 > x > -a \end{cases}$$

5. (a) Show that if the Fourier transform of a function $\psi(x)$ is $\phi(k)$, then the Fourier transform of $\psi(x - a)$ is $e^{-ika}\phi(k)$.

   (b) Use this result to find the Fourier transform of the function plotted in Fig. 2.7 from the Fourier transform of the function $g_b(x)$ defined by Eq. (30).

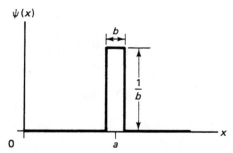

Fig. 2.7

6. Find the Fourier transform of the triangle function $g_b(x)$ plotted in Fig. 2.8. Construct a representation of the δ function by taking the limit $b \to 0$, and show that, in this limit, the Fourier transform becomes $1/\sqrt{2\pi}$.

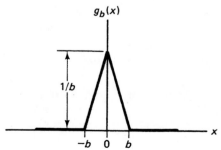

Fig. 2.8

7. Find the Fourier transform of $\delta'(x)$, the derivative of the δ function.

8. Show that $\int f(x)\delta''(x)dx = f''(0)$.

9. Show that $\int_{-\infty}^{\infty} \psi_1(x)\psi_2^*(x)\, dx = \int_{-\infty}^{\infty} \phi_1(k)\phi_2^*(k)\, dk$.

10. Show that $\int_{-\infty}^{\infty} e^{-ikx}\psi_1(x)\psi_2^*(x)\, dx = \int_{-\infty}^{\infty} \phi_1(k')\phi_2^*(k' - k)\, dk'$.

11. (a) Suppose that in Eq. (61), instead of $\omega = p^2/2m\hbar = \hbar k^2/2m$, we have some general function $\omega = \omega(k)$. Show that the group velocity in this case is

    $$v_g = \frac{d\omega}{dk}$$

    [Hint: Use the Taylor-series expansion $\omega(k) = \omega_0 + (k - k_0)\, d\omega/dk + \frac{1}{2}(k - k_0)^2 d^2\omega/dk^2$.]

    (b) For a relativistic particle with $\omega = E/\hbar = \sqrt{\hbar^2 k^2 c^2 + m^2 c^4}/\hbar$, find the group velocity and compare with the particle velocity.

12. Evaluate the uncertainties Δp and Δx for the distributions of momentum and position given by Eqs. (52) and (53), according to the precise definitions of Δp and Δx as rms deviations from the mean. How do the precise values compare with the rough estimate given in the qualitative discussion following Eq. (53)?

13. Perform the integral in Eq. (73). (Hint: Complete the square on p in the exponential.)

14. Obtain Eq. (75) from Eq. (74).

15. (a) Evaluate the expectation value of the kinetic energy $p^2/2m$ for the Gaussian wave packet given in Eq. (70).

    (b) If $p_0 = 0$, this Gaussian wave packet represents a particle at rest. Is the expectation value of the kinetic energy zero if $p_0 = 0$? Explain.

16. Suppose that at $t = 0$, the wavefunction of a free particle is

    $$\psi(x, 0) = \sqrt{b}\, e^{-|x|b + ip_0 x/\hbar}$$

    (a) Show that this wavefunction is correctly normalized.

    (b) What is the momentum amplitude for this wavefunction?

    (c) What is the probability for finding the momentum between p and $p + dp$?

    (d) What are the values of the uncertainties Δx and Δp at $t = 0$?

    (e) What is $\psi(x, t)$?

17. At a given time, an electron has equal probability amplitude for being either at $x = a$ or at $x = -a$. The probability amplitude is zero everywhere else.

(a) What is the wavefunction for the electron? Can this wavefunction be normalized?

(b) What is the momentum amplitude for this electron?

(c) What is the probability for finding the momentum in an interval dp centered around $\frac{1}{2}\pi\hbar/a$?

18. Assume that an electron of 20 keV in the beam of a TV picture tube is described by a Gaussian wave packet. Assume that the initial size of the packet, at the electron gun, is 10^{-7} m measured along the direction of the beam. How much does such a packet spread while the electron moves from the gun to the picture screen, a distance of about 30 cm?

19. A free particle moving in one dimension is known to be at the point x_1 at time t_1. Assume that the wavefunction at this time is $\psi(x, t_1) = \delta(x - x_1)$. Find the wavefunction $\psi(x, t_2)$ at some later time t_2.

20. Show that for a free particle, the uncertainty Δp of the momentum is constant, that is,

$$\frac{d}{dt}\Delta p = 0$$

21. At $t = 0$, the wavefunction of a free particle is

$$\psi(x, 0) = \begin{cases} \sqrt{\dfrac{b}{2\pi}}\ \sin bx & \text{for } |x| < \dfrac{2\pi}{b} \\[2ex] 0 & \text{for } |x| \geq \dfrac{2\pi}{b} \end{cases}$$

(a) What is the probability for finding the particle in the interval $0 \leq x \leq \pi/2b$?

(b) What is the momentum amplitude for this wavefunction?

(c) For what value of the momentum is the momentum probability maximum?

(d) What is the probability for finding the momentum in an interval dp around $\hbar b$?

22. Consider the wavefunction $\psi(x, 0)$ specified in the preceding problem. Evaluate the uncertainties Δx and Δp for this wavefunction.

23. Show that if $\psi(x, t)$ is the wavefunction of a free particle, then the first derivative of the expectation value of x equals the expectation value of the momentum divided by m,

$$\frac{d}{dt}\langle x \rangle = \frac{1}{m}\langle p \rangle$$

that is,

$$\frac{d}{dt} \int \psi^*(x, t) x \psi(x, t) \, dx = \frac{1}{m} \int \psi^*(x, t) p_{\text{op}} \psi(x, t) \, dx$$

(Hint: Use the Schrödinger equation to evaluate the time derivatives of ψ^* and of ψ.)

24. Show that if $\psi(x, t)$ is the wavefunction of a free particle, then the second derivative of the expectation value of x is zero,

$$\frac{d^2}{dt^2} \langle x \rangle = 0$$

that is,

$$\frac{d^2}{dt^2} \int \psi^*(x, t) x \psi(x, t) \, dx = 0$$

To what law of classical physics does this correspond?
(Hint: Use the Schrödinger equation to evaluate the time derivatives of ψ^* and of ψ.)

25. Evaluate $\left[x_{\text{op}}, \dfrac{1}{p_{\text{op}}} \right]$.

26. Evaluate $[x_{\text{op}}, \cos b p_{\text{op}}]$.

27. The operator $H = p_{\text{op}}^2/2m$ represents the kinetic energy of a particle. Show that $[H, x] = \hbar p_{\text{op}}/im$.

28. Consider the operators

$$\alpha = \frac{1}{\sqrt{2}} \left(\frac{x_{\text{op}}}{b} + \frac{ib p_{\text{op}}}{\hbar} \right)$$

and

$$\beta = \frac{1}{\sqrt{2}} \left(\frac{x_{\text{op}}}{b} - \frac{ib p_{\text{op}}}{\hbar} \right)$$

where b is a constant. Evaluate the commutator $[\alpha, \beta]$ of these two operators.

29. Consider the operator $e^{ia p_{\text{op}}/\hbar}$, where a is a constant. Show that when this operator acts on a function $f(x)$, the result is $f(x + a)$, that is,

$$e^{ia p_{\text{op}}/\hbar} f(x) = f(x + a)$$

3

Particles in Potentials

In this chapter we will consider the effects of an external force acting on a quantum-mechanical particle. The force is assumed to be derivable from some potential energy $V(x)$. The behavior of the wavefunction is then determined by the general Schrödinger wave equation, which includes the potential. This general Schrödinger wave equation with potential has two kinds of solutions, representing bound states and unbound states. Bound states occur if the potential is attractive and the particle has insufficient energy to escape from the potential well. The particle, or its wavefunction, then remains confined within finite limits that roughly correspond to the classical turning points of the motion. The bound states are stationary states; that is, their probability densities are constant in time. As we will see, it is a consequence of the Schrödinger equation that the energy of the bound states is quantized: the values of the energy are restricted to a *discrete set*.

Unbound states occur if the particle has sufficient energy to pass through the region of the potential and escape to infinity. Thus, the unbound states are, in essence, the free-particle states modified by the potential. The energy of the unbound states, like the energy of free-particle states, is not restricted by any quantization condition. The energy can assume any value in a *continuum* ranging from $E = 0$ to $E = \infty$.

3.1 *The Schrödinger Equation with Potential; Stationary States*

The Schrödinger wave equation with potential is a simple generalization of the Schrödinger wave equation for a free particle. Consider an initially free particle of definite, or nearly definite, momentum, and let this particle approach a region in which there acts some potential. For instance, consider a high-energy electron approaching and passing between the plates of a capacitor, where there is an electric field, constant in time. From our experience with electromagnetic waves and with sound waves, we know that if the characteristics of the medium are time independent, then the frequency of the wave necessarily remains constant as it propagates through the medium. This preservation of the frequency is a direct consequence of the basic mechanism of wave propagation: the oscillations of each portion of the wave generate the next adjoining portion of the wave, and therefore the frequencies of all the portions of the wave coincide. For our quantum-mechanical wave approaching and passing through a potential, the initial relation between the frequency and the energy is

$$\omega = \frac{E}{\hbar} \tag{1}$$

and since both the frequency and the energy are constants of the motion, we can conclude that this relation remains valid at all times.

We cannot draw any such general conclusion for the relation between the wavelength and the energy. In fact, the wavelength is not even well defined, since the separation between the wave peaks and troughs varies from place to place. However, if the potential changes slowly, with only a small fractional change of potential over a distance of several wave peaks, then we can treat the particle as approximately free over this distance, and we can use the de Boglie relation to calculate an approximate "local" wavelength. If the local value of the potential is $V(x)$, then the momentum is $p = \sqrt{2m[E - V(x)]}$, and the local wavelength is

$$\lambda = \frac{h}{p} = \frac{h}{\sqrt{2m[E - V(x)]}} \tag{2}$$

With these expressions for the frequency and the wavelength, we can construct an approximate wavefunction

$$\psi(x, t) = e^{\pm ikx - i\omega t} = e^{\pm i\sqrt{2m[E-V(x)]}x/\hbar - iEt/\hbar} \tag{3}$$

Given this wavefunction, it is then a simple matter to find the wave equation that it satisfies. This wave equation is

$$-\frac{\hbar^2}{2m}\frac{\partial^2}{\partial x^2}\psi(x, t) + V(x)\psi(x, t) = i\hbar\frac{\partial}{\partial t}\psi(x, t) \tag{4}$$

Exercise 1. Verify that Eq. (4) is the lowest-order differential equation satisfied by the wavefunction (3). Assume that the derivative of the potential $V(x)$ can be neglected, that is, the potential is approximately constant.

We will now postulate that the wave equation (4) remains valid even when the potential is *not* approximately constant and the approximate wavefunction (3) is *not* valid. Equation (4) is Schrödinger's wave equation for a particle interacting with a potential. This equation is the correct generalization of the free-particle wave equation. Equation (4) serves as the basic equation of motion in quantum mechanics; it determines the wavefunction at any later time from its initial condition at some initial time.

As we saw in Chapter 2, the free-particle Schrödinger equation can be written in terms of the kinetic-energy operator $p_{op}^2/2m$. Likewise, the Schrödinger equation for a particle interacting with a potential can be written in terms of an operator representing the total energy, that is, the sum of the kinetic and the potential energies,

$$\frac{p_{op}^2}{2m} + V(x) \tag{5}$$

In terms of this energy operator, the Schrödinger equation becomes

$$\left[\frac{p_{op}^2}{2m} + V(x)\right] \psi = i\hbar \frac{\partial \psi}{\partial t} \tag{6}$$

For a given potential $V(x)$, the Schrödinger equation has a large variety of possible solutions. But among all these solutions, those that are eigenfunctions of the energy operator play an eminent role. A wavefunction is an eigenfunction of the energy operator if

$$\left[\frac{p_{op}^2}{2m} + V(x)\right] \psi = E\psi \tag{7}$$

Here E is the energy eigenvalue. When a wavefunction is an eigenfunction of the energy operator, the energy is sharply defined, with no uncertainty.

The time dependence of the eigenstates of energy is always of the form $e^{-iEt/\hbar}$. This is not surprising, since we have already established this form of the time dependence in the case of unbound states [see Eq. (1)]. To establish it in general, note that for an eigenfunction of energy, the Schrödinger equation implies that

$$E\psi(x, t) = i\hbar \frac{\partial}{\partial t} \psi(x, t) \tag{8}$$

or

$$\frac{1}{\psi(x, t)} \frac{\partial}{\partial t} \psi(x, t) = -\frac{i}{\hbar} E \tag{9}$$

This equation can be readily integrated with respect to t, at a fixed value of x:

$$\ln \psi(x, t) = \frac{-iEt}{\hbar} + \text{constant} \tag{10}$$

where the constant of integration cannot depend on t, but it can depend on x. If we write this constant in the convenient form $\ln \psi(x)$, where $\psi(x)$ is the wavefunction at $t = 0$, we obtain from Eq. (10)

$$\psi(x, t) = e^{-iEt/\hbar}\psi(x) \tag{11}$$

Thus, in an eigenstate of energy, the wavefunction is the product of a function of x and a function of t, and the latter is of the form $e^{-iEt/\hbar}$.

Note that for a wavefunction with the time dependence (11), the probability density is time independent,

$$|\psi(x, t)|^2 = |e^{-iEt/\hbar}\psi(x)|^2 = |\psi(x)|^2 \tag{12}$$

Because of this time independence of the probability density, the energy eigenstates are called *stationary states*.[1]

For a wavefunction with the time dependence $e^{-iEt/\hbar}$, the Schrödinger equation (4) reduces to

$$\left[-\frac{\hbar^2}{2m}\frac{d^2}{dx^2} + V(x)\right]\psi(x)e^{-iEt/\hbar} = E\psi(x)e^{-iEt/\hbar} \tag{13}$$

or, upon canceling the exponential functions on both sides,

---

[1] Equation (12) establishes that in an energy eigenstate, the probability density is time independent. But, as we will see in Section 3.2, the converse is also true: if the probability density is time independent, then the state is necessarily an energy eigenstate.

$$\left[-\frac{\hbar^2}{2m}\frac{d^2}{dx^2} + V(x)\right]\psi(x) = E\psi(x) \tag{14}$$

This is called the *time-independent Schrödinger equation*. Obviously, this equation coincides with the eigenvalue equation for the energy [see Eq. (7)].

In the next sections we will examine some simple examples of solutions of the Schrödinger equation. These simple examples involve potentials that are piecewise constant, that is, the potentials are constant except for discontinuous increases at given points. Figure 3.1 illustrates such a potential, with a single discontinuity.

To proceed with the solution of the Schrödinger equation for such a potential, we need to know the boundary conditions that are to be imposed on the wavefunction ψ at the points of discontinuity of the potential. These boundary conditions are simply that the wavefunction ψ and its first derivative $d\psi/dx$ must remain continuous. We can justify these boundary conditions as follows: Suppose that the potential has a (finite) discontinuity at $x = a$, and integrate Eq. (14) across the discontinuity, from $x = a - \varepsilon$ to $x = a + \varepsilon$:

$$-\frac{\hbar^2}{2m}\int_{a-\varepsilon}^{a+\varepsilon}\frac{d^2\psi}{dx^2}\,dx = \int_{a-\varepsilon}^{a+\varepsilon}(E - V)\psi(x)\,dx \tag{15}$$

that is,

$$\frac{d\psi}{dx}\bigg|_{a+\varepsilon} - \frac{d\psi}{dx}\bigg|_{a-\varepsilon} = -\frac{2m}{\hbar^2}\int_{a-\varepsilon}^{a+\varepsilon}(E - V)\psi(x)\,dx \tag{16}$$

For any finite potential V, the integral on the right side will tend to zero as $\varepsilon \to 0$. Hence, $d\psi/dx$ is continuous at $x = a$. To see that ψ must also be continuous, we first write the Schrödinger equation

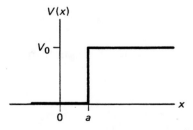

Fig. 3.1   This potential has a discontinuous step at the point $x = a$.

(14) as a differential-integral equation, by integrating it once with respect to x:

$$-\frac{\hbar^2}{2m}\frac{d}{dx}\,\psi(x) = -\frac{\hbar^2}{2m}\frac{d\psi}{dx}\bigg|_{x_0} + \int_{x_0}^{x}(E - V)\psi(x')\,dx' \tag{17}$$

where x_0 is some fixed point. Then we integrate Eq. (17) from $x = a - \varepsilon$ to $x = a + \varepsilon$:

$$\psi\bigg|_{a+\varepsilon} - \psi\bigg|_{a-\varepsilon} = \int_{a-\varepsilon}^{a+\varepsilon}\frac{d\psi}{dx}\bigg|_{x_0}dx - \frac{2m}{\hbar^2}\int_{a-\varepsilon}^{a+\varepsilon}\int_{x_0}^{x}(E - V)\psi(x')\,dx'\,dx$$

$$\tag{18}$$

The integrals approach zero as $\varepsilon \to 0$, and therefore ψ is continuous.

3.2 *The Infinite Square Well*

As the first and simplest example of the solution of the Schrödinger equation for a piecewise constant potential with stationary states, we will consider the potential

$$V(x) = \begin{cases} 0 & \text{for } 0 < x < L \\ \infty & \text{for } x < 0 \quad \text{and} \quad x > L \end{cases} \tag{19}$$

This potential looks like an infinitely deep square well (see Fig. 3.2). Physically, the potential represents a one-dimensional box, with rigid, elastic walls at $x = 0$ and at $x = L$, which reflect and confine the particle. Obviously, this system has no unbound states.

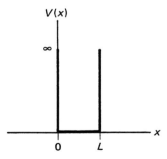

Fig. 3.2 The infinite square potential well.

To find the solution for the bound states, we note that in the region $0 < x < L$, the Schrödinger equation is the same as that for a free particle,

$$-\frac{\hbar^2}{2m}\frac{\partial^2}{\partial x^2}\psi(x, t) = i\hbar\frac{\partial}{\partial t}\psi(x, t) \qquad (20)$$

The solution of the Schrödinger equation is therefore trivial: the energy eigenstates are superpositions of the harmonic waves

$$e^{i\sqrt{2mE}\, x/\hbar - iEt/\hbar} \qquad \text{and} \qquad e^{-i\sqrt{2mE}\, x/\hbar - iEt/\hbar} \qquad (21)$$

However, the solution we seek must satisfy not only the Schrödinger equation, but also the boundary conditions at $x = 0$ and $x = L$. This means we must choose a special superposition of the two wavefunctions (21). Since the wavefunction is zero for $x < 0$ and $x > L$, the continuity of the wavefunction at $x = 0$ and at $x = L$ demands

$$\psi(0, t) = 0 \qquad \text{and} \qquad \psi(L, t) = 0 \qquad (22)$$

These are the only boundary conditions that the solution are required to satisfy. The extra boundary conditions for the derivative $d\psi/dx$ are not applicable when the potential has an *infinite* discontinuity. We can recognize this from Eq. (16), since, with $V = \infty$ and $\psi = 0$, the integral on the right side of the equation becomes undefined and imposes no restriction on the magnitude of the change of the derivative across the boundary.

The boundary condition (22) at $x = 0$ is satisfied by the following superposition:

$$\psi(x, t) \propto e^{i\sqrt{2mE}\, x/\hbar - iEt/\hbar} - e^{-i\sqrt{2mE}\, x/\hbar - iEt/\hbar} \qquad (23)$$

We can also write this as

$$\psi(x, t) = Ae^{-iEt/\hbar}\sin\frac{\sqrt{2mE}\, x}{\hbar} \qquad (24)$$

where A is some constant. The boundary condition at $x = L$ imposes the requirement

$$\sqrt{2mE}\,\frac{L}{\hbar} = n\pi \qquad (25)$$

or

$$E_n = \frac{n^2\hbar^2\pi^2}{2mL^2} \qquad n = 1, 2, 3, \ldots \qquad (26)$$

This is the quantization condition for the energy. The permitted values of the energy form a discrete set; Fig. 3.3 displays these permitted values of the energy on an energy-level diagram. The state of lowest energy is called the ground state, the next state is the first excited state, the next is the second excited state, and so on.

We will hereafter designate the wavefunction corresponding to the energy E_n by $\psi_n(x, t)$. Figure 3.4 shows plots of the first few of these wavefunctions at $t = 0$. As is obvious from these plots, the quantization condition corresponds to the requirement that some integer number of half wavelengths fit exactly in the width of the potential well.

The wavefunctions plotted in Fig. 3.4 are reminiscent of the wavefunctions for a standing wave on a string, say, a violin string. The discrete frequencies for the (quantum-mechanical) particle in the infinite potential well and for the (classical) violin string both arise in the same way, from the boundary conditions. However, the "quantization" of the frequency of the violin string does not imply a quantization of its energy. The energy of the classical string depends on the amplitude of oscillation, which is a freely adjustable parameter. In contrast, the amplitude of oscillation of the wavefunction of the quantum-mechanical particle in the potential well is fixed by the normalization condition, and the energy is therefore not adjustable.

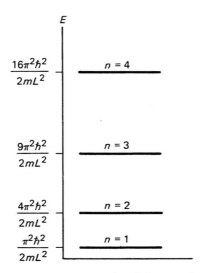

Fig. 3.3 Energy-level diagram for the infinite square well.

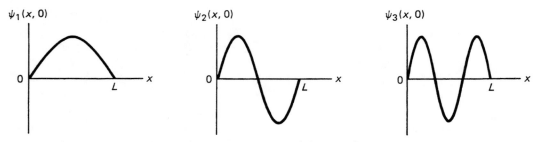

Fig. 3.4   Space dependence of the wavefunctions ψ_n.

The energy of the ground state, $E_1 = \hbar^2\pi^2/2mL^2$, is roughly the minimum possible energy compatible with the uncertainty principle. For an estimate of this minimum energy required for compatibility with the uncertainty principle, we exploit a general rule, valid for any (bound) stationary state: The expectation value of the square of the momentum in a stationary state equals the square of the uncertainty of the momentum,

$$\langle p^2 \rangle = (\Delta p)^2 \tag{27}$$

To establish this rule, we note that for a stationary state the expectation value of the momentum is zero, $\langle p \rangle = 0$.[2] According to the definition of Δp as a root-mean-square deviation from the mean, we then immediately obtain $(\Delta p)^2 = \langle p^2 - \langle p \rangle^2 \rangle = \langle p^2 \rangle$. This rule permits us to express the expectation value of the kinetic energy in terms of the uncertainty in the momentum:

$$\langle K \rangle = \frac{\langle p^2 \rangle}{2m} = \frac{(\Delta p)^2}{2m} \tag{28}$$

For the particle in the infinite square well, the uncertainty in x is roughly equal to $L/2$, half the width of the well. According to the uncertainty principle, the uncertainty in p is then $\Delta p \geq \hbar/2\Delta x \simeq \hbar/L$, and thus

$$\langle K \rangle \geq \frac{\hbar^2}{2mL^2} \tag{29}$$

The actual kinetic energy of the ground state, $\hbar^2\pi^2/2mL^2$, is consistent with this estimate.

---

[2] This is intuitively obvious, since the stationary state has no average translational motion. For a rigorous proof, we can use $\langle p \rangle = m(d/dt)\langle x \rangle$, which is one of Ehrenfest's equations (see Section 5.2); since $\langle x \rangle$ is necessarily time independent in a stationary state, it follows that $\langle p \rangle = 0$.

We still have to normalize the wavefunctions given by Eq. (24). The constant A in the wavefunction is fixed by the normalization condition:

$$1 = \int_0^L |\psi_n|^2 \, dx = \int_0^L A^2 \sin^2 \frac{\sqrt{2mE_n}\, x}{\hbar} \, dx$$

$$= \int_0^L A^2 \sin^2 \frac{n\pi x}{L} \, dx = A^2 \frac{L}{2} \tag{30}$$

From this

$$A = \sqrt{\frac{2}{L}}$$

and therefore

$$\psi_n(x, t) = e^{-iE_n t/\hbar} \sqrt{\frac{2}{L}} \sin \frac{n\pi x}{L} \tag{31}$$

It is convenient to write this as

$$\psi_n(x, t) = e^{-iE_n t/\hbar} \psi_n(x) \tag{32}$$

where the function $\psi_n(x)$ depends on x only:

$$\psi_n(x) = \sqrt{\frac{2}{L}} \sin \frac{n\pi x}{L} \tag{33}$$

The functions $\psi_n(x)$ have several interesting properties. First of all, the functions $\psi_n(x)$ are, of course, eigenfunctions of the energy operator with eigenvalue E_n:

$$\frac{p_{op}^2}{2m} \psi_n(x) = E_n \psi_n(x) \tag{34}$$

Further, these eigenfunctions are *orthonormal*, that is, the integral of the product $\psi_m{}^*(x)\psi_n(x)$ equals zero if $m \neq n$ and it equals 1 if $m = n$,

$$\int_0^L \psi_m{}^*(x)\psi_n(x) \, dx = \delta_{mn} \tag{35}$$

Here the quantity δ_{mn}, called the *Kronecker delta*, is defined as

$$\delta_{mn} = \begin{cases} 1 & \text{if } n = m \\ 0 & \text{if } n \neq m \end{cases} \tag{36}$$

In Eq. (35), the complex conjugation of the function $\psi_m{}^*$ is super-fluous, but it has been inserted for the sake of generality [with the complex conjugation included, Eq. (35) is also valid for the time-dependent eigenfunctions $\psi_n(x, t)$].

Exercise 2. Show that Eq. (35) holds.

Exercise 3. Show that the functions $\psi_n(x, t)$ also are eigenfunctions of the energy operator, and show that they also are orthonormal.

Finally, the eigenfunctions $\psi_n(x)$ form a *complete set*. This means that if $\psi(x)$ is any arbitrary function of x defined in the interval $x = 0$ to $x = L$ (not necessarily a solution of Schrödinger's equation), then it can be written as a superposition of the functions $\psi_n(x)$:

$$\psi(x) = \sum_n a_n\psi_n(x) \tag{37}$$

To see that this is true, we use the explicit form for $\psi_n(x)$ and write Eq. (37) as

$$\psi(x) = \sum_n a_n \sqrt{\frac{2}{L}} \sin \frac{n\pi x}{L} \tag{38}$$

This simply expresses ψ as a Fourier series, and it is well known that any function defined in the interval $0 < x < L$ can be expanded in such a series.[3]

The coefficients a_n in Eq. (37) can be determined as follows: Multiply the equation by $\psi_m{}^*(x)$, integrate over x, and use the orthonormality relation

$$\int_0^L \psi_m{}^*(x)\psi(x)\, dx = \sum_n a_n \int_0^L \psi_m{}^*(x)\psi_n(x)\, dx$$

$$= \sum_n a_n\delta_{mn}$$

$$= a_m \tag{39}$$

---

[3] In Chapter 2 we used a Fourier series with complex exponentials, which is equivalent to a Fourier series with sines and cosines. Here we use sines only. This is possible because the function $\psi(x)$ is now specified for positive values of x only, and for the purpose of comparison with the Fourier series of Chapter 2, we can pretend that the function is odd, with $\psi(-x) = -\psi(x)$, which requires a Fourier series with sines only.

By means of Eq. (39), we can then express Eq. (37) as

$$\psi(x) = \sum_n a_n\psi_n(x)$$

$$= \sum_n \int_0^L \psi_n{}^*(x')\psi(x') \, dx' \; \psi_n(x)$$

$$= \int_0^L \psi(x') \left[\sum_n \psi_n{}^*(x')\psi_n(x) \right] dx' \tag{40}$$

Since $\psi(x')$ is an arbitrary function, this shows that the quantity in brackets acts as a delta function, that is,

$$\sum_n \psi_n{}^*(x')\psi_n(x) = \delta(x - x') \tag{41}$$

This equation is called the *completeness* or the *closure relation*. In general, whenever a set of functions $\psi_n(x)$ satisfies the completeness relation, then these functions form a complete set. To see that Eq. (41) implies that an arbitrary function can be written as a superposition of the functions $\psi_n(x)$, we multiply both sides of Eq. (41) by the arbitrary function $\psi(x')$ and integrate over x'; we then immediately recover the superposition (40).[4]

If the function ψ in Eq. (37) depends on time, then the coefficients a_n will be different at different times and we must write

$$\psi(x, t) = \sum_n a_n(t)\psi_n(x) \tag{42}$$

In particular, (42) is a solution of the Schrödinger equation (20) if and only if the coefficients $a_n(t)$ have the form

$$a_n(t) = a_n(0)e^{-iE_nt/\hbar} \tag{43}$$

Exercise 4. Show this by substituting (42) into the Schrödinger equation (20) and solving for $a_n(t)$.

---

[4] The expression (41) for the delta function in terms of the eigenfunctions of the infinite square well,

$$\delta(x - x') = \sum \left(\frac{2}{L}\right) \sin \frac{n\pi x}{L} \sin \frac{n\pi x'}{L}$$

is valid only as long as we are dealing with functions defined in the interval $0 < x < L$. The expression given in Chapter 2 is not subject to this restriction.

From these results, we can conclude that the most general solution of the Schrödinger equation for a particle in the infinite square well is the superposition

$$\psi(x, t) = \sum_n a_n(0)e^{-iE_n t/\hbar}\psi_n(x) \tag{44}$$

$$= \sum_n a_n(0)\psi_n(x, t) \tag{45}$$

with arbitrary coefficients $a_n(0)$ [however, the coefficients $a_n(0)$ are subject to a normalization condition; see below]. This says that the most general state for the particle in the infinite square well is a superposition of stationary states.

Note that any superposition of two or more stationary states is *not* a stationary state. For instance, consider a superposition of the ground state and the first excited state, with $a_1(0) = \sqrt{3}/2$ and $a_2(0) = \frac{1}{2}$,

$$\psi(x, t) = \frac{\sqrt{3}}{2} e^{-iE_1 t/\hbar} \sqrt{\frac{2}{L}} \sin\frac{\pi x}{L} + \frac{1}{2} e^{-iE_2 t/\hbar} \sqrt{\frac{2}{L}} \sin\frac{2\pi x}{L} \tag{46}$$

The corresponding probability density is

$$|\psi(x, t)|^2 = \frac{3}{2L} \sin^2\frac{\pi x}{L} + \frac{1}{2L} \sin^2\frac{2\pi x}{L}$$

$$+ \frac{\sqrt{3}}{L} \sin\frac{\pi x}{L} \sin\frac{2\pi x}{L} \cos\frac{(E_2 - E_1)t}{\hbar} \tag{47}$$

The first two terms on the right side of the equation are independent of time, but the last term oscillates with a frequency $\omega = (E_2 - E_1)/\hbar$. Such oscillations in the probability density of superpositions of different stationary states play a crucial role in generating radiation in atomic and molecular systems. The oscillating probability density of an electron amounts to an oscillating charge distribution, which radiates electromagnetic waves, that is, it radiates photons.

Since a superposition, such as (45), of two or more eigenstates of different energies is not a stationary state, it follows that the individual eigenstates of energy are the *only* stationary states. Thus, a state is stationary if and only if it is an energy eigenstate.

While the probability density corresponding to a superposition of energy eigenstates oscillates, the total integrated probability remains constant. This preservation of the integrated probability is, of course, required on physical grounds. We can prove it mathematically by means of the following identity, which is a consequence of the Schrödinger equation (4):

$$\frac{\partial}{\partial t}(\psi^*\psi) = -\frac{\partial}{\partial x}\left[\frac{\hbar}{2im}\left(\psi^*\frac{\partial\psi}{\partial x} - \psi\frac{\partial\psi^*}{\partial x}\right)\right] \qquad (48)$$

Exercise 5. Prove Eq. (48). (Hint: Multiply the Schrödinger equation by ψ^*, multiply the complex conjugate of the Schrödinger equation by ψ, and take the difference between the resulting equations.)

If we multiply Eq. (48) by dx, we recognize that it can be interpreted as a conservation equation for the probability: it asserts that the change of the probability in the interval dx equals the difference between the currents of probability entering at one end of the interval and leaving at the other,

$$\frac{\partial}{\partial t}(\psi\psi^*)\,dx = -d\left[\frac{\hbar}{2im}\left(\psi^*\frac{\partial\psi}{\partial x} - \psi\frac{\partial\psi^*}{\partial x}\right)\right] \qquad (49)$$

According to this interpretation of Eq. (48), the quantity $\hbar/2im$ $(\psi^*\,\partial\psi/\partial x - \psi\,\partial\psi^*/\partial x)$ must be the probability current.

If we integrate Eq. (48) from $x = 0$ to $x = L$, we find that

$$\frac{d}{dt}\int_0^L \psi^*\psi\,dx = -\frac{\hbar}{2im}\left(\psi^*\frac{\partial\psi}{\partial x} - \psi\frac{\partial\psi^*}{\partial x}\right)\Bigg|_{x=L}$$

$$+\frac{\hbar}{2im}\left(\psi^*\frac{\partial\psi}{\partial x} - \psi\frac{\partial\psi^*}{\partial x}\right)\Bigg|_{x=0} \qquad (50)$$

The right side of this equation is zero, since ψ and ψ^* are zero at $x = 0$ and at $x = L$. Thus, the integrated probability density is constant.

Exercise 6. Verify explicitly that for a stationary state, each side of Eq. (48) is identically zero.

3.3 *The Energy Representation*

We have shown that any solution of the Schrödinger equation for the infinite square well can be expressed as a superposition of the eigenfunctions $\psi_n(x)$ with coefficients $a_n(t)$:

$$\psi(x, t) = \sum_n a_n(t)\psi_n(x) \qquad (51)$$

The time dependence of these coefficients $a_n(t)$ is given by Eq. (43). To specify the state of a particle in the infinite square well, we can therefore use the set of coefficients $a_n(t)$, $n = 1, 2, 3, \ldots$ instead of using $\psi(x, t)$. The description of the state by the infinite set of coefficients $a_n(t)$ is called the *energy representation*. We can write the coefficients $a_n(t)$ in the form of column vector of an infinite number of components:[5]

$$
\begin{pmatrix} a_1 \\ a_2 \\ a_3 \\ \vdots \end{pmatrix}
\tag{52}
$$

This column vector is called the *state vector*. The components of the state vector must satisfy the condition

$$
\sum_{n=1}^{\infty} |a_n|^2 = 1
\tag{53}
$$

To show this, we use Eq. (39):

$$
a_n = \int \psi_n{}^*(x)\psi(x, t)\, dx
\tag{54}
$$

which yields

$$
\begin{aligned}
\sum_{n=1}^{\infty} a_n{}^*a_n &= \sum_{n=1}^{\infty} \left[\int \psi_n{}^*(x)\psi(x, t)\, dx\right]^* \left[\int \psi_n{}^*(x')\psi(x', t)\, dx'\right] \\
&= \int dx\, dx' \left[\sum_{n=1}^{\infty} \psi_n{}^*(x')\psi_n(x)\right] [\psi^*(x, t)\psi(x', t)] \\
&= \int dx\, dx'\, \delta(x - x')\psi^*(x, t)\psi(x', t) \\
&= \int dx\, |\psi(x, t)|^2 = 1
\end{aligned}
\tag{55}
$$

The last integral equals 1 by the normalization of $\psi(x, t)$.

The meaning of the coefficients a_n is this: $|a_n|^2$ gives the probability that a measurement of energy results in the value E_n for the energy. To see how this probability interpretation of the coeffi-

---

[5] The time dependence of the a_n is left understood.

cients a_n arises, let us calculate the expectation value of the energy operator, $p_{op}^2/2m$:

$$\left\langle \frac{p_{op}^2}{2m} \right\rangle = \int \psi^*(x, t) \frac{p_{op}^2}{2m} \psi(x, t) \, dx$$

$$= \int \left[\sum_n a_n\psi_n(x) \right]^* \frac{p_{op}^2}{2m} \left[\sum_k a_k\psi_k(x) \right] dx$$

$$= \int \left[\sum_n a_n\psi_n(x) \right]^* \left[\sum_k a_k \frac{p_{op}^2}{2m} \psi_k(x) \right] dx \qquad (56)$$

But we know that $\psi_k(x)$ is an eigenfunction of energy, with $(p_{op}^2/2m)\psi_k(x) = E_k\psi_k(x)$. Hence we obtain

$$\left\langle \frac{p_{op}^2}{2m} \right\rangle = \int \left[\sum_n a_n\psi_n(x) \right]^* \left[\sum_k a_k E_k\psi_k(x) \right] dx$$

$$= \sum_n \sum_k a_n^* a_k E_k \delta_{nk}$$

$$= \sum_n |a_n|^2 E_n \qquad (57)$$

Exercise 7. More generally, show that

$$\left\langle \left(\frac{p_{op}^2}{2m} \right)^l \right\rangle = \sum_n |a_n|^2 (E_n)^l \qquad (58)$$

where l is any positive integer.

Equation (57) establishes that the coefficients $|a_n|^2$ determine the probability distribution of the energy values. Since the $|a_n|^2$ are normalized [see Eq. (53)], we can identify $|a_n|^2$ as the probability for the particle having energy E_n. Note that if only one of the a_n is different from zero, then there is certainty that only the value of the energy corresponding to this nonzero a_n can result when a measurement is performed. The energy of such a state therefore shows no uncertainty. If more than one a_n is different from zero, then the energy is not sharply defined, and we have an uncertainty in energy.

In the energy representation, the state is described by an (infinite) column vector. Hence linear operators will be described by *matrices.* As a simple example, let us look at the energy operator $p_{op}^2/2m$. Since

$$\frac{p_{op}^2}{2m} \psi(x, t) = \sum_n a_n \frac{p_{op}^2}{2m} \psi_n(x)$$

$$= \sum_n E_n a_n \psi_n(x) \tag{59}$$

this operator acting on an arbitrary wave function ψ has the effect of multiplying each a_n by E_n. The matrix that multiplies each a_n by E_n when applied to the column vector (52) obviously is

$$\begin{pmatrix} E_1 & 0 & 0 & 0 & . & . \\ 0 & E_2 & 0 & 0 & . & . \\ 0 & 0 & E_3 & 0 & . & . \\ 0 & 0 & 0 & E_4 & . & . \\ \vdots & \vdots & \vdots & \vdots & \vdots & \vdots \end{pmatrix} \tag{60}$$

This is the energy operator in the energy representation. It is a general result that the energy operator is *diagonal* in the energy representation.

How do multiplication by coordinate and by momentum look in the energy representation? We have

$$x\psi(x, t) = \sum_n x a_n \psi_n(x) \tag{61}$$

The product $x\psi_n(x)$ can be expressed as a superposition of the eigenfunctions $\psi_m(x)$:

$$x\psi_n(x) = \sum_m x_{mn} \psi_m(x) \tag{62}$$

The coefficients x_{mn} of this expansion of the function $x\psi_n$ in terms of the eigenfunctions ψ_m can be found in the usual way. These coefficients are given by the integral

$$x_{mn} = \int_0^L \psi_m^*(x) x \psi_n(x) \, dx \tag{63}$$

The evaluation of this integral is straightforward. For $m \neq n$, the result is

$$x_{mn} = \int_0^L \frac{2}{L} \left(\sin \frac{m\pi x}{L} \right) x \left(\sin \frac{n\pi x}{L} \right) dx$$

$$= \frac{L}{\pi^2} \left[\frac{4mn}{(m^2 - n^2)^2} \right] [(-1)^{(m-n)} - 1] \tag{64}$$

and for $m = n$, the result is $x_{mn} = L/2$.

If we rearrange Eq. (61) slightly, we obtain

$$x\psi(x,\, t) = \sum_n a_n \sum_m x_{mn}\psi_m(x)$$

$$= \sum_m \left(\sum_n x_{mn}a_n \right) \psi_m(x)$$

$$= \sum_m a_m{}'\psi_m(x) \tag{65}$$

Hence, under multiplication by the coordinate, the expansion coefficients a_n are changed into new coefficients $a_m{}'$ related to the old ones according to

$$a_m{}' = \sum_n x_{mn}a_n \tag{66}$$

This shows that, in the energy representation, multiplication by the coordinate is *matrix multiplication.* With the standard row-by-column rule for matrix multiplication, Eq. (66) is equivalent to

$$
\begin{pmatrix} a_1' \\ a_2' \\ a_3' \\ \vdots \end{pmatrix} =
\begin{pmatrix}
x_{11} & x_{12} & x_{13} & \cdot & \cdot \\
x_{21} & x_{22} & x_{23} & \cdot & \cdot \\
x_{31} & x_{32} & x_{33} & \cdot & \cdot \\
\vdots & \vdots & \vdots & \vdots & \vdots
\end{pmatrix}
\begin{pmatrix} a_1 \\ a_2 \\ a_3 \\ \vdots \end{pmatrix}
$$

$$
=
\begin{pmatrix}
L/2 & -16L/9\pi^2 & 0 & \cdot & \cdot \\
-16L/9\pi^2 & L/2 & -48L/25\pi^2 & \cdot & \cdot \\
0 & -48L/25\pi^2 & L/2 & \cdot & \cdot \\
\vdots & \vdots & \vdots & \vdots & \vdots
\end{pmatrix}
\begin{pmatrix} a_1 \\ a_2 \\ a_3 \\ \vdots \end{pmatrix} \tag{67}
$$

Note that the matrix representing multiplication by the coordinate is symmetric, $x_{mn} = x_{nm}$. This is somewhat of an accident. If we had chosen our eigenfunctions $\psi_n(x)$ with different phase factors instead of choosing them all real, we would not have obtained a real symmetric matrix, but rather a complex *hermitian* matrix, that is, a matrix such that its complex conjugate equals its transpose:

$$x_{mn} = (x_{mn})^* \tag{68}$$

In the same way we can find what happens when we multiply by momentum:

$$p_{op}\psi(x, t) = \frac{\hbar}{i}\frac{d}{dx}\psi(x, t) = \sum_n a_n \frac{\hbar}{i}\frac{d}{dx}\psi_n(x) \tag{69}$$

Here, the product $d/dx\, \psi_n(x)$ can be expressed as a superposition of the eigenfunctions $\psi_m(x)$:

$$\frac{\hbar}{i}\frac{d}{dx}\psi_n(x) = \sum_m p_{mn}\psi_m(x) \tag{70}$$

The coefficients p_{mn} are given by the integral

$$p_{mn} = \int \psi_m^*(x)\left(\frac{\hbar}{i}\frac{d}{dx}\right)\psi_n(x)\, dx \tag{71}$$

For $m \neq n$, the result of this integration is

$$p_{mn} = \int_0^L \frac{2}{L}\sin\frac{m\pi x}{L}\left(\frac{\hbar}{i}\frac{d}{dx}\right)\sin\frac{nx\pi}{L}\, dx$$

$$= \frac{\hbar}{iL}\left[\frac{2mn}{m^2 - n^2}\right][1 - (-1)^{m-n}] \tag{72}$$

and for $m = n$, the result is $p_{mn} = 0$.

We then obtain

$$p_{op}\psi(x, t) = \sum_n a_n \sum_m p_{mn}\psi_m(x)$$

$$= \sum_m \left(\sum_n p_{mn}a_n\right)\psi_m(x)$$

$$= \sum_m a'_m\psi_m(x)$$

that is,

$$a'_m = \sum_n p_{mn}a_n$$

This shows that multiplication by momentum is, again, a row-by-column multiplication by a matrix:

$$
\begin{pmatrix}
p_{11} & p_{12} & p_{13} & \cdot & \cdot \\
p_{21} & p_{22} & p_{23} & \cdot & \cdot \\
p_{31} & p_{32} & p_{33} & \cdot & \cdot \\
\vdots & \vdots & \vdots & \vdots & \vdots
\end{pmatrix}
=
\begin{pmatrix}
0 & 8i\hbar/3L & 0 & \cdot & \cdot \\
-8i\hbar/3L & 0 & 24i\hbar/5L & \cdot & \cdot \\
0 & -24i\hbar/5L & 0 & \cdot & \cdot \\
\vdots & \vdots & \vdots & \vdots & \vdots
\end{pmatrix}
\tag{74}
$$

Note that this matrix is, again, hermitian:

$$
p_{mn} = (p_{mn})^{*}
\tag{75}
$$

We could next use the explicit forms for x_{mn} and p_{mn} given by Eqs. (64) and (72) to check that the commutator of the matrices x_{mn} and p_{mn} is $i\hbar$:

$$
\sum_{k} (x_{mk}p_{kn} - p_{mk}x_{kn}) = i\hbar\delta_{mn}
\tag{76}
$$

This says that the matrix product of the x matrix by the p matrix minus the product of the p matrix by the x matrix equals $i\hbar$ times the unit matrix. Equation (76) involves doing some complicated infinite sums, and it is easier to check it by using the expressions (63) and (71) and manipulating the eigenfunctions by integration by parts and the completeness relation.

We conclude that there are several ways of representing the state of a particle: by means of the wavefunction $\psi(x)$, the amplitude in momentum space $\phi(p)$, or the state vector $(a_1, a_2, a_3, \ldots)$. Since a function of a continuous variable can be regarded as a vector with a continuously infinite number of components,[6] we can say that all these descriptions represent the state by a vector. Position and momentum are, in general, represented by operators. Further, the representations are equivalent in the sense that the operators have the same expectation values in any given physical state, and they have the same commutation relations. In Chapter 4 we will consider those properties of state vectors and operators that all representations have in common, that is, we will look at quantum theory abstractly. By stripping the theory down to its essentials we can better appreciate its structure.

---

[6] $\psi(1)$ is component number 1, $\psi(1.2)$ is component "number" 1.2, $\ldots$, and in general, $\psi(x)$ is component number x. The total number of such components is the same as the total number of points on the real line; that is, the number of components is an uncountable infinity.

3.4 *The Finite Square Well*

The solution of the Schrödinger equation for the infinite square well was trivial, since the wavefunctions of the stationary states are simply superpositions of free-particle wavefunctions. A somewhat more complicated potential with stationary states is the finite square well:

$$V(x) = \begin{cases} -V_0 & \text{for } |x| < L \\ 0 & \text{for } |x| > L \end{cases} \tag{77}$$

This potential is plotted in Fig. 3.5.

In the region $|x| < L$, the Schrödinger equation is

$$-\frac{\hbar^2}{2m}\frac{\partial^2}{\partial x^2}\psi(x,\,t) - V_0\psi(x,\,t) = i\hbar\frac{\partial}{\partial t}\psi(x,\,t) \tag{78}$$

This differs from the equation for a free particle only in the extra term $-V_0\psi(x,\,t)$. The bound states have negative values of the energy; thus, the time dependence of the energy eigenstates is

$$\psi(x,\,t) = e^{-iEt/\hbar}\psi(x) = e^{+i|E|t/\hbar}\psi(x) \tag{79}$$

and the time-independent Schrödinger equation is

$$-\frac{\hbar^2}{2m}\frac{d^2}{dx^2}\psi(x) - V_0\psi(x) = -|E|\psi(x) \tag{80}$$

The solutions of this equation are harmonic waves,

$$e^{-i\sqrt{2m(V_0-|E|)}\,x/\hbar} \quad \text{and} \quad e^{+i\sqrt{2m(V_0-|E|)}\,x/\hbar} \quad \text{for } |x| < L \tag{81}$$

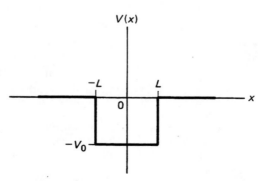

Fig. 3.5 A finite square potential well, of width $2L$ and depth $-V_0$.

These resemble the familiar harmonic waves for a free particle, but the momentum $\sqrt{2mE}$ for a free particle has been replaced by $\sqrt{2m(V_0 - |E|)}$; this is as expected, since the kinetic energy of a free particle is E, whereas the kinetic energy of the particle confined in a bound state in the square-well potential is $E + V_0 = -|E| + V_0$.

In the region $|x| > L$, the time-independent Schrödinger equation is

$$-\frac{\hbar^2}{2m}\frac{d^2}{dx^2}\psi(x) = -|E|\psi(x) \tag{82}$$

This differential equation has the solutions

$$e^{\sqrt{2m|E|}\,x/\hbar} \quad \text{and} \quad e^{-\sqrt{2m|E|}\,x/\hbar} \quad \text{for } |x| > L \tag{83}$$

These solutions are exponential functions rather than oscillatory functions.

Exercise 8. Verify that these exponential functions are solutions of the differential equation (82).

One of the exponential functions listed in (83) is an increasing function, the other is decreasing. Since the wavefunction must remain finite as $x \to \pm\infty$, we must select the first of the functions listed in (83) for $x < -L$, and the second for $x > L$. Furthermore, since both the functions are real and they must be matched to the functions (81) by boundary conditions, we must select a real superposition of the functions (81). There are two possible independent real superpositions of the functions (81):

$$\cos\sqrt{2m(V_0 - |E|)}\,x/\hbar \quad \text{and} \quad \sin\sqrt{2m(V_0 - |E|)}\,x/\hbar$$

$$\text{for } |x| < L \tag{84}$$

First, we will deal with the cosine function in (84). We can summarize the wavefunctions in the regions inside and outside the potential well as follows:

$$\psi(x) = \begin{cases} A\cos\sqrt{2m(V_0 - |E|)}\,x/\hbar & \text{for } |x| < L \\ Be^{-\sqrt{2m|E|}\,x/\hbar} & \text{for } x > L \\ Be^{\sqrt{2m|E|}\,x/\hbar} & \text{for } x < -L \end{cases} \tag{85}$$

Here, we have included equal constants of proportionality B for the solutions to the left and the right of the well. This is required by the symmetry of the potential well and the symmetry of the cosine function. Since the cosine function has equal values at $x = L$ and $x = -L$, the outside wavefunctions that match up with the cosine function at these points must also have equal values (if instead of the cosine function, we use the sine function in Eq. (84), then we need opposite constants B and $-B$ for the solutions to the left and the right of the well).

Since the choice of constants of proportionality in Eq. (85) takes full account of the symmetry of the wavefunction, it will be sufficient to examine the boundary conditions at just one of the points of discontinuity of the potential, say, at $x = L$. According to the general discussion in Section 3.1, the boundary conditions are the continuity of ψ and the continuity of $d\psi/dx$, that is,

$$A \cos \sqrt{2m(V_0 - |E|)} \, L/\hbar = Be^{-\sqrt{2m|E|} \, L/\hbar} \tag{86}$$

and

$$-A \frac{\sqrt{2m(V_0 - |E|}}{\hbar} \sin \sqrt{2m(V_0 - |E|)} \, L/\hbar = -B \frac{\sqrt{2m|E|}}{\hbar} e^{-\sqrt{2m|E|} \, L/\hbar} \tag{87}$$

If we divide Eq. (86) by Eq. (87), we obtain

$$\cot \frac{\sqrt{2m(V_0 - |E|)} \, L}{\hbar} = \frac{\sqrt{V_0 - |E|}}{\sqrt{|E|}} \tag{88}$$

This equation determines the energy eigenvalues. The equation has a discrete set of solutions, and it therefore constitutes a quantization condition for the energy.

If instead of the cosine function in Eq. (84), we use the sine function, we obtain another equation for the energy eigenvalues, with another discrete set of solutions:

$$\tan \frac{\sqrt{2m(V_0 - |E|)} \, L}{\hbar} = -\frac{\sqrt{V_0 - |E|}}{\sqrt{|E|}} \tag{89}$$

Exercise 9. Derive Eq. (89).

Unfortunately, Eqs. (88) and (89) cannot be solved explicitly for $|E|$; they must be solved numerically or graphically. For the graphical solution, it is convenient to introduce the wave number

$$k = \frac{\sqrt{2m(V_0 - |E|)}}{\hbar} \tag{90}$$

and to express these equations as

$$\cot kL = \frac{kL}{\sqrt{2mL^2V_0/\hbar^2 - k^2L^2}} \tag{91}$$

and

$$\tan kL = -\frac{kL}{\sqrt{2mL^2V_0/\hbar^2 - k^2L^2}} \tag{92}$$

Figure 3.6 shows plots of the left sides and the right sides of Eqs. (91) and (92) vs. kL. The intersections of these plots determine the solutions of these equations. With the choice of width and depth of the potential used in Fig. 3.6, there are two intersections in the first plot and one intersection in the second plot, that is, there are three possible solutions altogether. In general, the number of solutions, or the number of bound states, depends on the value of the parameter $2mL^2V_0/\hbar^2$. If $2mL^2V_0/\hbar^2 < \pi^2/4$, there is only one bound state; if $\pi^2/4 < 2mL^2V_0/\hbar^2 < \pi^2$, there are two bound states; if $\pi^2 < 2mL^2V_0/\hbar^2 < 9\pi^2/4$, there are three bound states; and so on.

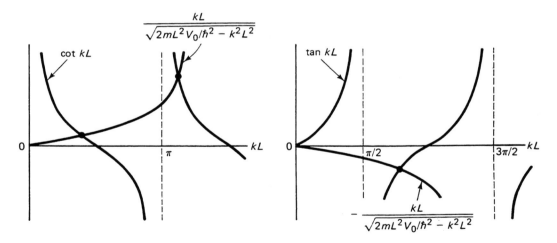

Fig. 3.6 (a) Plots of $\cot kL$ and of $kL/\sqrt{2mL^2V_0/\hbar^2 - k^2L^2}$ vs. kL; (b) plots of $\tan kL$ and of $-kL/\sqrt{2mL^2V_0/\hbar^2 - k^2L^2}$ vs. kL. In these plots, a value $2mL^2V_0/\hbar^2 = 1.5\pi^2$ has been assumed.

Exercise 10. Show that Eq. (91) has N solutions if

$$\pi^2(N-1)^2 < \frac{2mL^2}{\hbar^2} V_0 < \pi^2 N^2 \tag{93}$$

[Hint: Plot the left side and the right side of (91) as a function of kL and count the number of intersections of the two curves.]

Exercise 11. Show that Eq. (92) has N solutions if

$$\pi^2\left(N-\frac{1}{2}\right)^2 < \frac{2mL^2}{\hbar^2} V_0 < \pi^2\left(N+\frac{1}{2}\right)^2 \tag{94}$$

Figure 3.7 gives a plot of the wavefunctions for the ground state and the first and second excited states. Note that according to these plots, the particle has a finite probability to be found in the regions $x > L$ and $x < -L$, beyond the edges of the potential well. Classically, the points $x = \pm L$ are turning points of the motion for a bound particle, and the regions beyond these points are

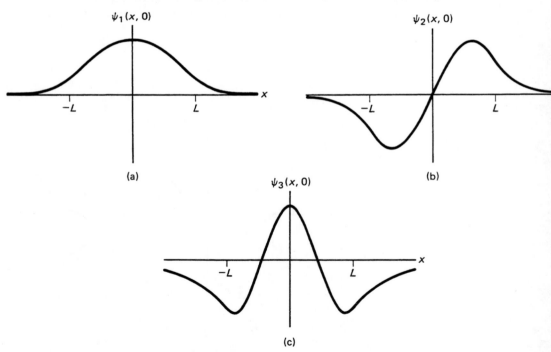

Fig. 3.7   The wavefunctions $\psi(x)$ for the ground state and the first and second excited states of the finite square well with $2mL^2V_0/\hbar^2 = 1.5\pi^2$.

forbidden regions. We can see this by examining the kinetic energy; since the potential is zero in these regions, the kinetic energy equals the total energy, $K = E = -|E|$. This means that if we ever find the particle in this region, we will find it with a negative kinetic energy! However, it turns out that Heisenberg's uncertainty relation saves us from this disaster. To see what role uncertainties play in the measurement, note that according to Eq. (83), the probability density of the wavefunction decreases by a factor e^{-1} when the distance x increases by $\hbar/2\sqrt{2m|E|}$. Because of this rapid decrease of the wavefunction, we can regard $\hbar/2\sqrt{2m|E|}$ as the typical penetration distance of the particle into the forbidden region. If we want to detect the particle in this region, we must therefore measure its position with an uncertainty less than the penetration distance:

$$\Delta x < \frac{\hbar}{2\sqrt{2m|E|}} \tag{95}$$

By the Heisenberg relation, the uncertainty in the momentum will then be at least

$$\Delta p \geq \frac{\hbar}{2\,\Delta x} > \sqrt{2m|E|} \tag{96}$$

The uncertainty of kinetic energy associated with this uncertainty in momentum is

$$\Delta K = \frac{(\Delta p)^2}{2m} > |E| \tag{97}$$

From this we see that after the position measurement, the uncertainty in the kinetic energy is larger than the magnitude of the nominal negative kinetic energy. Hence the negative kinetic energy is hidden by the uncertainty; the negative kinetic energy is an unobservable, virtual kinetic energy.

In the above calculations, we have dealt only with the solutions with negative energies. Of course, the Schrödinger equation for the finite square well has not only solutions with negative energies, but also solutions with positive energies. The latter are unbound states. We can think of these unbound states as the free-particle states modified by the potential; the possible values of the energies for these unbound states are not subject to any quantization condition.[7]

---

[7] Note that for every positive value of the energy, there are two independent unbound states, of positive and of negative momentum.

The bound states in conjunction with the unbound states form a complete set of states, that is, an arbitrary function of x can be expressed as a superposition of the bound states and the unbound states. Since the unbound states are characterized by a continuous parameter (the energy or, better, the momentum), the superposition of these states actually involves an integral, akin to the Fourier integral. Thus, for any arbitrary function of x,

$$\psi(x) = \sum_n a_n\psi_n(x) + \frac{1}{\sqrt{2\pi\hbar}} \int \phi(p)\psi_p(x)\, dp \tag{98}$$

where the eigenfunctions ψ_n are the solutions for the bound states and the eigenfunctions ψ_p are the solutions for the unbound states. Likewise, all our formulas in Section 3.3 for the expectation values, the completeness relation, and so on, now involve both a summation over the discrete set of bound states and an integration over the continuous set of unbound states. However, for the sake of notational convenience, we will usually write all such formulas with summations only, and pretend that the integration over unbound states is included in the summation.

3.5 *Barrier Penetration*

A potential barrier is a region of high potential energy. Figure 3.8 shows a square potential barrier of height V_0 extending from $x = -L$ to $x = +L$ on the x axis. The potential is

$$V(x) = \begin{cases} V_0 & \text{for } |x| < L \\ 0 & \text{for } |x| > L \end{cases}$$

The energy level of a particle approaching this barrier from, say, the left may be either above or below the height of the barrier. In

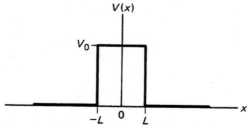

Fig. 3.8 A square potential barrier.

either case, the particle is in an unbound state, and the energy is not restricted by any quantization condition. For a classical particle, the outcome of the encounter with the barrier is entirely determined by the value of the energy. If $E > V_0$, the particle will always pass beyond the barrier, but if $E < V_0$, the particle will always be reflected at the barrier. If $E < V_0$, the point $x = -L$ is a turning point for the motion of a classical particle, and the region $-L < x < L$ is forbidden. However, as we saw in the preceding section, a quantum-mechanical particle can move beyond the classical turning point and penetrate the forbidden region. After traversing this region, it can then emerge on the far side. This is sometimes called *tunneling* through the barrier.

In the region $x < -L$, the wavefunction is a superposition of an incident wave traveling toward the right and a reflected wave traveling toward the left. We assume that the amplitude of the incident wave is $A = 1$, and we designate the amplitude of the reflected wave by R. The time-independent part of the wavefunction is then

$$\psi(x) = e^{ikx} + Re^{-ikx} \qquad \text{for } x < -L \tag{99}$$

where $k = \sqrt{2mE}/\hbar$.

In the region $x > L$, the wavefunction consists of a transmitted wave, traveling toward the right. We designate the amplitude of this transmitted wave by T:

$$\psi(x) = Te^{ikx} \qquad \text{for } x > L \tag{100}$$

In the region $-L < x < L$, within the barrier, the time-independent Schrödinger equation is

$$-\frac{\hbar^2}{2m}\frac{d^2}{dx^2}\psi(x) = -(V_0 - E)\psi(x) \tag{101}$$

If $E < V_0$, the solutions of this equation are increasing or decreasing exponentials, $e^{\pm\sqrt{2m(V_0-E)}\,x/\hbar}$. For convenience, we write these solutions as $e^{\pm\kappa x}$, where $\kappa = \sqrt{2m(V_0 - E)}/\hbar$. In contrast to the case of the square well, where only the decreasing exponential was acceptable, now both kinds of exponential are acceptable. Thus, the wavefunction is a superposition, with coefficients C and D:

$$\psi = Ce^{-\kappa x} + De^{\kappa x} \qquad \text{for } -L < x < L \tag{102}$$

The coefficients R, C, D, and T will have to be evaluated by means of the boundary conditions for ψ and $d\psi/dx$. At $x = -L$, these boundary conditions read

$$e^{-ikL} + Re^{ikL} = Ce^{\kappa L} + De^{-\kappa L} \tag{103}$$

$$ike^{-ikL} - ikRe^{ikL} = -\kappa Ce^{\kappa L} + \kappa De^{-\kappa L} \tag{104}$$

and at $x = L$,

$$Ce^{-\kappa L} + De^{\kappa L} = Te^{ikL} \tag{105}$$

$$-\kappa Ce^{-\kappa L} + \kappa De^{\kappa L} = ikTe^{ikL} \tag{106}$$

Equations (103)–(106) are four equations for the four unknown amplitudes R, C, D, and T. With some manipulation, the solution of these equations yields

$$T = \frac{-4ik\kappa e^{-2\kappa L}e^{-2ikL}}{(\kappa - ik)^2 - (\kappa + ik)^2\, e^{-4\kappa L}} \tag{107}$$

This is the amplitude of the transmitted wave.

Exercise 12. Derive the result (107) from Eqs. (103)–(106).

Similar expressions can be obtained for the other amplitudes. Figure 3.9 shows the real part of the wavefunction ($Re\ \psi$) at one instant of time. Note that the amplitude within the barrier decreases from left to right, that is, the decreasing exponential $e^{-\kappa x}$ dominates over the increasing exponential $e^{\kappa x}$.

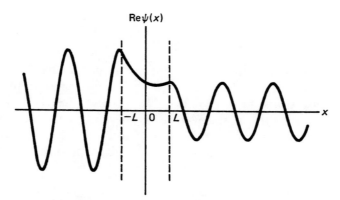

Fig. 3.9   The real part of the wavefunction for the square potential barrier at one instant of time.

The probability for finding a particle in an interval dx is $|e^{-ikx}|^2 \, dx = 1 \times dx$ for the incident wave, and it is $|Te^{-ikx}|^2 \, dx = |T|^2 \, dx$ for the transmitted wave. Hence $|T|^2$ tells us the factor by which the probability for the transmitted wave is reduced compared with the incident wave, and thus $|T|^2$ is the probability for transmission:

$$P = |T|^2 = \frac{16k^2\kappa^2}{|(\kappa - ik)^2 - (\kappa + ik)^2 e^{-4\kappa L}|^2} \, e^{-4\kappa L} \tag{108}$$

If the barrier is fairly thick, then $e^{-4\kappa L} \ll 1$, and the second term in the denominator can be neglected compared with the first, which leads to

$$P \simeq \frac{16k^2\kappa^2}{|\kappa - ik|^4} \, e^{-4\kappa L} = \frac{16k^2\kappa^2}{(k^2 + \kappa^2)^2} \, e^{-4\kappa L} \tag{109}$$

With $k = \sqrt{2mE}/\hbar$ and $\kappa = \sqrt{2m(V_0 - E)}/\hbar$, this reduces to

$$P \simeq 16 \frac{E}{V_0} \left(1 - \frac{E}{V_0}\right) e^{-4\kappa L} \tag{110}$$

Figure 3.10 is a plot of the probability for transmission through a square barrier as a function of the energy E of the incident particle.

From our result for a square barrier, we can deduce an approximate formula for the probability for transmission through a barrier of arbitrary shape. We can regard such a barrier as a succession of adjacent thin square barriers of width $\Delta x_i = 2L$ (see Fig. 3.11). For a thin barrier, we can approximate $e^{-4\kappa L} \simeq 1$ in the *denominator* of Eq. (108). This denominator then reduces to

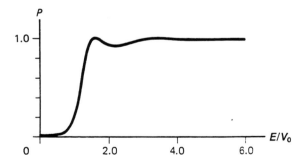

Fig. 3.10 Probability of transmission through a square barrier vs. E. This plot includes both energies smaller than V_0 and energies larger than V_0. For this plot, a value $2mL^2V_0/\hbar^2 = 4$ has been assumed.

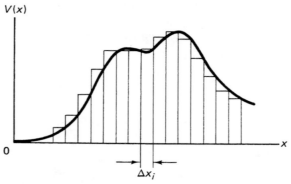

Fig. 3.11   A barrier of arbitrary shape can be approximates as a "stack" of adjacent thin barrier of with Δx_i.

$$|(\kappa - ik)^2 - (\kappa + ik)^2|^2 = 16k^2\kappa^2 \tag{111}$$

and Eq. (108) reduces to $P \simeq e^{-4\kappa L}$. Thus, the transmission probabilities for the individual thin barriers in Fig. 3.11 are $P_i \simeq e^{-2\kappa \Delta x_i}$. According to the usual rule for combining the probabilities of successive independent events, the overall probability for transmission through the entire barrier is then

$$P = P_1 P_2 P_2 \cdots \simeq e^{-2 \sum \kappa \Delta x_i} \tag{112}$$

If the intervals Δx_i are small, we can replace the sum by an integral, and we obtain

$$P \simeq e^{-2 \int \kappa dx} \tag{113}$$

or, with $\kappa = \sqrt{2m(V_0 - E)}/\hbar$,

$$P \simeq e^{-2 \int \sqrt{2m(V_0-E)}/\hbar \, dx} \tag{114}$$

where the integration extends through the barrier, from one classical turning point to the other.

Equation (114) is only an approximation. The probability for transmission through the succession of adjacent thin barriers in Fig. 3.11 is not simply the product of the individual probabilities, because a particle attempting to pass through this succession of barriers may suffer multiple reflections that send it back and forth between several of these barriers. Our simple calculation ignores these complications; our calculation assumes that if the particle is reflected once, by one of the thin barriers, it has not further chance of penetrating through the whole barrier. However, a more careful calculation, based on an approximate solution of the Schrödinger equation, shows that Eq. (114) is a good approximation whenever the change of potential is gradual, that is, whenever the fractional change in the potential is small within a distance $1/\kappa$.

Equation (114) can be applied to the theory of alpha decay of a nucleus. This is a case of barrier penetration. Before the decay, the alpha particle is confined within the nucleus by a high potential barrier. The alpha particle bounces back and forth within the nucleus, striking the barrier repeatedly, until it succeeds in penetrating through it. The alpha particle finally emerges from the barrier with an energy much below that required for passing over it (typical energies of alpha particles produced in nuclear decay are a few MeV, in contrast to the 30 or 40 MeV required for passing over the top of the barrier of a heavy nucleus).

Figure 3.12 shows the potential barrier holding the alpha particle in the nucleus. The sloping right side of this barrier is simply the Coulomb potential energy corresponding to the repulsive electric force between the alpha particle (charge $2e$) and the nucleus (charge Ze; this is the charge remaining in the nucleus *after* the alpha particle has been emitted). In SI units, this Coulomb potential energy is

$$V(r) = \frac{1}{4\pi\varepsilon_0} \frac{2Ze^2}{r} \tag{115}$$

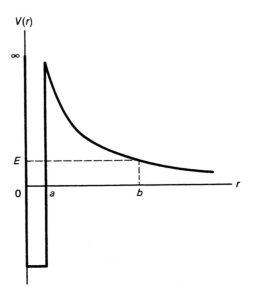

Fig. 3.12 Potential barrier for an alpha particle in and near a nucleus.

The steep left side of the barrier arises from the strongly attractive nuclear binding force. The dashed line shows a typical energy for an alpha particle. Obviously, the left side of the barrier (or the nuclear radius $r = a$) is a classical turning point.

To escape from the nucleus, the alpha particle must tunnel through the barrier. In the calculation of the probability for tunneling, we can treat the radial motion of the alpha particle as one-dimensional motion, equivalent to motion along the x axis. Equation (114) then gives us a probability

$$P \simeq \exp\left[-\frac{2}{\hbar}\int_a^b \sqrt{2m_\alpha\left(\frac{2Ze^2}{4\pi\varepsilon_0 r} - E\right)}\, dr\right] \qquad (116)$$

At the point b, the Coulomb potential energy is equal to the total energy,

$$\frac{2Ze^2}{4\pi\varepsilon_0 b} = E \qquad (117)$$

We can then rewrite the integral in (116) as

$$P \simeq \exp\left[-\frac{2}{\hbar}\sqrt{2m_\alpha E}\int_a^b \sqrt{\frac{b}{r} - 1}\, dr\right] \qquad (118)$$

For the evaluation of this integral, it is convenient to change the variable of integration to an angular variable θ defined by

$$r = b\,\sin^2\theta \qquad (119)$$

This leads to

$$P \simeq \exp\left[-\frac{2}{\hbar}\sqrt{2m_\alpha E}\, b\int_{\theta_a}^{\theta_b} 2\cos^2\theta\, d\theta\right]$$

$$= \exp\left[-\frac{2}{\hbar}\sqrt{2m_\alpha E}\, b\left(\theta + \frac{1}{2}\sin 2\theta\right)\Big|_{\theta_a}^{\theta_b}\right] \qquad (120)$$

In this expression, the upper limit of the integral is $\theta_b = \pi/2$ and the lower limit is $\theta_a = \sin^{-1}\sqrt{a/b}$. The distance b at which the alpha particle emerges from the barrier is much larger than the nuclear radius a; thus, $b \gg a$, and $\theta_a + \frac{1}{2}\sin 2\theta_a \simeq 2\theta_a \simeq 2\sqrt{a/b}$. This leads to

$$P \simeq \exp\left[-\frac{2}{\hbar}\sqrt{2m_\alpha E}\, b\,\frac{\pi}{2} + \frac{4}{\hbar}\sqrt{2m_\alpha E}\,\sqrt{ab}\right]$$

$$\simeq \exp\left[-\frac{4\pi}{\hbar}\frac{Ze^2}{4\pi\varepsilon_0}\frac{1}{v_\alpha} + \frac{4}{\hbar}\sqrt{\frac{4Ze^2 m_\alpha a}{4\pi\varepsilon_0}}\right] \qquad (121)$$

Here, v_α is the speed of the alpha particle, $v_\alpha = \sqrt{2E/m_\alpha}$.
The first factor in Eq. (121),

$$\exp\left[-\frac{4\pi}{\hbar} \frac{Ze^2}{4\pi\varepsilon_0} \frac{1}{v_\alpha} \right] \tag{122}$$

is called the *Gamow factor;* it determines the dependence of the
tunneling probability on the speed of the emitted alpha particle.
The half-life of a nucleus is inversely proportional to the tunneling
probability P. For nuclei of the same, or nearly the same, radius a
and atomic number Z, the second factor in Eq. (121) is nearly
constant, and the Gamow factor then gives a prediction for the
dependence of the half-life on the speed of the emitted alpha
particles:

$$\ln T_{1/2} \propto -\ln P \propto \frac{1}{v_\alpha} \tag{123}$$

This prediction is in good agreement with the observed half-lives
of isotopes of heavy nuclei.

PROBLEMS   **1.** A particle of mass m moving in one dimension is subject to the following potential (see Fig. 3.13):

$$V(x) = \begin{cases} 0 & \text{for } x > 0 \\ \infty & \text{for } x < 0 \end{cases}$$

Find the wavefunctions of the stationary states and the energy eigenvalues.

Fig. 3.13

2. A particle of mass m, energy E is moving in the potential shown in
Fig. 3.14. The energy E, indicated by the dashed line in Fig. 3.14, is
less than the height V_0 of the potential step. Find the solution of the
Schrödinger equation in the regions $x < 0$ and $x > 0$.

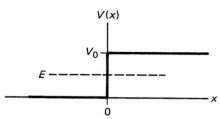

Fig. 3.14

3. Find the Fourier transform of the wavefunction $\psi_1(x) = \sqrt{2/L} \sin \pi x/L$ for the ground state of the infinite square well. What is the probability that a measurement of momentum yields a value between p and $p + dp$?

4. The potential for a particle of mass m moving in one dimension is (see Fig. 3.15)

$$V(x) = \begin{cases} \infty & \text{for } x < 0 \\ 0 & \text{for } 0 < x < L \\ V_0 & \text{for } L < x < 2L \\ \infty & \text{for } x > 2L \end{cases}$$

Assume that the energy of the particle is in the range $0 < E < V_0$. Find the energy eigenfunctions and the equation that determines the energy eigenvalues. You need not normalize the eigenfunctions.

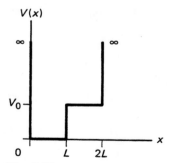

Fig. 3.15

5. A particle of mass m is within an infinite square-well potential of width L. Its wavefunction at $t = 0$ is

$$\psi(x, 0) = \frac{1}{\sqrt{3}} \sqrt{\frac{2}{L}} \sin \frac{2\pi x}{L} + \sqrt{\frac{2}{3}} \sqrt{\frac{2}{L}} \sin \frac{3\pi x}{L}$$

(a) Is this a stationary state?

(b) What will be the wavefunction at a later time t?

(c) What is the probability that a measurement of the energy at time t yields the value
$\hbar^2\pi^2/2mL^2$? $4\hbar^2\pi^2/2mL^2$? $9\hbar^2\pi^2/2mL^2$? $16\hbar^2\pi^2/2mL^2$?

(d) What is the expectation value of x at the time t?

(e) What is the expectation value of p at the time t?

6. For the ground state and the first excited state of the infinite square-well potential, evaluate Δx and Δp exactly, according to their definitions as rms deviations from the mean. What is the product $\Delta x\ \Delta p$ in each case?

7. Suppose that a particle in the infinite square well has the initial wavefunction

$$\psi(x) = \frac{\sqrt{30}}{L^{5/2}}\,(x^2 - Lx)$$

(a) What are the expectation values of the position and of the energy for this particle?

(b) What is the uncertainty Δx in the position of the particle?

8. Figure 3.16 shows a square well with the potential

$$V(x) = \begin{cases} \infty & \text{for } x < 0 \\ 0 & \text{for } 0 < x < L \\ V_0 & \text{for } x > L \end{cases}$$

For this well, what is the equation that determines the energy eigenvalues? Show that if the parameter $2mL^2V_0/\hbar^2$ is smaller than a critical magnitude, there are no bound states. What is this critical magnitude?

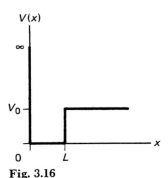

Fig. 3.16

9. A particle in an infinitely deep square-well potential of width L is in the first excited state. What are the x representation, the p representation, and the energy representation for this state?

10. Consider a finite square well (see Fig. 3.5) with $2mL^2V_0/\hbar^2 = 3\pi^2/4$. Find the energies of the ground state and the first and the second excited states by numerical methods (successive approximation) to three significant figures.

11. Prove that the quantization condition [Eq. (26)] for the infinite square well can be derived from the quantization condition Eq. (89) for the finite square well by taking a suitable limit $V_0 \to \infty$.

12. The energy of the ground state of a finite square well with $2mL^2V_0/\hbar^2 = \pi^2/2$ is

$$E = -\frac{3.792\hbar^2}{2mL^2}$$

Normalize the wavefunction (85) for this ground state. Calculate the probability for finding the particle outside the potential well, that is, in the region $|x| > L$. Give a numerical answer.

13. At time $t = 0$, a particle of mass m in an infinite square potential well of width L has a wavefunction .

$$\psi(x, 0) = \begin{cases} 1/\sqrt{L} & \text{for } 0 < x < L \\ 0 & \text{for } x < 0 \text{ and } x > L \end{cases}$$

(a) Find the wavefunction at a time $t > 0$.

(b) What is the probability that an energy measurement performed at some time $t > 0$ results in $9\pi^2\hbar^2/2mL^2$?

14. Consider the wavefunction given in Eq. (46).

(a) What is the mean value of the energy, or the expectation value of the energy?

(b) What is the uncertainty ΔE in the energy, defined as a rms deviation from the mean?

(c) What is the period of oscillation T of the probability density $|\psi|^2$? What is the product $T \, \Delta E$ of this period and ΔE? According to one form of the time–energy uncertainty relation (which we will examine in Chapter 5), the product $T \, \Delta E$ should be larger than $\hbar/2$. Is this the case here?

15. Consider a superposition of the ground state, the first excited state, and the third excited state of the infinite square well,

$$\psi(x, t) = a_1(0)\psi_1(x, t) + a_2(0)\psi_2(x, t) + a_3(0)\psi_3(x, t)$$

What are the frequencies found in the oscillations of the probability density? What is the periodicity of this superposition, that is, after what time interval does the probability density return to its initial value?

16. Evaluate the integral for x_{mn} in Eq. (64).

17. Evaluate the integral for p_{mn} in Eq. (72).

18. Consider a particle of mass m in the infinite square potential well of width L. At $t = 0$, the wavefunction of the particle is

$$\psi(x, 0) = \frac{\sqrt{3}}{2} \psi_1(x) - \frac{1}{2} \psi_3(x)$$

where ψ_1 and ψ_2 are the wavefunctions for the ground state and the second excited state.

   (a) What is the probability that the particle be found in an interval dx and $x = L/2$ when a position measurement is performed?

   (b) Suppose that the particle is found at $x = L/2$. What is the wavefunction immediately after this position measurement?

   (c) Immediately after the position measurement, an energy measurement is performed. Calculate the value of the ratio $P(E_n)/P(E_1)$, where $P(E_n)$ is the probability for finding the energy E_n in the energy measurement. [Hint: Express $\delta(x - L/2)$ as a superposition of the energy eigenfunctions $\psi_n(x)$.]

19. Substitute the expressions (63) and (71) into Eq. (76), and show that the result is an identity.

20. Find the solution of Schrödinger equation for a particle of mass m and energy E incident from the left on the potential step shown in Fig. 3.14. Assume that $E > V_0$. Compare the probability currents in the reflected wave and in the transmitted wave with the probability current in the incident wave, and deduce the probabilities of reflection and transmission.

21. Repeat the preceding problem for a particle incident from the *right*.

22. From the equations given in Section 3.5, solve for the coefficient R representing the amplitude of the wave reflected by the square potential barrier. Verify that $|R|^2 + |T|^2 = 1$. What is the physical interpretation of this equation?

23. A particle of energy E and mass m is incident (from the left) on the square potential barrier shown in Fig. 3.8. Suppose that the energy is such that $E > V_0$.

   (a) Write down the solution of Schrödinger's equation (including the time dependence) for each of the regions $x < -L$, $-L < x < L$, and $x > L$. The solution should contain several unknown constants. Be sure to express all the parameters of the solution in terms of E, m, V_0, and L.

   (b) State the boundary conditions that apply to this problem and use them to obtain a sufficient number of equations so all the unknown constants can be determined.

(c) Solve for the transmission amplitude.

(d) Show that the probability of transmission equals 1 when $2k'L = n\pi$, where $k' = \sqrt{2m(E - V_0)}/\hbar$ is the wave number in the region of the barrier. Compare the condition $2k'L = n\pi$ with the condition for energy eigenstates in an infinitely deep potential well of width $2L$.

(e) Show that the probability of transmission approaches 1 for $E \gg V_0$.

24. A particle of *positive* energy approaches the square well illustrated in Fig. 3.5 from the left. Some of the wave will be transmitted, and some will be reflected.

(a) Calculate the transmission amplitude.

(b) Show that the transmission amplitude equals 1 when $2k'L = n\pi$, where $k' = \sqrt{2m(E + V_0)}/\hbar$ is the wave number in the region of the well. Thus, the potential well becomes transparent at a discrete set of energies; this is called the *Ramsauer effect*.

(c) What is the transmission probability if the energy of the particle is much larger than V_0?

25. Suppose that you bombard a sample of uranium with alpha particles. According to the calculations of Section 3.5, what is the probability that an alpha particle of 6.0 MeV succeeds in penetrating into a uranium nucleus in a head-on collision? For uranium, $Z = 92$ and $a = 7.4 \times 10^{-15}$ m.

26. Use the approximate formula given in Section 3.5 to calculate the transmission probability for a particle of energy E incident on the potential barrier:

$$V(x) = \begin{cases} 0 & \text{for } x < 0 \\ A - Bx & \text{for } 0 < x < A/B \\ 0 & \text{for } x > A/B \end{cases}$$

Assume that $E < A$.

27. Use the approximate formula given in Section 3.5 to calculate the transmission probability for a particle of energy E incident on the potential barrier:

$$V(x) = \begin{cases} 0 & \text{for } |x| > \sqrt{A/B} \\ A - Bx^2 & \text{for } |x| < \sqrt{A/B} \end{cases}$$

Assume that $E < A$.

4

Axiomatic Formulation of Quantum Mechanics

Whereas in the preceding chapters we tried to make quantum mechanics plausible, and we tried to justify the use of wavefunctions, momentum operators, energy operators, and so on by physical arguments, in the present chapter we adopt an axiomatic approach, that is, we lay down a set of axioms and study their consequences. These axioms will be justified a posteriori, by the success of the theory. The advantage of this approach is that we can better appreciate the logical and mathematical structure of quantum mechanics, and we can better see what is and what is not essential. Of course, the motivation for our choices of axioms arises from the physical arguments of the preceding chapters, and we will occasionally refer to these chapters for guidance on how to proceed with the axiomatic construction.

In the logical structure of quantum mechanics, the axioms play a role similar to that of Newton's laws in classical mechanics. However, several of the axioms of quantum mechanics are concerned with just how the "state" of a system is to be defined, whereas Newton did not bother to postulate explicitly that the "state" of a classical particle is defined by giving position and velocity. In contrast to the definition of the state in classical mechanics, the definition of the state of a system in quantum mechanics is far from trivial.

4.1 *Vector Spaces*

Before we set down the axioms of quantum mechanics, we need to introduce the abstract concept of a state vector. We will use the Dirac *ket notation* $|\alpha\rangle$ for a vector. The symbol $|\ \rangle$ is called a "ket" because it is the second half of a bracket, $\langle\ \rangle$. The label α in the "ket" | ⟩ will serve to distinguish different vectors. The ket symbol | ⟩ is here used in much the same way as the arrow → is used in elementary vector analysis; we could, but will not, use $\vec{\alpha}$ instead of |α⟩. The abstract definition of vectors given by linear algebra is the following: *A vector is a mathematical object for which there exists an operation of vector addition and an operation of multiplication by a real or complex number such that*

(i) $|\alpha\rangle + |\beta\rangle = |\beta\rangle + |\alpha\rangle$ (commutative law for addition)

(ii) $|\alpha\rangle + (|\beta\rangle + |\gamma\rangle) = (|\alpha\rangle + |\beta\rangle) + |\gamma\rangle$ (associative law for addition)

(iii) There exists a "zero vector" such that its addition to any vector leaves that vector unchanged, $|\alpha\rangle + 0 = |\alpha\rangle$.

(iv) For any vector $|\alpha\rangle$ there exists an *inverse vector* $-|\alpha\rangle$ such that $|\alpha\rangle + (-|\alpha\rangle) = 0$.

(v) $1\,|\alpha\rangle = |\alpha\rangle$

(vi) $(ab)|\alpha\rangle = a(b|\alpha\rangle)$, where a and b are any real or complex numbers   (associative law for multiplication)

(vii) $a(|\alpha\rangle + |\beta\rangle) = a|\alpha\rangle + a|\beta\rangle$ (distributive law for multiplication)

(viii) $(a + b)|\alpha\rangle = a|\alpha\rangle + b|\alpha\rangle$ (distributive law for multiplication)

A set of vectors with operations of addition and multiplication satisfying the conditions (i)–(viii) is called a *vector space*.

We have defined vectors by their properties. Such a definition is much more general (and useful) than to assert, for instance, a vector is a triplet of numbers. Obviously, any set of n-tuples with the usual rules for addition and multiplication by a scalar will form a vector space. The set of all functions of x also forms a vector space. Each function $\psi(x)$ can be regarded as a vector, the functions can be added, multiplied by scalars, etc. This space has all of the foregoing properties in common with the space of n-tuples; that is why it is useful to regard both as vector spaces. We now recognize that the different representations, given in Chapter 3, of the state of a particle by a function $\psi(x)$, or $\phi(p)$, or an n-tuple (with $n = \infty$ for the case of our column vectors in the energy representation), all have this in common: They all represent the state of the system by a *vector*.

In an abstract vector space we can, again abstractly, define a *scalar product* of two vectors. The scalar product (or inner product) of two vectors $|\alpha\rangle$ and $|\beta\rangle$ is a real or complex number. We will use the bracket notation $\langle\alpha|\beta\rangle$ for the scalar product; this is the analog of $\vec{\alpha} \cdot \vec{\beta}$ in elementary vector analysis. The symbol $\langle \; | \; \rangle$ is called Dirac's bracket. For our purposes, this bracket plays the same role as a multiplication sign for the vectors. (It is also possible to give a separate meaning to the front half $\langle \; |$ of the bracket, which is then called a "bra," or a dual vector; but we will have no occasion to use bras by themselves.) The scalar product has the following properties:

(i) $\langle\alpha|(|\beta\rangle + |\gamma\rangle) = \langle\alpha|\beta\rangle + \langle\alpha|\gamma\rangle$

(ii) $\langle\alpha|(a|\beta\rangle) = a\langle\alpha|\beta\rangle$

(iii) $\langle\alpha|\beta\rangle = \langle\beta|\alpha\rangle^*$

(iv) $\langle\alpha|\alpha\rangle \geq 0$, and $\langle\alpha|\alpha\rangle = 0$ if and only if $|\alpha\rangle = 0$

The notation in (i) and (ii) is somewhat clumsy. The left side of (i) stands for the scalar product of $|\alpha\rangle$ and $|\beta\rangle + |\gamma\rangle$. The left side of (ii) stands for the scalar product of $|\alpha\rangle$ and $a|\beta\rangle$.

Exercise 1. Show that $(\langle\alpha|a) \; |\beta\rangle = a^*\langle\alpha|\beta\rangle$. (The left side stands for the scalar product of $a|\alpha\rangle$ and $|\beta\rangle$.)

In the space of *n*-tuples of complex numbers, the scalar product of two vectors a_k and b_k has the familiar form of a sum of products of the components:

$$\sum_{k=1}^{n} a_k^* b_k$$

In the space of functions of x, the scalar product of two functions $\psi(x)$ and $\chi(x)$ is an integral:

$$\int_{-\infty}^{\infty} \chi^*(x)\psi(x) \; dx$$

We have seen examples of such integrals in Chapter 3, for instance in Eq. (3.35).

In terms of the scalar product, we can define the *magnitude*, or the *norm*, of a vector $|\alpha\rangle$ as $(\langle\alpha|\alpha\rangle)^{1/2}$.

Note that we have not found it necessary to say anything about components of vectors; in general, vectors are just abstract "objects." Of course, as soon as we want to do a numerical calculation, we will need to introduce the components of the vector with respect to some basis and give numerical values, but general theorems on vectors can be proved without using components.

Exercise 2. Prove the Schwartz inequality:

$$\langle \alpha | \alpha \rangle \langle \beta | \beta \rangle \geq |\langle \alpha | \beta \rangle|^2 \tag{1}$$

(Hint: Start with $\langle \gamma | \gamma \rangle \geq 0$, where

$$|\gamma\rangle = |\beta\rangle - \frac{\langle\beta|\alpha\rangle}{\langle\alpha|\alpha\rangle}|\alpha\rangle)$$

We need another concept, that of a *Hilbert space*. A Hilbert space is (i) a vector space in which (ii) there is defined a scalar product and a norm in terms of this scalar product, and (iii) the vector space is *complete*. In brief, a Hilbert space is a normed complete vector space. The completeness requirement will be of little concern to us, although it would be crucial if we wanted to give rigorous proofs of theorems about vectors in Hilbert space. This requirement amounts to the following: Suppose that we have an infinite sequence of vectors $|\alpha_1\rangle, |\alpha_2\rangle, \ldots, |\alpha_n\rangle, \ldots$ such that the magnitude of $|\alpha_n\rangle - |\alpha_m\rangle$ approaches zero as $n, \ m \to \infty$.[1] Then there exists a limit for the sequence, that is, there exists a vector $|\alpha\rangle$ such that the magnitude of $|\alpha\rangle - |\alpha_n\rangle$ approaches zero as $n \to \infty$.

The space of n-tuples with

$$\sum_{k=1}^{n} |a_k|^2 = 1$$

is a Hilbert space. This also remains true for $n = \infty$. Furthermore, the space of square-integrable functions, that is, functions with $\int |\psi(x)|^2 \, dx < \infty$, is a Hilbert space.

4.2   *The Axioms of Quantum Mechanics*

The connection between all these purely mathematical concepts and physics is given by the following three fundamental axioms of quantum mechanics:

---

[1] Such a sequence is called a Cauchy sequence.

I. *To each state of a physical system there corresponds a "state vector" in a quantum-mechanical Hilbert space $\mathcal{H}$. The state vector has unit magnitude. Thus,*

$$\text{physical state} \leftrightarrows \text{state vector } |\alpha\rangle \qquad (2)$$

$$\langle \alpha | \alpha \rangle = 1 \qquad (3)$$

For example, suppose that the physical state consists of an electron in the ground state of an infinite square well. Using a convenient, self-explanatory notation, we can designate the corresponding state vector by $|\text{ground}\rangle$. Similarly, if the electron is in the first excited state, we can designate its state vector by $|\text{first excited}\rangle$.[2] The significance of axiom I then lies in this: Given that $|\text{ground}\rangle$ and $|\text{first excited}\rangle$ are possible state vectors, it follows that $a|\text{ground}\rangle + b|\text{first excited}\rangle$ is also a possible state vector. The (complex) coefficients a and b are arbitrary, except for the normalization requirement $|a|^2 + |b|^2 = 1$. The physical interpretation of such a superposition of state vectors will emerge with axiom II. Note that in this example we are treating the state vectors in the quantum-mechanical Hilbert space $\mathcal{H}$ abstractly. We are not saying anything about components or numerical values of vectors.

Axiom I applies not only to physical systems of one particle, but also to systems of two or more particles. For example, if we have a two-particle system consisting of one electron with some given momentum and one proton with some given momentum, then there is a corresponding state vector $|\text{electron;proton}\rangle$. It is obvious that the dimension of the quantum-mechanical Hilbert space $\mathcal{H}$ must be infinite, since it must contain a state vector for every conceivable state of every conceivable physical system.

Here, an excuse becomes necessary. We know that states of definite momentum are not normalizable. Strictly speaking, such states are therefore not permitted by axiom I. The excuse for nevertheless using such states in the preceding (and many following) examples is that this normalization problem can be handled by a formal trick (see Section 4.3). It is then possible to pretend that the momentum states are "good" states.

There are two difficulties with our formulation of axiom I. First, for a given physical state the state vector is not uniquely determined, since, if $|\alpha\rangle$ is an acceptable vector, so is any vector $e^{i\theta} |\alpha\rangle$ which differs from the former only by a phase factor. To

---

[2] A more elegant notation can be devised. But we want to emphasize that it is irrelevant what labels are used inside the ket $|\ \rangle$ as long as they are unambiguous.

avoid this ambiguity we will always make some particular choice of phase.

The second difficulty is more serious. Suppose that $|\text{electron}\rangle$ is the state vector for an electron of some given momentum and that $|\text{proton}\rangle$ is the state vector for a proton of some given momentum. Then the superposition $a|\text{electron}\rangle + b|\text{proton}\rangle$ is presumably a possible state vector.[3] However, for technical reasons having to do with some fundamental symmetries obeyed by quantum-mechanical states, such a superposition of electron and proton states never seems to occur in nature. The rule that the superposition of electron and proton states is forbidden is a special case of a general rule that forbids the superposition of states of different electric charge. This is called the *superselection rule* for electric charge. Other superselection rules that nature seems to obey are those for baryon number, lepton number, and statistics which assert, respectively, that superpositions of states of different numbers of baryons, different numbers of leptons, or different statistics (Bose–Einstein or Fermi–Dirac) do not occur. For the purpose of axiom I it is therefore necessary to keep in mind that some of the vectors in $\mathcal{H}$ do not correspond to any physically realizable state.

The next axiom provides the connection between the abstract states and the probabilities measured in experiments:

> II. *If the state of the system is $|\alpha\rangle$, then the probability that a measurement finds the system in a state $|\beta\rangle$ is $|\langle\beta|\alpha\rangle|^2$.*

In this context, to "observe" means to carry out whatever measurements are required to decide whether the system is in the state $|\beta\rangle$. We now can give the physical interpretation of a superposition of state vectors. Suppose that our system consists of an electron in the state[4] $a|\text{ground}\rangle + b|\text{first excited}\rangle$, as in Eq. (3). According to axiom II, the probability for the electron to be found in the ground state is then

$$|\langle\text{ground}|(a|\text{ground}\rangle + b|\text{first excited}\rangle)|^2 \qquad (4)$$

---

[3] Note the distinction between this state vector and $|\text{electron;proton}\rangle$. The latter vector describes a system of two particles, one electron *and* one proton. The former vector describes a system consisting of a single particle, one electron *or* one proton.

[4] For the sake of brevity we say "the system is in the state $|\alpha\rangle$" or "the state of the system is $|\alpha\rangle$" instead of the more accurate "the system is in the state corresponding to the state vector $|\alpha\rangle$."

If we anticipate the result that the scalar product ⟨ground|first excited⟩ is zero (or remember this result from the preceding chapter), then Eq. (4) simplifies to

$$|\langle \text{ground}|(a|\text{ground}\rangle)\rangle|^2 = |a\langle \text{ground}|\text{ground}\rangle|^2 = |a|^2 \qquad (5)$$

Likewise, the probability for the electron to be found in the first excited state is $|b|^2$. This supplies us with the physical interpretation of the coefficients a and b in the superposition; their absolute squares are probabilities.

It is important to note that axiom II applies no matter what the states $|\alpha\rangle$ and $|\beta\rangle$ are. Thus, suppose that $|x\rangle$ describes an electron localized at a definite position x and $|p\rangle$ describes an electron of some given momentum p, then $|\langle x|\text{ground}\rangle|^2$ gives the probability for an electron that is in the ground state to be at x; $|\langle p|\text{ground}\rangle|^2$ gives the probability for an electron that is in the ground state to have momentum p; and $|\langle x|p\rangle|^2$ gives the probability for an electron that is in a state of momentum p to be found at x.

One more remark: The probability given by axiom II is symmetric in $|\alpha\rangle$ and $|\beta\rangle$; that is, $|\langle \beta|\alpha\rangle|^2 = |\langle \alpha|\beta\rangle|^2$. This means that the probability that the system be found in state $|\beta\rangle$ if it is initially in state $|\alpha\rangle$ is exactly the same as the probability that the system be found in state $|\alpha\rangle$ if it is initially in state $|\beta\rangle$.

Before we set down the next axiom, we need to introduce the concept of a *physical observable*. By a physical observable is meant any numerical quantity that can be measured about a system. Examples of observables are position, momentum, energy, mass, angular momentum, spin, electric charge, magnetic dipole moment, baryon number, parity, and so on.

> III. *To every physical observable there corresponds an operator in the Hilbert space $\mathcal{H}$. This operator is linear and hermitian. The only possible results of measurement are the eigenvalues of the corresponding operator. These possible values will occur with definite probabilities that depend on the state being measured; if the state is an eigenstate of the observable, then, and only then, the result is certain to be the eigenvalue belonging to the state.*

We have already encountered hermitian operators, or hermitian matrices, in the preceding chapter, in the special case of the position and the momentum operators in the energy representation. We will deal with the general definition of a hermitian operator in the next section.

4.3 *Operators and Eigenvectors*

We begin with the general definition of an operator. An *operator* acting on a vector in a vector space transforms it into another vector,

$$Q|\alpha\rangle = |\beta\rangle \tag{6}$$

The operator is said to be *linear* if

$$Q(a|\alpha\rangle + b|\gamma\rangle) = aQ|\alpha\rangle + bQ|\gamma\rangle \tag{7}$$

for any two vectors $|\alpha\rangle$ and $|\gamma\rangle$. For example, in the space of n-tuples, the $n \times n$ matrices multiplying vectors are linear operators. Obviously, the standard row-by-column multiplication of a column vector by a matrix satisfies the conditions for a linear operation.

To lead up to the general definition of a hermitian operator, let us first define the *hermitian conjugate,* or the *adjoint,* of an operator. We say that $Q^\dagger$ is the hermitian conjugate of Q if for any vector $|\alpha\rangle$ and any vector $|\beta\rangle$ we have

$$\langle\alpha|(Q|\beta\rangle) = (\langle\alpha|Q^\dagger)|\beta\rangle \tag{8}$$

The expression on the left side represents the scalar product of $|\alpha\rangle$ and $Q|\beta\rangle$; that on the right side represents the scalar product of $Q^\dagger|\alpha\rangle$ and $|\beta\rangle$. We will use the convention that an operator "sandwiched" between two vectors always acts to the right, unless parentheses indicate otherwise.[5] Hence (8) can also be written as

$$\langle\alpha|Q|\beta\rangle = (\langle\alpha|Q^\dagger)|\beta\rangle \tag{9}$$

For example, in the space of n-tuples, a linear operator Q is a matrix Q_{mn}. Equation (8) then reads

$$\sum_{k,m} a_k^* Q_{km} b_m = \sum_{k,m} (Q_{mk}^\dagger a_k)^* b_m \tag{10}$$

This can be true for arbitrary vectors a_k and b_m only if

$$Q_{km} = (Q_{mk}^\dagger)^*$$

or

$$Q_{mk}^\dagger = (Q_{km})^* \tag{11}$$

---

[5] This convention will also apply to multiplication by a complex number. Thus, $\langle\alpha|a|\beta\rangle \equiv \langle\alpha|(a|\beta\rangle)$.

Hence the hermitian conjugate of a matrix Q_{km} is the transpose of the complex conjugate of that matrix.

Exercise 3. Show that for any operators A and B,

$$(A^\dagger)^\dagger = A \tag{12}$$

$$(A + B)^\dagger = A^\dagger + B^\dagger \tag{13}$$

$$(aA)^\dagger = a^*A^\dagger \tag{14}$$

$$(AB)^\dagger = B^\dagger A^\dagger \tag{15}$$

In the space of square-integrable functions, Eq. (8) reads

$$\int_{-\infty}^{\infty} \chi^*(x) Q \psi(x) \, dx = \int_{-\infty}^{\infty} [Q^\dagger \chi(x)]^* \psi(x) \, dx \tag{16}$$

Exercise 4. Show that in the space of square-integrable functions, the hermitian conjugate of the operator d/dx is $-d/dx$, that is,

$$\int_{-\infty}^{\infty} \chi^*(x) \frac{d}{dx} \psi(x) \, dx = -\int_{-\infty}^{\infty} \left(\frac{d}{dx} \chi(x)\right)^* \psi(x) \, dx \tag{17}$$

We now define an operator to be *hermitian* if $Q = Q^\dagger$. Note that for a matrix operator, this means that

$$(Q_{mk})^* = Q_{km} \tag{18}$$

As we already found in Chapter 3 [see Eqs. (3.68) and (3.75)], the matrices x_{mk} and p_{mk} representing the position and the momentum in the energy representation satisfy the condition (18); that is, the operators x_{op} and p_{op} are hermitian in this representation, just as they are supposed to be according to axiom III. It is easy to check that p_{op} and x_{op} are also hermitian in the position representation and in the momentum representation.

Exercise 5. Show that $(\hbar/i)d/dx$ is hermitian in the space of square-integrable functions. Show that x is also hermitian.

In general, for any operator Q, the vector $|\alpha\rangle$ is said to be an *eigenvector* of the operator if

$$Q|\alpha\rangle = q|\alpha\rangle$$

where q is a real or complex number, called the *eigenvalue.* Since the result of any measurement is necessarily some *real* number, axiom III makes sense only provided that we can show that the eigenvalues that are supposed to be the results of measurements are always real. That this is so is the content of the following theorem:

The eigenvalues of an hermitian operator are real.

The proof is easy. Suppose that $Q|\alpha\rangle = q|\alpha\rangle$, so $|\alpha\rangle$ is an eigenvector of Q with eigenvalue q. Then

$$q\langle\alpha|\alpha\rangle = \langle\alpha|(Q|\alpha\rangle) = (\langle\alpha|Q^\dagger)|\alpha\rangle = (\langle\alpha|Q)|\alpha\rangle$$

$$= (\langle\alpha|q)|\alpha\rangle = q^*\langle\alpha|\alpha\rangle \tag{19}$$

This implies that $q = q^*$, and thus q is real.

Another important theorem about the eigenvectors of hermitian operators is the following:

The eigenvectors belonging to different eigenvalues of a hermitian operator are orthogonal.

For a proof, suppose that $Q|\alpha_1\rangle = q_1|\alpha_1\rangle$ and $Q|\alpha_2\rangle = q_2|\alpha_2\rangle$. Then

$$q_1\langle\alpha_2|\alpha_1\rangle = \langle\alpha_2|(q_1|\alpha_1\rangle) = \langle\alpha_2|(Q|\alpha_1\rangle)$$

$$= (\langle\alpha_2|Q^\dagger)|\alpha_1\rangle = (\langle\alpha_2|Q)|\alpha_1\rangle = (\langle\alpha_2|q_2)|\alpha_1\rangle$$

$$= q_2^*\langle\alpha_2|\alpha_1\rangle$$

that is,

$$0 = (q_1 - q_2^*)\,\langle\alpha_2|\alpha_1\rangle$$

But we already know that the eigenvalues are real, and therefore

$$0 = (q_1 - q_2)\,\langle\alpha_2|\alpha_1\rangle$$

This implies that $\langle\alpha_2|\alpha_1\rangle = 0$ if $q_1 \neq q_2$.

Sometimes it happens that there exist several distinct eigenvectors with the *same* eigenvalue; we would like to have orthogonality even in this case. This is of particular interest in the case of eigenstates of energy, where there are often several different eigenstates of the same energy. Different eigenstates that belong to the same eigenvalue are said to be *degenerate.* The number of such degenerate eigenstates is usually called the *degeneracy* of the eigenvalue. For example, the following exercise shows that the degeneracy of the eigenvalue $6\pi^2\hbar^2/3mL^2$ is three for the case of a particle in a three-dimensional infinite square well.

Exercise 6. Show that the energy eigenvalues for a particle of mass m in a three-dimensional infinite square well of dimensions $L \times L \times L$, with a potential of zero in the region $0 < x < L$, $0 < y < L$, $0 < z < L$ and an infinite potential elsewhere, are

$$E_{n_1, n_2, n_3} = \frac{\pi^2\hbar^2}{2mL^2} (n_1{}^2 + n_2{}^2 + n_3{}^2) \qquad n_1, n_2, n_3 = 1, 2, 3, 4, \ldots \qquad (20)$$

Show that there are three states of energy $6\pi^2\hbar^2/2mL^2$.

We would like the degenerate eigenvectors to be orthogonal just as nondegenerate ones are. This is a matter of convenience. We could use eigenvectors that are not orthogonal, but we want to use eigenvectors to construct a basis in our Hilbert space $\mathcal{H}$, and a nonorthogonal basis is a nuisance. To get what we want, we can use the *Schmidt orthogonalization procedure.* Suppose that $|\alpha_1\rangle$, $|\alpha_2\rangle, |\alpha_3\rangle, \ldots, |\alpha_n\rangle$ are linearly independent vectors, but not necessarily orthogonal. (By *linearly independent* we mean that there exists no choice of constants $a_1, a_2, \ldots, a_n$ that will give $\Sigma\, a_k |\alpha_k\rangle = 0$, that is, none of the vectors can be expressed as a superposition of the others.) The orthogonalization procedure is the following. We take $|\alpha_1'\rangle = |\alpha_1\rangle$; we assume that this vector is normalized. Then take

$$|\alpha_2'\rangle = |\alpha_2\rangle - (\langle \alpha_1'|\alpha_2\rangle)\, |\alpha_1'\rangle$$

Obviously, $\langle \alpha_1'|\alpha_2'\rangle = 0$. Furthermore, we can normalize the vector $|\alpha_2'\rangle$. For simplicity, we continue to designate the normalized vector by the same symbol $|\alpha_2'\rangle$. Next, we take

$$|\alpha_3'\rangle = |\alpha_3\rangle - (\langle \alpha_2'|\alpha_3\rangle)\, |\alpha_2'\rangle - (\langle \alpha_1'|\alpha_3\rangle)\, |\alpha_1'\rangle$$

This vector is orthogonal to $|\alpha_2'\rangle$ and $|\alpha_1'\rangle$. We can normalize this vector, and so on. In the end, we obtain a set of n orthonormal eigenvectors.

Without loss of generality, we will from now on assume that all the eigenvectors of an observable are orthonormal. For each observable we then have an *orthonormal set of eigenvectors.*

A concise notation, due to Dirac, for the eigenvectors of an operator is the following: If the eigenvalues are $q_1, q_2, q_3, \ldots,$ then the eigenvectors are designated by the kets $|q_1\rangle, |q_2\rangle, |q_3\rangle,$

..., respectively. This means that we are using the eigenvalue as a label to distinguish the different eigenvectors. (This notation runs into a difficulty when some of the eigenvalues are degenerate. For example, if there are two eigenvectors with the eigenvalue q_3, then "$|q_3\rangle$" becomes ambiguous since it is not clear which of the two eigenvectors is meant. We take it for granted that whenever the need arises enough *extra* labels are to be inserted into the ket to remove any ambiguity.) In this notation the eigenvalue equation becomes

$$Q|q_n\rangle = q_n |q_n\rangle \tag{21}$$

and the orthonormality condition reads

$$\langle q_m|q_n\rangle = \delta_{mn} \tag{22}$$

Using this concise notation, we can now say that according to axiom III the result of a measurement of the observable Q must always be one or another of the numbers $q_1, q_2, q_3, \ldots$. Furthermore, if the state of the system is $|q_n\rangle$, then a measurement of Q gives the result q_n with a probability equal to 1 (the result $q_m \neq q_n$ has, of course, a probability equal to zero).

The obvious question to ask next is this: If the system is in some state $|\psi\rangle$ that is not one of the eigenstates $|q_n\rangle$, then what are the probabilities for the different possible outcomes of the measurement? Suppose that for the state $|\psi\rangle$, we measure the observable Q, and we obtain the result q_n. This means that immediately *after* this measurement the value of the observable is certain to be q_n (in the sense that an immediate repetition of the measurement will again yield q_n). By axiom III, the state of the system immediately after measurement must then be $|q_n\rangle$. Thus, we see that the probability for obtaining the result q_n in the measurement is simply equal to the probability for finding the system in the state $|q_n\rangle$, which is $|\langle q_n|\psi\rangle|^2$, by axiom II:

$$[\text{probability for } q_n] = |\langle q_n|\psi\rangle|^2 \tag{23}$$

The important role played by the measurement process should be noted: The measurement of the observable Q changes the state of the system from $|\psi\rangle$ to one of the eigenstates $|q_n\rangle$ of the observable. This is a discontinuous, unpredictable change; it is a change of the second kind, as discussed in Section 2.8.

Since the sum of the probabilities for all the possible outcomes must be equal to 1, Eq. (23) implies that

$$1 = \sum_n |\langle q_n|\psi\rangle|^2 \tag{24}$$

It is a consequence of Eq. (24) that the set of eigenvectors of an observable is *complete,* that is, any arbitrary vector can be written as a superposition of eigenvectors. We have already encountered an example of such a complete set in the case of the eigenfunctions of the energy operator for the particle in the infinite square well discussed in Chapter 3. The general completeness theorem asserts:

The set of eigenvectors of any observable is complete.

Suppose that $|\psi\rangle$ is an arbitrary vector; we will prove that $|\psi\rangle$ can be written as a superposition of the eigenvectors $|q_n\rangle$. Define a vector $|\psi'\rangle$ by

$$|\psi'\rangle = |\psi\rangle - \sum_n (\langle q_n|\psi\rangle) |q_n\rangle \tag{25}$$

The magnitude squared of this vector is

$$\langle \psi'|\psi'\rangle = \langle \psi|\psi\rangle - \sum_n \langle q_n|\psi\rangle \langle \psi|q_n\rangle - \sum_n \langle q_n|\psi\rangle^* \langle q_n|\psi\rangle$$
$$+ \sum_{m,n} \langle q_n|q_m\rangle \langle q_m|\psi\rangle \langle \psi|q_n\rangle \tag{26}$$

The last term on the right side reduces to

$$\sum_{m,n} \delta_{mn} \langle q_m|\psi\rangle \langle \psi|q_n\rangle = \sum_n |\langle q_n|\psi\rangle|^2 \tag{27}$$

which cancels the next to the last term. We are then left with

$$\langle \psi'|\psi'\rangle = \langle \psi|\psi\rangle - \sum_n |\langle q_n|\psi\rangle|^2 \tag{28}$$

Here, the right side is zero because of Eq. (24) and the normalization of $|\psi\rangle$. Since only the zero vector can have zero magnitude, we conclude that $|\psi'\rangle = 0$, which is equivalent to

$$|\psi\rangle = \sum_n (\langle q_n|\psi\rangle) |q_n\rangle \tag{29}$$

This expresses $|\psi\rangle$ as a superposition of the eigenvectors $|q_n\rangle$ and establishes the theorem.

We can also state our results in the following way: Any arbitrary vector $|\psi\rangle$ can always be written in the form

$$|\psi\rangle = \sum_n a_n|q_n\rangle \tag{30}$$

where the expansion coefficients a_n are given explicitly by

$$a_n = \langle q_n|\psi\rangle \tag{31}$$

This means that the expansion coefficients are probability amplitudes, that is, $|a_n|^2$ gives the probability for the measurements of the observable Q to result in q_n. Note that the a_n satisfy the normalization condition [Eq. (24)]:

$$\sum_n |a_n|^2 = 1 \tag{32}$$

It is convenient to write the coefficient $\langle q_n | \psi \rangle$ in Eq. (29) *after* the vector $|q_n\rangle$,

$$|\psi\rangle = \sum_n |q_n\rangle \langle q_n | \psi \rangle \tag{33}$$

In this expression we can regard $\sum_n |q_n\rangle \langle q_n|$ as an operator acting on $|\psi\rangle$.[6] Equation (33) asserts that this operator is the *unit* operator, that is,

$$\sum_n |q_n\rangle \langle q_n| = \boldsymbol{1} \tag{34}$$

This is the *completeness* or the *closure relation*. It is equivalent to the statement that the vectors $|q_n\rangle$ form a complete set, since if both the right and left sides of this equation are allowed to act on an arbitrary $|\psi\rangle$, we recover the superposition (33).

Next we derive an important expression for the expectation value of an observable when the system is in some arbitrary state $|\psi\rangle$. Consider the observable Q with eigenvectors $|q_n\rangle$ and eigenvalues q_n. The expectation value, or the mean value, of this observable is

$$\langle Q \rangle = \sum_n q_n \times (\text{probability for finding the system in state } n)$$

$$= \sum_n q_n |\langle \psi | q_n \rangle|^2$$

$$= \sum_n q_n \langle \psi | q_n \rangle \langle q_n | \psi \rangle = \sum_n \langle \psi | Q | q_n \rangle \langle q_n | \psi \rangle$$

$$= \langle \psi | Q \left(\sum_n |q_n\rangle \langle q_n| \right) |\psi\rangle = \langle \psi | Q(\boldsymbol{1}) |\psi\rangle = \langle \psi | Q |\psi\rangle \tag{35}$$

The result (35) means that the expectation value of an observable is equal to the scalar product of $|\psi\rangle$ with $Q|\psi\rangle$. We have, of course,

---

[6] In general, the object $|\alpha\rangle \langle \beta|$ can be regarded as an operator, in the sense that multiplication of a vector $|\psi\rangle$ by $|\alpha\rangle \langle \beta|$ gives $|\alpha\rangle \langle \beta|\psi\rangle$, that is, it gives the vector $|\alpha\rangle$ multiplied by the scalar $\langle \beta|\psi\rangle$.

already seen examples of this: to obtain the expectation value of momentum we took the scalar product of $\psi(x)$ and $p_{op}\psi(x)$, that is, we took the integral

$$\int \psi^* \frac{\hbar}{i} \frac{d}{dx} \psi \, dx$$

Let us review the eigenstates of the momentum, position, and energy of a particle in the context of our abstract Hilbert-space description. In accord with the notation of earlier chapters, we designate the momentum operator by p_{op}. An eigenstate of the momentum operator is a state $|\psi_p\rangle$ such that

$$p_{op} |\psi_p\rangle = p|\psi_p\rangle \tag{36}$$

where the real number p is the eigenvalue of the momentum. The state $|\psi_p\rangle$ represents a particle of definite momentum p, without uncertainty. We introduce the concise Dirac notation by omitting the ψ in Eq. (36):

$$p_{op} |p\rangle = p|p\rangle \tag{37}$$

This means that we use the momentum eigenvalue as a label to distinguish different momentum eigenvectors. For example, $|2.15\rangle$ might denote the state in which the particle has a momentum of 2.15 MeV/c toward the right (we are dealing with one dimension only; in three dimensions, we would need three eigenvalues, one for each component of momentum).

Likewise, the position operator is designated by x_{op}, and an eigenstate of position is a state such that

$$x_{op} |x\rangle = x|x\rangle \tag{38}$$

Here $|x\rangle$ represents a state in which the particle is sharply localized at x, without uncertainty.

The other operator we have used in earlier chapters is the energy operator. We will hereafter designate this operator by H; it is often called the *Hamiltonian* operator, for reasons to be discussed in the next chapter. For a free particle, H is simply $p_{op}^2/2m$; but for a particle with interactions, H will be more complicated. The eigenstates of H, or the eigenstates of energy, will be labeled by the corresponding energy eigenvalue:

$$H|E_n\rangle = E_n |E_n\rangle \tag{39}$$

Here $|E_n\rangle$ represents a state in which the particle has a sharply defined energy.

Consider now an arbitrary state $|\psi\rangle$. By our second axiom, $|\langle x|\psi\rangle|^2$ must be the probability that the particle be found at x, and thus $\langle x|\psi\rangle$ is the probability amplitude for finding the particle at x. But we have used the symbol $\psi(x)$ for this amplitude in earlier chapters. Hence

$$\psi(x) = \langle x|\psi\rangle \tag{40}$$

This supplies us with the connection between the abstract formalism of state vectors and the wavefunction: the wavefunction is the scalar product of $|x\rangle$ with $|\psi\rangle$.

Exercise 7. Show by a similar argument that $\langle p|\psi\rangle = \phi(p)$ and $\langle E_n|\psi\rangle = a_n$, where a_n is the coefficient used in Eq. (3.51).

We have said earlier that the set of coefficients a_n describes the state in the "energy representation." We can express this more precisely as follows. The energy eigenvectors form a complete orthonormal set. We can use these eigenvectors to describe any state vector by giving the *components* of the vector with respect to all the energy eigenvectors. Since the energy eigenvectors are unit vectors, the component of a vector $|\psi\rangle$ with respect to the energy eigenvector $|E_n\rangle$ is simply $\langle E_n|\psi\rangle$.

Similar formulas are valid in the "position representation" and in the "momentum representation." In these representations, instead of using the energy eigenvectors as a basis, we use the position eigenvectors $|x\rangle$ or the momentum eigenvectors $|p\rangle$ as a basis, and we use the components with respect to these position eigenvectors or momentum eigenvectors to describe any state vector. Thus, the component of $|\psi\rangle$ along $|x\rangle$ is $\langle x|\psi\rangle$, *provided we normalize $|x\rangle$ correctly*. The difficulty that occurs here is this: the vector $|x\rangle$ cannot be normalized to unit magnitude, that is, $\langle x|x\rangle \neq 1$ and $\langle x|x'\rangle \neq \delta_{xx'}$. The best we can do is to normalize the position eigenvectors $|x\rangle$ according to what is called delta-function normalization:[7]

$$\langle x|x'\rangle = \delta(x - x') \tag{41}$$

---

[7] A corresponding delta-function normalization must also be adopted for the energy eigenvectors of the unbound eigenstates.

Of course, the fact that $|x\rangle$ cannot be normalized contradicts our first axiom. This means $|x\rangle$ is actually not a vector in the Hilbert space $\mathcal{H}$. However, it is convenient for many calculations to pretend that $|x\rangle$ is an acceptable vector. If we are careful, this will not cause any trouble. Incidentally, Eq. (41) says that the wavefunction $\psi_{x'}(x)$ for a particle localized at x' is

$$\psi_{x'}(x) = \delta(x - x') \tag{42}$$

This, of course, is in agreement with the picture of the delta function as a sharp spike at $x = x'$.

For the momentum eigenvectors $|p\rangle$, we use a similar delta-function normalization:

$$\langle p'|p\rangle = \delta(p' - p) \tag{43}$$

Finally, we want to write down the completeness relation in terms of the basis vectors $|E_n\rangle$, $|x\rangle$, and $|p\rangle$. Obviously, for the basis vectors $|E_n\rangle$,

$$I = \sum_n |E_n\rangle \langle E_n| \tag{44}$$

Note that both the bound energy eigenstates and unbound eigenstates must be included in the sum on the right side of this equation. Since the unbound states form a continuum, the summation over such states is an integral, or a "continuous sum." But, according to the convention adopted at the end of Section 3.4, we do not bother to make this distinction between summation and integration for the energy eigenvectors.

With the basis vectors $|x\rangle$, the completeness relation takes the form of an integral over all the possible values of x,

$$I = \int dx \, |x\rangle \langle x| \tag{45}$$

Here we need an integral, instead of a discrete sum, because x can take on a continuum of values. To verify Eq. (45), multiply from the left by $|x'\rangle$ and from the right by $|x''\rangle$,

$$\langle x'|x''\rangle = \int dx \, \langle x'|x\rangle \langle x|x''\rangle$$

Using Eq. (41), we obtain

$$\delta(x' - x'') = \int dx \, \delta(x' - x) \, \delta(x - x'')$$

which is true, at lest in a formal sense.

For the basis vectors $|p\rangle$, the completeness relation takes the form

$$1 = \int dp \; |p\rangle \langle p| \tag{46}$$

If we multiply this from the left by $|x'\rangle$ and from the right by $|x\rangle$, we obtain

$$\langle x'|x\rangle = \int dp \; \langle x'|p\rangle \langle p|x\rangle \tag{47}$$

Here, the left side of the equation is the delta function, $\delta(x' - x)$; and the right side is an integral over a product of momentum wavefunctions $\langle x|p\rangle^*$ and $\langle x'|p\rangle$. But we know, from Chapter 2, that the delta function has the following expression as an integral over a product of momentum wavefunctions:

$$\delta(x' - x) = \frac{1}{2\pi\hbar} \int e^{ipx'/\hbar} e^{-ipx/\hbar} \; dp$$

Comparison of this expression with Eq. (47) tells us that the delta-function normalization of the position and momentum eigenvectors requires that

$$\psi_p(x) = \langle x|p\rangle = \frac{1}{\sqrt{2\pi\hbar}} \; e^{ipx/\hbar} \tag{48}$$

Table 4.1 on Dirac's bracket notation summarizes some of the preceding equations.

Exercise 8. Check the other entries in Table 4.1.

Now that we recognize that $\psi(x)$, $\phi(p)$, and a_n (which describe a state in the position, momentum, and energy representations, respectively) are nothing but the components of the *same* vector $|\psi\rangle$ taken with respect to different sets of basis vectors, we can easily understand why the commutation relations between x_{op} and p_{op} are the same in all these representations. It is simply that the multiplication table for the operators cannot depend on what basis we use to describe the vectors, since operators give *relationships* between vectors. If an operator transforms one vector into another, this relationship between the two is the same no matter what basis we use to specify components.

The invariance of the commutation relation with respect to changes in the representation permits us to introduce the following axiom:

TABLE 4.1 Dirac's Bracket Notation

	Operator		
	H	x_{op}	p_{op}
Eigenvector	$\lvert E_n \rangle$	$\lvert x \rangle$	$\lvert p \rangle$
Component of arbitrary $\lvert \psi \rangle$ w.r.t. eigenvector	$\langle E_n \lvert \psi \rangle \equiv A_n$	$\langle x \lvert \psi \rangle \equiv \psi(x)$	$\langle p \lvert \psi \rangle \equiv \phi(p)$
Component of eigenvector w.r.t. $\lvert E_m \rangle$	$\langle E_m \lvert E_n \rangle = \delta_{mn}$	$\langle E_n \lvert x \rangle = \psi^*_{E_n}(x)$	$\langle E_n \lvert p \rangle = \phi^*_{E_n}(p)$
$\lvert x' \rangle$	$\langle x' \lvert E_n \rangle = \psi_{E_n}(x')$	$\langle x' \lvert x \rangle = \delta(x' - x)$	$\langle x' \lvert p \rangle = \dfrac{e^{ipx'/\hbar}}{\sqrt{2\pi\hbar}}$
$\lvert p' \rangle$	$\langle p' \lvert E_n \rangle = \phi_{E_n}(p')$	$\langle p' \lvert x \rangle = \dfrac{e^{-ip'x/\hbar}}{\sqrt{2\pi\hbar}}$	$\langle p' \lvert p \rangle = \delta(p' - p)$
Completeness of eigenvectors	$1 = \displaystyle\sum_n \lvert E_n \rangle\langle E_n \rvert$	$1 = \displaystyle\int dx \lvert x \rangle\langle x \rvert$	$1 = \displaystyle\int dp \lvert p \rangle\langle p \rvert$
Completeness in position representation	$\langle x' \lvert x \rangle = \displaystyle\sum_n \langle x' \lvert E_n \rangle\langle E_n \lvert x \rangle$ i.e., $\delta(x' - x) = \displaystyle\sum_n \psi_{E_n}(x')\psi^*_{E_n}(x)$	$\langle x' \lvert x \rangle = \displaystyle\int dx'' \langle x' \lvert x'' \rangle\langle x'' \lvert x \rangle$ $\delta(x' - x) = \displaystyle\int dx'' \delta(x' - x'')\delta(x'' - x)$	$\langle x' \lvert x \rangle = \displaystyle\int dp \langle x' \lvert p \rangle\langle p \lvert x \rangle$ $\delta(x' - x) = \dfrac{1}{2\pi\hbar}\displaystyle\int dp\, e^{ip(x'-x)/\hbar}$
Expansion of arbitrary $\lvert \psi \rangle$	$\lvert \psi \rangle = \displaystyle\sum_n \lvert E_n \rangle\langle E_n \lvert \psi \rangle$	$\lvert \psi \rangle = \displaystyle\int dx' \lvert x' \rangle\langle x' \lvert \psi \rangle$	$\lvert \psi \rangle = \displaystyle\int dp \lvert p \rangle\langle p \lvert \psi \rangle$
Expansion in position representation	$\langle x \lvert \psi \rangle = \displaystyle\sum_n \langle x \lvert E_n \rangle\langle E_n \lvert \psi \rangle$ i.e., $\psi(x) = \displaystyle\sum_n \psi_{E_n}(x)A_n$	$\langle x \lvert \psi \rangle = \displaystyle\int dx' \langle x \lvert x' \rangle\langle x' \lvert \psi \rangle$ $\psi(x) = \displaystyle\int dx' \delta(x - x')\psi(x')$	$\langle x \lvert \psi \rangle = \displaystyle\int dp \langle x \lvert p \rangle\langle p \lvert \psi \rangle$ $\psi(x) = \displaystyle\int dp\, \dfrac{e^{ipx/\hbar}}{\sqrt{2\pi\hbar}} \phi(p)$

IV. *The operator representing the coordinate and the operator representing the corresponding momentum obey the "canonical" commutation relation*

$$[x_{op}, p_{op}] = i\hbar \tag{49}$$

This statement is somewhat more general than it looks. The "coordinate" need not be the position coordinate of a particle. It could be a generalized coordinate in the sense of Lagrangian theory. If we interpret (49) this way, it applies not only to the quantum theory of particles, but also to that of fields.

4.4 *Compatible Observables*

The eigenvectors of most operators representing observables are highly degenerate. For example, suppose we know that the energy of a free electron is 3.15 GeV. Since the direction of the momentum does not affect the value of the energy, there are many states of this energy—one state for every conceivable direction of the momentum (in three dimensions). Further, the electron spin can have different orientations without changing the energy; this introduces even more degeneracy. The situation gets far worse if we are dealing with a system of more than one particle, since a given amount of total energy can then be distributed in diverse ways among the particles. So if we want to specify a quantum-mechanical state uniquely, it is not sufficient to say that it is an eigenstate of one particular observable with some eigenvalue. Usually, there are many eigenstates with the same eigenvalue. What we must do is to specify *simultaneously* the eigenvalues of several distinct observables.

In terms of measurements, we can describe the situation as follows. We know that the measurement of any observable puts the system into an eigenstate of that observable. Thus, immediately after an energy measurement, the system is in an eigenstate of energy with some particular value of the energy. If this eigenvalue is nondegenerate, then we have thrown the system into one definite state. But if the eigenvalue is degenerate, then there are many eigenfunctions of this energy, and the system is not yet in one definite state.

Let us take a system consisting of an atom of hydrogen as an example. Suppose we measure the energy and find that, apart from translational kinetic energy of the entire atom and apart from rest-mass energies, the energy is -3.40 eV. This corresponds to the second energy level of the atom, an energy level which is degenerate. For these degenerate states, the magnitude squared of the angular momentum can be either 0 or $2\hbar^2$.[8] We therefore next measure this magnitude squared of the angular momentum (represented in the Hilbert space $\mathcal{H}$ by an operator L^2). Suppose that the result of the measurement is $2\hbar^2$. Even this is not enough

---

[8] According to Bohr's semiclassical theory, described in Chapter 1, the possible values of the angular momentum for the first excited state are $\hbar$ and $2\hbar$. But according to the fully quantum-mechanical treatment of the angular momentum, given in Chapter 7, the possible values of the angular momentum are 0 and $\sqrt{2}\,\hbar$. For the purposes of the present discussion, the exact values of the angular momentum are not crucial; what matters is that there are *several* values.

to define the state uniquely. If $\mathbf{L}^2$ has the eigenvalues $2\hbar^2$, there are three states of different orientation of the angular momentum, that is, the z component of the angular momentum (represented by an operator L_z) can take on the alternative values $-\hbar$, 0, or $+\hbar$. So we must measure this z component. Suppose that the result of the measurement is $-\hbar$. Then we finally have defined the state of the system uniquely (provided we ignore spin). If we use the notation $|E_n,\lambda,m_l\rangle$ for the states (where E_n is the energy eigenvalue, λ is the eigenvalue of $\mathbf{L}^2$, and m_l the eigenvalue of L_z), then our state would be $|-3.4, 2\hbar^2, -\hbar\rangle$. In this example, we see that to specify the state uniquely we must give the eigenvalues of several operators; the state is a *simultaneous eigenstate* of the operators H, $\mathbf{L}^2$, L_z:

$$H\,|E_n,\lambda,m_1\rangle = E_n\,|E_n,\lambda,m_1\rangle$$

$$\mathbf{L}^2\,|E_n,\lambda,m_1\rangle = \lambda\,|E_n,\lambda,m_1\rangle \tag{50}$$

$$L_z\,|E_n,\lambda,m_1\rangle = m_l\,|E_n,\lambda,m_1\rangle$$

When we need to measure *several* observables, which should we choose to measure? First of all, there must be enough of them so that no ambiguity remains in the state. Furthermore, if measurement of the observables is to yield always a simultaneous eigenvector of all of them, then these simultaneous eigenvectors must form a complete set, so any arbitrary state vector $|\psi\rangle$ can be written as a superposition of the simultaneous eigenvectors. Let us focus on two of the observables, and write the state $|\psi\rangle$ as a superposition of their simultaneous eigenvectors:

$$|\psi\rangle = \sum_{a,b} c_{ab}\,|a,b\rangle \tag{51}$$

Here $|a,b\rangle$ is a simultaneous eigenvector of the two observables A and B with eigenvalues a and b, respectively, and $\Sigma_{a,b}$ means summation over all these eigenvectors. We can then check that the operator AB acting on $|\psi\rangle$ produces the same result as the operator BA:

$$AB|\psi\rangle = \sum_{a,b} c_{ab}AB|a,b\rangle = \sum_{a,b} c_{ab}Ab|a,b\rangle$$

$$= \sum_{a,b} c_{ab}ab|a,b\rangle = \sum_{a,b} ba|a,b\rangle$$

$$= \sum_{a,b} Ba|a,b\rangle = \sum_{a,b} BA|a,b\rangle$$

$$= BA|\psi\rangle \tag{52}$$

Since $|\psi\rangle$ is arbitrary, this shows that $AB = BA$, that is, the operators A and B commute.

The requirements that (i) the simultaneous eigenvectors of a number of observables form a complete set and (ii) the observables commute are actually equivalent. We have shown that the first requirement implies the second. The converse is also true. For simplicity let us deal with only two observables and assume that they commute. If $|a\rangle$ is an eigenvector of A, then

$$A(B|a\rangle) = B(A|a\rangle) = B(a|a\rangle) = a(B|a\rangle) \tag{53}$$

This shows that $B|a\rangle$ is an eigenvector of A with eigenvalue a. In general, there will be several eigenvectors of A with this eigenvalue. We can then say only that B operating on one of these eigenvectors gives some superposition of them,

$$B|(k)a\rangle = \sum_n d_{kna} |(n)a\rangle \tag{54}$$

Here an extra label k has been inserted in the ket in order to distinguish between the different eigenvectors of A that have the same eigenvalue: $|(k)a\rangle$ stands for the kth eigenvector of A with the eigenvalue a. The set of eigenvectors $|(k)a\rangle$ (including all possible values of a and k) may be assumed to form a complete orthonormal set, so any arbitrary $|\psi\rangle$ can be written as

$$|\psi\rangle = \sum_{a,k} c_{ak} |(k)a\rangle \tag{55}$$

In particular, suppose that $|\psi\rangle$ is an eigenvector of B with eigenvalue b. Then we have

$$B|\psi\rangle = b|\psi\rangle = b \sum_{a,k} c_{ak} |(k)a\rangle \tag{56}$$

But according to Eq. (54),

$$B|\psi\rangle = \sum_{a,k} c_{ak}B |(k)a\rangle = \sum_{a,k,n} c_{ak}d_{kna} |(n)a\rangle \tag{57}$$

We therefore have the equality

$$\sum_{a,k,n} c_{ak}d_{kna} |(n)a\rangle = b \sum_{a,k} c_{ak} |(k)a\rangle \tag{58}$$

and if we multiply both sides by the vector $|(k')a'\rangle$ and use the orthonormality condition

$$\langle (k')a'|(n)a\rangle = \delta_{aa'}\delta_{nk'} \tag{59}$$

we find that

$$\sum_k c_{a'k} d_{kk'a'} = b c_{a'k'} \tag{60}$$

or, omitting the prime on a,

$$\sum_k c_{ak} d_{kk'a} = b c_{ak'} \tag{61}$$

The possible eigenvectors of B with eigenvalue b are obtained by finding all possible solutions of Eq. (61) for the unknowns c_{ak}. These equations do not mix different eigenvalues of A (there is no summation over a), that is, the equations do not impose any restrictions at all between two coefficients c_{ak} if their a values are different. This means that we can get all possible solutions of (61) by the following procedure: Take $c_{ak} = 0$ for all values of a *except one*. Most of the equations (61) will then reduce to $0 = 0$; the equations corresponding to the one chosen value of a remain to be solved. The solution always exists, because the number of equations matches the number of unknown coefficients. For instance, if there are three eigenvectors for a given value of a, then there are three relevant equations in Eq. (61), with $k' = 1, 2, 3$, and there are three unknowns c_{a1}, c_{a2}, c_{a3}. The resulting expression for $|\psi\rangle$ [Eq. (55)] will then not involve any summation over a; it will contain only the one chosen value of a. We have therefore obtained a simultaneous eigenvector of A and B. Next we choose a different value of a and go through the same procedure, and so on. The resulting B eigenvectors will all be simultaneous eigenvectors of A. Since the eigenvectors of B form a complete set and, as we have shown, all these eigenvectors are simultaneous eigenvectors of A, it follows that the simultaneous eigenvectors form a complete set.

We summarize our conclusions this way: The observables used to specify uniquely the state of a system must be a "complete" set of commuting observables. The word "complete" in this context simply means that the observables specify the state completely, that is, uniquely. Such a set of observables is also called a "complete" set of *compatible* observables. This refers to the fact that observables that do commute can be specified simultaneously, whereas observables that do not commute cannot be specified simultaneously. For example, if we first measure the position of a particle, we throw the system into an eigenstate of x_{op}. If we immediately afterward measure the momentum, we throw the system into an eigenstate of p_{op}. But it now will not be in an eigenstate of x_{op} any more, that is, x_{op} and p_{op} are not compatible observables.

4.5   *The Uncertainty Relations*

If two observables commute, then they can be measured simultaneously with arbitrary precision. But if they do not commute, then there is a limit on how precisely their values can be defined. As an indication of the uncertainty of an observable in some state $|\psi\rangle$, we use the root-mean-square deviation from the mean:

$$\Delta A = [\langle\psi|(A - \langle\psi|A|\psi\rangle)^2|\psi\rangle]^{1/2} \tag{62}$$

We will show that, if A and B are two observables, then they obey the general uncertainty relation

$$\Delta A \ \Delta B \geq \tfrac{1}{2} |\langle\psi|[A, B]|\psi\rangle| \tag{63}$$

We will first deal with the special case in which both A and B have zero expectation value, $\langle\psi|A|\psi\rangle = \langle\psi|B|\psi\rangle = 0$, so Eq. (63) reduces to

$$\langle\psi|A^2|\psi\rangle \ \langle\psi|B^2|\psi\rangle \geq \tfrac{1}{4} |\langle\psi|[A, B]|\psi\rangle|^2$$

For the proof, we need the following results:

$$\langle\psi|(Q|\psi\rangle) + \langle\psi|(Q^\dagger \ |\psi\rangle) \qquad \text{is real} \tag{64}$$

$$\langle\psi|(Q|\psi\rangle) - \langle\psi|(Q^\dagger \ |\psi\rangle) \qquad \text{is imaginary} \tag{65}$$

Exercise 9. Show that these statements are true for any operator Q (not necessarily hermitian).

We begin by exploiting the hermitian property of A and of B:

$$\langle\psi|A^2|\psi\rangle \ \langle\psi|B^2|\psi\rangle = ((\langle\psi|A) \ (A|\psi\rangle)) \ ((\langle\psi|B) \ (B|\psi\rangle)) \tag{66}$$

We then use the Schwartz inequality, Eq. (1):

$$\langle\alpha|\alpha\rangle \ \langle\beta|\beta\rangle \geq |\langle\alpha|\beta\rangle|^2 \tag{67}$$

With $|\alpha\rangle = A|\psi\rangle$, $\beta = B|\psi\rangle$, this yields

$$\langle\psi|A^2|\psi\rangle \ \langle\psi|B^2|\psi\rangle = ((\langle\psi|A) \ (A|\psi\rangle)) \ ((\langle\psi|B) \ (B|\psi\rangle))$$

$$\geq |\langle\psi|AB|\psi\rangle|^2$$

$$\geq |\tfrac{1}{2} \langle\psi|AB + BA|\psi\rangle + \tfrac{1}{2} \langle\psi|AB - BA|\psi\rangle|^2$$

$$\geq |\tfrac{1}{2} \langle\psi|AB + BA|\psi\rangle + \tfrac{1}{2} \langle\psi|[A, B]|\psi\rangle|^2 \tag{68}$$

But, according to the result stated in Eqs. (64) and (65), the quantity $\langle\psi|AB|\psi\rangle + \langle\psi|BA|\psi\rangle = \langle\psi|AB|\psi\rangle + \langle\psi|(AB)^\dagger|\psi\rangle$ is real and the quantity $\langle\psi|[A, B]|\psi\rangle$ is imaginary. Hence the first of the two terms between the absolute-value signs on the right side of Eq. (68) is real and the second is imaginary. The right side of Eq. (68) is therefore necessarily larger than the square of the imaginary term:

$$\langle\psi|A^2|\psi\rangle \langle\psi|B^2|\psi\rangle \geq \tfrac{1}{4} |\langle\psi|[A, B]|\psi\rangle|^2 \tag{69}$$

which is what we wanted to prove.

If the average values of A and B are not zero, we can introduce the operators

$$A' = A - \langle\psi|A|\psi\rangle \tag{70}$$

$$B' = B - \langle\psi|B|\psi\rangle \tag{71}$$

These operators have zero expectation value. It is easy to see that

$$\Delta A = \Delta A' \quad\text{and}\quad \Delta B = \Delta B' \tag{72}$$

and

$$[A', B'] = [A, B] \tag{73}$$

Our proof obviously applies to A' and B', and it then follows that the inequality (63) holds for A and B.

Exercise 10. Show that Eqs. (72) and (73) hold.

Let us now apply Eq. (63) to the position and momentum operators x_{op}, p_{op}. This gives us the uncertainty relation:

$$\Delta p\,\Delta x \geq \tfrac{1}{2}|\langle\psi|[p_{op}, x_{op}]|\psi\rangle|$$

that is,

$$\Delta p\,\Delta x \geq \tfrac{1}{2}\hbar \tag{74}$$

which is the precise version of our earlier approximate statement of the Heisenberg uncertainty relation. It can be shown that the *only* wavefunction that satisfies the *equality* in Eq. (74) is the Gaussian packet of Chapter 2.

PROBLEMS   1. Prove that, for any two vectors $|\alpha\rangle$ *and* $|\beta\rangle$,

$$(\langle\alpha|\alpha\rangle^{1/2} + \langle\beta|\beta\rangle^{1/2})^2 \geq (\langle\alpha| + \langle\beta|)(|\alpha\rangle + |\beta\rangle)$$

(Hint: Use the Schwartz inequality.)

2. Show that the sum of two linear operators is a linear operator. Show that the product of two linear operators is a linear operator.

3. Suppose that A is a linear operator. Show that e^A is a linear operator. In general, show that if f is any function that can be expressed as a power series, then $f(A)$ is a linear operator.

4. Show that if $|\psi\rangle$ is an eigenvector of A with eigenvalue a, then it is an eigenvector of $f(A)$ with eigenvalue $f(a)$,

$$f(A)|\psi\rangle = f(a)|\psi\rangle$$

   Here $f(A)$ is any function that can be expressed as a power series.

5. Suppose that B is the inverse of A, that is, $BA = 1$. Show that if $|\psi\rangle$ is an eigenvector of A with eigenvalue $a \neq 0$, then $|\psi\rangle$ is an eigenvector of B with eigenvalue $1/a$.

6. Show that $(AB)^\dagger = B^\dagger A^\dagger$ and that, in general, $(ABC \ldots)^\dagger = \ldots C^\dagger B^\dagger A^\dagger$.

7. Show that if A and B are hermitian operators, then all the following operators are hermitian: $A + B$, $AB + BA$, $i(AB - BA)$, $A^n B^m + B^m A^n$.

8. Consider the following operators constructed out of products of x_{op} and p_{op}: $p_{op}x_{op}$, $p_{op}x_{op} + x_{op}p_{op}$, $x_{op}{}^2 p_{op}$, $x_{op}p_{op}x_{op}$, $i(x_{op}{}^2 p_{op} - p_{op}x_{op}{}^2)$. Which of these are hermitian?

9. Show that $AA^\dagger$ is hermitian, even if A is not hermitian.

10. Show that if A is hermitian, then the expectation value of A^2 is nonnegative (that is, $\langle\psi|A^2|\psi\rangle \geq 0$), and show that all the eigenvalues of A^2 are nonnegative.

11. Suppose that there exists a linear operator A that has an eigenvector $|\psi\rangle$ with the eigenvalue a,

$$A|\psi\rangle = a|\psi\rangle$$

   Suppose that there also exists an operator B such that

$$[A, B] = B + 2BA^2$$

   Show that the vector $B|\psi\rangle$ is an eigenvector of A and find the eigenvalue.

12. Prove the following identity for two operators A and B:

$$e^A B e^{-A} = B + [A, B] + \frac{1}{2!}[A, [A, B]] + \frac{1}{3!}[A, [A, [A, B]]] + \cdots$$

   (Hint: Consider the operator $g(\lambda) = e^{\lambda A} B e^{-\lambda A}$, where λ is a numerical parameter. Show that, at $\lambda = 0$, $dg/d\lambda = [A, B]$, $d^2g/d\lambda^2 = [A, [A, B]], \ldots$. Then develop $g(\lambda)$ in a Taylor series in λ about $\lambda = 0$.)

13. In the abstract Hilbert space $\mathcal{H}$, the energy eigenvectors of a particle in an infinite square well of width L are designated by $|E_1\rangle, |E_2\rangle, |E_3\rangle,$ ..., for the ground state, the first excited state, the second excited state, and so on. Suppose that at a given time a particle is in the state

$$|\psi\rangle = \frac{1}{2}|E_1\rangle - \frac{\sqrt{3}}{2}|E_3\rangle$$

(a) Express this state in the x representation.

(b) Express this state in the p representation.

(c) Express this state in the energy representation.

(d) Write down the energy operator in each of these three representations.

(e) Calculate the expectation value of the energy. Do this calculation three times, once in each of the representations.

(f) Using whichever representation you like best, find the rms deviation of the energy from the mean.

14. An operator Q has eigenvectors $|q_n\rangle$,

$$Q|q_n\rangle = q_n|q_n\rangle \qquad n = 1, 2, 3, \ldots$$

Suppose that these eigenvectors $|q_n\rangle$ form a complete set. Show that in this case the operator Q can be written as

$$Q = \sum_n q_n |q_n\rangle \langle q_n|$$

15. Suppose that three normalized state vectors $|\alpha\rangle, |\beta\rangle,$ and $|\gamma\rangle$, have the following multiplication table:

$$\langle\alpha|\beta\rangle = \langle\beta|\alpha\rangle = 0.3 \qquad \langle\alpha|\gamma\rangle = \langle\gamma|\alpha\rangle = 0.2 \qquad \langle\beta|\gamma\rangle = \langle\gamma|\beta\rangle = 0.8$$

Use the Schmidt procedure to construct two new vectors $|\beta'\rangle$ and $|\gamma'\rangle$ orthogonal to $|\alpha\rangle$ and to each other.

16. Verify that the wavefunction $\langle x|p\rangle = (1/\sqrt{2\pi\hbar})\, e^{ipx/\hbar}$ has delta-function normalization, that is,

$$\langle p'|p\rangle = \int \langle p'|x\rangle \langle x|p\rangle\, dx = \delta(p - p')$$

17. Suppose that A is a hermitian operator. Show that

$$|\langle\phi|A^2|\psi\rangle|^2 \leq \langle\phi|A^2|\phi\rangle \langle\psi|A^2|\psi\rangle$$

18. Suppose that two observables A and B commute. Can you conclude that $\Delta A\, \Delta B = 0$? Prove this if it is true and give an explicit counterexample if it is false.

19. Use the general uncertainty relation to prove that $\Delta E \,\Delta x \geq \hbar |\langle p \rangle|/2m$ for a free particle. Evaluate ΔE explicitly for the Gaussian packet of Section 2.5, and verify that this inequality is satisfied.

20. Show that with $A = x_{op} - \langle \psi | \, x_{op} \, | \psi \rangle$ and $B = x_{op}^2 - \langle \psi | \, x_{op}^2 \, | \psi \rangle$, Eq. (68) yields

$$\Delta(x)\,\Delta(x^2) \geq |\,\langle \psi | x_{op}^3 | \psi \rangle - \langle \psi | x_{op}^2 | \psi \rangle \,\langle \psi | x_{op} | \psi \rangle \,|$$

and verify this explicitly for the Gaussian packet of Eq. (2.74).

21. Prove the following generalization of the uncertainty relation, valid for any two hermitian operators A and B:

$$(\Delta A)^2 (\Delta B)^2 \geq |\tfrac{1}{2} \langle \psi | \,(AB - BA)\, | \psi \rangle \,|^2 + |\,\tfrac{1}{2} \langle \psi | \,(AB + BA)\, | \psi \rangle - \langle \psi | A | \psi \rangle \langle \psi | B | \psi \rangle \,|^2$$

5

The Evolution of States in Time

So far our axioms of quantum mechanics have dealt only with "kinematics," that is, the description of states and of observables. We now want to discuss "dynamics," or the evolution of states in time caused by interactions. For this, we need an equation of motion, that is, an equation that determines the time derivative of the state vector. This equation, which we will introduce in axiom V, is a general form of the Schrödinger equation. It plays the same role in quantum mechanics as Newton's equation of motion does in classical mechanics—it permits us to calculate the state at a later time from the given state at an initial time. Thus, the evolution of the quantum-mechanical state vector is deterministic; however, since the state vector provides information only about probabilities of the outcome of measurements, we are unable to make the firm predictions we are accustomed to from classical mechanics. But it must be emphasized that this lack of predictability is inherent in the concept of the state vector (see axiom II), and is not brought on by the equation of motion. The uncertainties in the predictions of quantum mechanics are to be blamed on the kinematics, not on the dynamics. No information is lost during the evolution of the state vector—both the initial state vector and the state vector at a later time provide the same kind of information about probabilities of the outcome of measurements. We can regard quantum mechanics as a deterministic theory because it makes definite predictions for the probabilities of the outcome of measurements at any later time. This is the most we can ask for, in the context of a description of states by state vectors.

5.1 *The Hamiltonian Operator*

Since the state vector contains *all* of the information that can be known about a system, the initial state at some time t is completely specified by the state vector $|\psi(t)\rangle$. This is in contrast to the specification of the initial state of a classical system, which requires both the position and the velocity at the initial time. Given the initial state vector $|\psi(t)\rangle$, the equation for the time evolution has to determine the state vector $|\psi(t + dt)\rangle$. Equivalently, the equation for the time evolution has to determine the derivative $d/dt \, |\psi(t)\rangle$. Thus, the derivative must somehow be related to the state vector. We proclaim this relationship in another axiom:

V. *The time evolution of the state vector describing a system is given by*

$$i\hbar \frac{d}{dt} |\psi(t)\rangle = H|\psi(t)\rangle \tag{1}$$

where H is the Hamiltonian operator. This operator is linear and hermitian.

Equation (1) is the *general Schrödinger equation*. The Hamiltonian operator H appearing in this equation is the energy operator of the system.[1] The precise form of this operator depends on the system. We can give a general prescription for how to find the operator from the *classical* Hamiltonian for the corresponding *classical* system: Simply take the classical Hamiltonian and replace the variables p and x appearing in it by the corresponding operators p_{op} and x_{op}. For instance, suppose our system consists of a particle moving in one dimension in a potential $V(x)$, with a classical Hamiltonian $H = p^2/2m + V(x)$; then the prescription tells us to replace p by p_{op} and x and x_{op}, with the result

$$H = \frac{p_{op}^2}{2m} + V(x_{op})$$

But this simple replacement prescription is subject to some ambiguities if the Hamiltonian contains products of positions and momenta. The order of factors in such a product does not affect the

---

[1] It is known from classical mechanics that the Hamiltonian of a classical system coincides with the energy, provided that there are no time-dependent constraints among the coordinates and there is no velocity-dependent potential. These provisos are satisfied by all the systems of interest in quantum mechanics.

classical Hamiltonian, but it does affect the quantum-mechanical Hamiltonian, since the position and momentum operators do not commute. We will resolve this factor-ordering ambiguity case by case, whenever it arises.

The last sentence in axiom V is actually redundant since we already have postulated that every observable is represented by a linear and hermitian operator. As a consequence of the linearity of H, the solutions of the differential equation (1) obey a superposition principle: If $|\psi_1(t)\rangle$ and $|\psi_2(t)\rangle$ are solutions of Eq. (1), then $a|\psi_1(t)\rangle + b|\psi_2(t)\rangle$ is also a solution.

The linearity of the Hamiltonian is required by our general axiom III, since the Hamiltonian corresponds to the energy, that is, it corresponds to a physical observable. From the point of view of dynamics, the linearity of the Hamiltonian, and the consequent superposition principle satisfied by the solutions of Eq. (1), means that the state vector does not interact with itself—all the interactions are contained in the Hamiltonian.

The importance of the hermiticity requirement for the Hamiltonian lies in the following: We must have $\langle\psi(t)|\psi(t)\rangle = 1$ for all times, and thus

$$0 = \frac{d}{dt}\langle\psi(t)|\psi(t)\rangle = \langle\psi(t)|\left(\frac{d}{dt}|\psi(t)\rangle\right) + \left(\langle\psi(t)|\frac{\overleftarrow{d}}{dt}\right)|\psi(t)\rangle \qquad (2)$$

Here the arrow on the differential operator indicates that it acts to the *left*. If we use Eq. (1), we obtain

$$0 = \langle\psi(t)|\left(\frac{1}{i\hbar}H|\psi(t)\rangle\right) + \left(\langle\psi(t)|\frac{1}{i\hbar}H\right)|\psi(t)\rangle$$

$$= \frac{1}{i\hbar}[\langle\psi(t)|(H|\psi(t)\rangle) - (\langle\psi(t)|H)|\psi(t)\rangle] \qquad (3)$$

Obviously, this requires that H be hermitian.

Keep in mind that Eq. (1) is an equation of motion, that is, a condition imposed on the state vector. It is not a condition imposed on the operator H. Thus, it would be wrong to write the operator equation $i\hbar\partial/\partial t = H$. [We can see that $i\hbar\,\partial/\partial t = H$ makes no sense as an operator equation by examining what happens if we multiply this putative equation by some other operator, say, x_{op}. The result is inconsistent, since the left side ($\partial/\partial t$) commutes with x_{op}, whereas the right side (H) does not!]

Let us check that, for a particle, Eq. (1) implies the usual Schrödinger wave equation. We know that the Schrödinger wavefunction $\psi(x, t)$ can be expressed as $\psi(x, t) = \langle x|\psi(t)\rangle$. To extract a differential equation for $\psi(x, t)$ from Eq. (1), multiply the latter by $\langle x|$ from the left and use the completeness relation $\int dx'|x'\rangle\langle x'| = \mathbf{1}$:

$$i\hbar \frac{\partial}{\partial t} \langle x|\psi(t)\rangle = \langle x|H|\psi(t)\rangle$$

$$= \int dx' \langle x|H|x'\rangle \langle x'|\psi(t)\rangle \tag{4}$$

We can rewrite this as

$$i\hbar \frac{\partial}{\partial t} \psi(x, t) = \int dx' \langle x|H|x'\rangle \psi(x', t) \tag{5}$$

On the right of this equation, there appears a complicated integral operator. To go further, we need to know more about the Hamiltonian operator. For instance, suppose that we are dealing with a particle moving in one dimension in a potential $V(x)$. As we saw above, the replacement prescription then tells us that the Hamiltonian operator is

$$H = \frac{p_{\mathrm{op}}^2}{2m} + V(x_{\mathrm{op}}) \tag{6}$$

With this, Eq. (5) becomes

$$i\hbar \frac{\partial}{\partial t} \psi(x, t) = \int dx' \langle x| \frac{p_{\mathrm{op}}^2}{2m} |x'\rangle \psi(x', t)$$

$$+ \int dx' \langle x| V(x_{\mathrm{op}}) |x'\rangle \psi(x', t) \tag{7}$$

Let us first evaluate the term involving the potential. Since $|x\rangle$ is an eigenstate of x_{op}, we have[2]

$$\int dx' \langle x|V(x_{\mathrm{op}})|x'\rangle \psi(x', t) = \int dx' \langle x|V(x')|x'\rangle \psi(x', t)$$

$$= \int dx' \langle x|x'\rangle V(x') \psi(x', t)$$

$$= \int dx' \delta(x - x') V(x') \psi(x', t)$$

$$= V(x) \psi(x, t) \tag{8}$$

---

[2] If $V(x_{\mathrm{op}})$ is a polynomial in x_{op}, then it is easy to show that $V(x_{\mathrm{op}})|x\rangle = V(x)|x\rangle$ (see Problem 2). If $V(x_{\mathrm{op}})$ is *not* a polynomial, then the latter equation serves as definition of $V(x_{\mathrm{op}})$.

The term involving $p_{op}^2/2m$ is harder to evaluate. We will not do it directly, but use a trick. Suppose that we look at Eq. (1) in the momentum representation. To find the equation in this representation, multiply Eq. (1) from the left by $|p\rangle$:

$$i\hbar \frac{\partial}{\partial t} \langle p|\psi(t)\rangle = \langle p| \frac{p_{op}^2}{2m} |\psi(t)\rangle + \langle p|V(x_{op})|\psi(t)\rangle$$

$$= \int dp' \langle p| \frac{p_{op}^2}{2m} |p'\rangle \langle p'|\psi(t)\rangle$$

$$+ \int dp' \langle p|V(x_{op})|p'\rangle \langle p'|\psi(t)\rangle \tag{9}$$

that is,

$$i\hbar \frac{\partial}{\partial t} \phi(p, t) = \int dp' \langle p| \frac{p_{op}^2}{2m} |p'\rangle \phi(p', t)$$

$$+ \int dp' \langle p|V(x_{op})|p'\rangle \phi(p', t) \tag{10}$$

The first term on the right is easy to evaluate because $|p'\rangle$ is an eigenvector of p_{op}:

$$\int dp' \langle p| \frac{p_{op}^2}{2m} |p'\rangle \phi(p', t) = \int dp' \langle p|p'\rangle \frac{p'^2}{2m} \phi(p', t)$$

$$= \int dp' \, \delta(p - p') \frac{p'^2}{2m} \phi(p', t)$$

$$= \frac{p^2}{2m} \phi(p, t) \tag{11}$$

The Schrödinger equation in the momentum representation is therefore

$$i\hbar \frac{\partial}{\partial t} \phi(p, t) = \frac{p^2}{2m} \phi(p, t) + \int dp' \langle p|V(x_{op})|p'\rangle \phi(p', t) \tag{12}$$

Let us now go back to the position representation. Comparing Eqs. (12) and (7), we see that the expression

$$\int dx' \langle x| \frac{p_{op}^2}{2m} |x'\rangle \psi(x', t)$$

must be such that it gives $(p^2/2m) \phi(p, t)$ when we transform it into the momentum representation, that is, when we take its Fourier transform. Obviously,

$$\frac{1}{2m} \left(\frac{\hbar}{i} \frac{\partial}{\partial x} \right)^2 \psi(x, t)$$

has this property. Hence the Schrödinger equation in the position representation is

$$i\hbar \frac{\partial}{\partial t} \psi(x, t) = -\frac{\hbar^2}{2m} \frac{\partial^2}{\partial x^2} \psi(x, t) + V(x)\psi(x, t) \tag{13}$$

We already became acquainted with this form of the Schrödinger equation in Chapter 3.

Exercise 1. By comparing Eqs. (12) and (8), show similarly that the Schrödinger equation in the momentum representation can be written as

$$i\hbar \frac{\partial}{\partial t} \phi(p, t) = \frac{p^2}{2m} \phi(p, t) + V \left(i\hbar \frac{\partial}{\partial p} \right) \phi(p, t) \tag{14}$$

This assumes that $V(x)$ is a polynomial, so $V(i\hbar\, \partial/\partial p)$ makes sense [see Section 5.4 for an alternative version of Eq. (14), which does not require a polynomial function $V(x)$].

From the general Schrödinger equation, we can obtain a useful expression for the time derivative of the expectation value of an operator. A straightforward evaluation of the time derivative yields[3]

$$\frac{d}{dt} \langle \psi | Q | \psi \rangle = \left(\langle \psi | \overset{\leftarrow}{\frac{d}{dt}} \right) Q |\psi\rangle + \langle \psi | \frac{\partial Q}{\partial t} |\psi\rangle + \langle \psi | Q \left(\frac{d}{dt} |\psi\rangle \right)$$

$$= \left(\langle \psi | \frac{H}{i\hbar} \right) Q |\psi\rangle + \langle \psi | \frac{\partial Q}{\partial t} |\psi\rangle + \langle \psi | Q \left(\frac{H}{i\hbar} |\psi\rangle \right)$$

$$= -\frac{1}{i\hbar} \langle \psi | HQ |\psi\rangle + \frac{1}{i\hbar} \langle \psi | QH |\psi\rangle + \langle \psi | \frac{\partial Q}{\partial t} |\psi\rangle$$

that is,

$$\frac{d}{dt} \langle \psi | Q | \psi \rangle = \frac{i}{\hbar} \langle \psi | [H, Q] |\psi\rangle + \langle \psi | \frac{\partial Q}{\partial t} |\psi\rangle \tag{15}$$

---

[3] We recall that the notation $\overset{\leftarrow}{d}/dt$ means that the time derivative acts on what stands to the *left*.

The last term in this equation takes into account the possibility that the operator Q is an explicit function of time.

If the operator Q commutes with the Hamiltonian and is not an explicit function of time, Eq. (15) tells us that

$$\frac{d}{dt} \langle \psi | Q | \psi \rangle = 0$$

Thus, the expectation value of Q is constant in time. This is the quantum-mechanical version of a conservation law. In later chapters we will examine divers observables—such as angular momentum, parity, or total momentum of an isolated system—represented by observables Q that commute with the Hamiltonian. The expectation values of these observables are constant.

Let us consider a special case of Eq. (15), with $Q = H$. For an isolated system—that is, a system not subject to any external disturbance—the Hamiltonian is independent of time. Since $[H, H] \equiv 0$, we immediately obtain

$$\frac{d}{dt} \langle \psi | H | \psi \rangle = 0 \tag{16}$$

which says that the expectation value of the energy is constant in time. This is the quantum-mechanical version of the conservation law for energy.

From earlier chapters we know that the energy eigenvectors play a central role in quantum theory. These energy eigenvectors satisfy the eigenvalue equation

$$H | \psi(t) \rangle = E | \psi(t) \rangle \tag{17}$$

Substituting this into the general Schrödinger equation (1) and integrating with respect to time, we find, as in Chapter 3 [see Eqs. (3.8)–(3.11)], that the energy eigenvectors have a time dependence

$$| \psi(t) \rangle = e^{-iEt/\hbar} | \psi(0) \rangle \tag{18}$$

Thus, *the eigenstates of the energy operator are stationary states,* a result we have already established in the special case of a single particle in Chapter 3. In accord with the terminology we used earlier, Eq. (17) is called the general time-independent Schrödinger equation.

We can best appreciate the stationary character of the eigenvectors of energy from the following result:

In an energy eigenstate, the expectation value of any (time-independent) operator is constant.

This is a direct consequence of the energy eigenvalue equation and of Eq. (15):

$$\frac{d}{dt} \langle \psi | Q | \psi \rangle = \frac{i}{\hbar} \langle \psi | [H, Q] | \psi \rangle = \frac{i}{\hbar} \langle \psi | (HQ - QH) | \psi \rangle$$

$$= \frac{i}{\hbar} [(\langle \psi | H) Q | \psi \rangle - \langle \psi | Q (H | \psi \rangle)]$$

$$= \frac{i}{\hbar} (E \langle \psi | Q | \psi \rangle - E \langle \psi | Q | \psi \rangle) = 0 \tag{19}$$

5.2 *Ehrenfest's Equations*

We will now show that classical mechanics is a limiting case of quantum mechanics. Consider a particle in a state $|\psi\rangle$ that represents some fairly narrow wave packet. We can then evaluate the rate of change of the expectation value of the position by taking $Q = x_{op}$ in Eq. (15). Since x_{op} has no explicit time dependence, we obtain

$$\frac{d}{dt} \langle \psi | x_{op} | \psi \rangle = \frac{i}{\hbar} \langle \psi | [H, x_{op}] | \psi \rangle \tag{20}$$

We recall from Eq. (2.103) that for any polynomial function $f(p_{op})$, the commutator with x_{op} is $[f(p_{op}), x_{op}] = \hbar/i \, \partial f/\partial p_{op}$. Since the Hamiltonian for the particle is of the form $H = p_{op}^2/2m + V(x_{op})$, we obtain

$$[H, x_{op}] = \frac{\hbar}{i} \frac{\partial H}{\partial p_{op}} = \frac{\hbar}{im} p_{op} \tag{21}$$

Equation (20) now becomes

$$\frac{d}{dt} \langle \psi | x_{op} | \psi \rangle = \frac{1}{m} \langle \psi | p_{op} | \psi \rangle \tag{22}$$

Thus, the rate of change of the average position equals the average momentum divided by the mass. This equality is in accord with the classical relation between velocity and momentum. Equation (22) tells us that whenever the uncertainty in position and momentum can be neglected, so we can speak of the average x and p as *the* x and p, then the relation between the two is the classical one.

Next, let us look at the rate of change of the expectation value of the momentum. With $Q = p_{op}$, Eq. (15) yields

$$\frac{d}{dt} \langle \psi | p_{op} | \psi \rangle = \frac{i}{\hbar} \langle \psi | [H, p_{op}] | \psi \rangle$$

$$= \frac{i}{\hbar} \langle \psi | \left(i\hbar \frac{\partial H}{\partial x_{op}} | \psi \rangle \right)$$

$$= - \langle \psi | \frac{\partial V(x_{op})}{\partial x_{op}} | \psi \rangle \qquad (23)$$

The classical equation corresponding to Eq. (23) is Newton's law of motion: the rate of change of momentum equals the force. We see therefore that in the classical limit (when uncertainties can be neglected), the average motion of the quantum-mechanical particle is along the classical orbit. Thus, narrow wave packets move like classical particles, and the correspondence principle is satisfied.

Equations (22) and (23) are called *Ehrenfest's equations.* Although in this discussion we have focused on wave-packet states, Ehrenfest's equations are valid for any arbitrary state $|\psi\rangle$. These equations have an interesting implication for stationary states. For such a state, the expectation value of x_{op} is constant [see Eq. (19)], and hence the first Ehrenfest equation implies that the expectation value of the momentum is zero, $\langle p \rangle = 0$. Note that this result is in conflict with the properties of the eigenstates of momentum for a free particle, which are stationary states of nonzero momentum. The reason the result $\langle p \rangle = 0$ does not apply to these states is that they are not normalizable, and the first Ehrenfest equation does not apply to them ($\langle \psi | x_{op} | \psi \rangle$ does not exist if $|\psi\rangle$ is an eigenstate of momentum).

5.3   *The Energy–Time "Uncertainty Relation"*

We know that the uncertainties of position and of momentum for the wave packet of a free particle obey the Heisenberg relation

$$\Delta x \, \Delta p \geq \frac{\hbar}{2} \qquad (24)$$

Let us suppose that we are dealing with a wave packet of fairly well defined momentum and velocity. If the velocity of the packet is approximately v, then the uncertainty in the energy of the packet is related to the uncertainty in the momentum by $\Delta E = \Delta(p^2/2m) \simeq (p/m)\, \Delta p \simeq v\, \Delta p$. Furthermore, the uncertainty Δx can be regarded as the approximate width of the wave packet. This width of the packet in space implies a width in time. If the velocity of the packet is v, it will take a time

$$\delta t \simeq \frac{\Delta x}{v} \tag{25}$$

to pass some given point. Expressing Eq. (24) in terms of these values of ΔE and δt, we find

$$v\, \delta t\, \frac{\Delta E}{v} \geq \frac{\hbar}{2}$$

or

$$\Delta E\, \delta t \geq \frac{\hbar}{2} \tag{26}$$

This equation is usually called the *energy–time uncertainty relation*. The time δt can be regarded as an uncertainty because, until the wave packet passes the given point, we cannot be certain that the particle has passed.

Note that with $E = \hbar\omega$, the energy–time uncertainty relation becomes a frequency–time uncertainty relation:

$$\Delta\omega\, \delta t \geq \frac{1}{2} \tag{27}$$

This is a familiar equation in radio engineering, where it is applied to pulses of radio waves. For a radio pulse, $\Delta\omega$ is called the bandwidth and δt the pulse width. The bandwidths of radio pulses are usually obtained directly from a Fourier analysis of the wave pulse with respect to the time variable t, rather than the space variable x (for radio waves, frequency and wave number are directly proportional, $k = \omega/c$, and hence Fourier analysis in time is equivalent to Fourier analysis in space).

Although in the special case of the moving wave packet, the energy–time uncertainty relation is a trivial consequence of the position–momentum uncertainty relation, in more general cases it is an independent statement. The mathematical form of the energy–time uncertainty relation is always as given by Eq. (26), but the meaning of δt is somewhat different in the case of a general quantum-mechanical system. As we will see, the time interval δt

must be given the meaning of a characteristic time interval for the evolution of the system, that is, the time interval required for a significant change to occur in the system.

Consider *any* observable B of the system, without explicit time dependence; for instance, position, or momentum, or dipole moment, and so on. According to the general uncertainty relation stated in Eq. (4.63),

$$\Delta A \; \Delta B \geq \frac{1}{2} \, |\langle\psi| \, [A, B] \, |\psi\rangle \tag{28}$$

If we take the Hamiltonian operator H for A, we obtain

$$\Delta E \; \Delta B \geq \frac{1}{2} \, | \, \langle\psi| \, [H, B] \, |\psi\rangle \, | = \frac{1}{2} \, | \, \frac{\hbar}{i} \frac{d}{dt} \langle\psi|B|\psi\rangle \, |$$

that is,

$$\Delta E \; \Delta B \geq \frac{\hbar}{2} \, | \, \frac{d}{dt} \langle\psi|B|\psi\rangle \, | \tag{29}$$

Suppose that we define a characteristic time δt by

$$\delta t = \frac{\Delta B}{| \, \frac{d}{dt} \langle\psi|B|\psi\rangle \, |} \tag{30}$$

If the time derivative appearing in this equation is approximately constant, then δt is the time required for the average value of B to change by ΔB. Thus, δt is roughly the time required for an *appreciable change* to occur in the value of the observable B. For different observables, the corresponding values of δt will, of course, be different. But if we choose that observable B which gives the smallest value of δt, then we can say that this δt is the time required for an appreciable change to occur in the physical properties of the system. With this definition of δt, we can write Eq. (29) as

$$\Delta E \; \delta t \geq \frac{\hbar}{2} \tag{31}$$

This is the general energy–time uncertainty relation.

Note that the time interval δt defined by Eq. (30) can be regarded as an uncertainty in time, since until the expectation value changes by an amount comparable to ΔB, we cannot be sure that any change has occurred. But this "uncertainty" in time means

something rather different from the other quantum-mechanical uncertainties we have encountered. The uncertainties in energy,
momentum, position, and so on, arise directly from the probability
distribution of these observables—the uncertainties are defined as
rms deviations from the mean. But the uncertainty δt is not an
uncertainty in the time intself, but an uncertainty in the time at
which something happens. If we were to ignore this subtle distinction and were to attempt to derive the energy–time uncertainty relation in the same way as the position–momentum uncertainty relation, we would not obtain Eq. (31). If we take the
Hamiltonian operator H for A and the time t for B in the general
uncertainty relation (28), we obtain

$$\Delta H \, \Delta t \geq \frac{1}{2} \, |\langle\psi| \, [H, \, t] \, |\psi\rangle \qquad (32)$$

Here the right side is zero, since t is a number, and any number
commutes with every operator. The left side, of course, is also
zero. In Eq. (32), Δt is defined as an rms deviation, and it is easy to
check that with this definition $\Delta t \equiv 0$, that is, there is no uncertainty in the time itself. The time in quantum mechanics, as in
classical mechanics, is a parameter; it is the time at which we
choose to consider a system. The time parameter is subject to
experimental error, but it has no quantum-mechanical fluctua-
tion. This is very different from what happens with the position
(x) variable. The position is, of course, subject to experimental
error, since the measuring apparatus is never perfectly constructed; but besides that, it is subject to quantum-mechanical
fluctuations which are an intrinsic property of the observed system. The position is "unsharp," and this is not the fault of measuring apparatus but of the quantum properties of the observed system. In contrast, the time has an uncertainty only indirectly. If
we ask when a quantum-mechanical observable reaches some
specified value or when it changes by some specified amount, the
answer is somewhat uncertain because the value of the observable
is uncertain, and we cannot be sure just when the specified change
has occurred. This gives rise to the uncertainty δt in Eq. (30).

   Of course, the general definition (30) for δt includes the simple example of the moving wave packet as a special case. To see
this, take $B = x_{\text{op}}$ in Eq. (30). Then

$$\delta t = \frac{\Delta x}{\left| \dfrac{d}{dt} \langle\psi|x_{\text{op}}|\psi\rangle \right|} = \frac{\Delta x}{\left| \langle\psi|p_{\text{op}}|\psi\rangle/m \right|} \qquad (33)$$

where the first Ehrenfest equation has been used. Since $\langle\psi|p_{\text{op}}|\psi\rangle/m$ is the average velocity of the wave packet, this expression for δt agrees with Eq. (25).

For another simple example, consider some quantum-mechanical system in a *stationary* state. In such a state, $\Delta E = 0$ and consequently the energy–time uncertainty relation yields $\delta t = \infty$. This is obviously correct, since in a stationary state the expectation values of all observables (assumed to have no explicit time dependence) remain constant.

As a more interesting example, consider an atom that is initially in an excited state and spontaneously decays to the ground state by emission of a photon. A typical lifetime for this process is $\tau \simeq 10^{-8}$ s, and this lifetime can be regarded as the characteristic time required for the relevant observables to change by an appreciable amount (in this context, a relevant observable is any observable that changes during the decay, for instance, $B = x_{\text{op}}{}^2$ or $B = p_{\text{op}}{}^2$). The energy–time uncertainty relation then demands that the decaying state have an uncertainty in its energy:

$$\Delta E \geq \frac{\hbar}{2\tau} \simeq \frac{6 \times 10^{-22}\ \text{MeV·s}}{2 \times 10^{-8}\ \text{s}} \simeq 10^{-7}\ \text{eV} \tag{34}$$

Note that we have ascribed this uncertainty to the initial (excited) state, not to the final (ground) state. After the atom reaches its ground state, no further decay process can occur. The ground state, and only the ground state, is an absolutely stable state; it is an *exact* stationary state, with $\Delta E = 0$ and $\delta t = \infty$.

This raises the question: What happens to the initial uncertainty of the energy? It seems unlikely that the uncertainty of the energy of an isolated system could decrease in time. In fact, we can prove a theorem about the uncertainty in the energy of an isolated system:

> *The uncertainty in energy of an isolated system remains constant in time.*

The proof is simple. Consider the energy uncertainty squared; it is equal to the expectation value of the operator $(H - \langle\psi|H|\psi\rangle)^2$,

$$(\Delta E)^2 = \langle\psi|(H - \langle\psi|H|\psi\rangle)^2|\psi\rangle$$

Since $H - \langle\psi|H|\psi\rangle$ is time independent, we obtain

$$\frac{d}{dt}(\Delta E)^2 = \frac{d}{dt}\langle\psi|(H - \langle\psi|H|\psi\rangle)^2|\psi\rangle$$

$$= \frac{i}{\hbar}\langle\psi|[H, (H - \langle\psi|H|\psi\rangle)^2]|\psi\rangle = 0 \tag{35}$$

If we apply this theorem to the decaying atom, we conclude that the final uncertainty should equal the initial uncertainty, since the decay is spontaneous, without any external disturbance. However, we must remember that the theorem applies only to a complete *isolated system*, that is, it applies to the system of atom plus emitted photon. Since the final uncertainty of the energy of the atom is zero, the photon must carry away all of the uncertainty. In our example, the energy of the emitted photon therefore has an uncertainty of 10^{-7} eV.[4] This uncertainty, or the corresponding uncertainty in the frequency, is often called the *natural linewidth* of emitted photon.

A similar situation obtains in a nucleus undergoing alpha or beta decay. The appropriate value for δt is then again the lifetime τ of the decaying nucleus, since this is the characteristic time for an appreciable change to occur in the nucleus. The energy–time uncertainty relation then requires an uncertainty in the energy of the decaying nucleus, and, later, an uncertainty in the energy of the emitted alpha or beta ray.

In particle physics, Eq. (31) is commonly used to determine the lifetime of a short-lived particle, or "resonance," from the measured value of the uncertainty in the formation energy (or the decay energy) of the particle. For instance, the measured rest-mass energy of the particle $N^*(1688)$ is known to have an intrinsic uncertainty $\Delta E \simeq 100$ MeV. Therefore,

$$\tau \simeq \frac{\hbar}{2\,\Delta E} = \frac{6 \times 10^{-22}\ \text{MeV·s}}{2 \times 100\ \text{MeV}} \simeq 10^{-24}\ \text{s} \tag{36}$$

A particle of such a short lifetime cannot be observed directly by a track in a bubble chamber, but its existence can be inferred from correlations observed in the momenta of decay products (pions and nucleons in this case).

It is always true that for a system subject to a decay process, the characteristic time δt is roughly equal to the lifetime. Obviously, in a decaying system there must exist at least one observable that changes by an appreciable amount within a lifetime. If there were no such observable, we could never know whether the decay has taken place. When we say that the system has decayed, we mean that something *has* changed.

---

[4] We neglect energy that might be transferred to the recoil of the atom.

5.4 *Representations of State Vectors and Their Transformation*

In Section 5.1 we formulated the equation for the time evolution of the state of a particle in the position and in the momentum representations. For this, we needed to express the state vector and the Hamiltonian operator in these representations. The theory of the representations of state vectors and operators and of the transformations between different representations is of great interest in quantum mechanics. Here, we will discuss representations in general. Many of the formulas in this section will look familiar because we have already encountered them in our discussion of the special case of the energy representation.

Suppose that we have a complete set of orthonormal states. We designate the vectors in this set by $|\psi_1\rangle, |\psi_2\rangle, |\psi_3\rangle, \ldots$. This set of vectors will usually be infinite, but for simplicity we assume that we are dealing with a discrete set. Any arbitrary state vector $|\chi\rangle$ can be written as a superposition

$$|\chi\rangle = \sum_n a_n |\psi_n\rangle \tag{37}$$

where

$$a_n = \langle \psi_n | \chi \rangle \tag{38}$$

We can regard the vectors $|\psi_1\rangle, |\psi_2\rangle, |\psi_3\rangle, \ldots$ as forming an orthogonal basis of unit vectors. The coefficient a_n can then be interpreted as giving the component of $|\chi\rangle$ along the unit vector $|\psi_n\rangle$. The set of coefficients $a_1, a_2, a_3, \ldots$ therefore describes the state vector by giving its components with respect to a basis. The set of these coefficients provides us with a representation of the state vector. We can arrange the coefficients as a column vector,

$$\mathbf{a} \equiv \begin{pmatrix} a_1 \\ a_2 \\ a_3 \\ \vdots \end{pmatrix} \tag{39}$$

Thus, any state vector is represented by a column vector. It is convenient to write the column vector (39) in vector notation simply as **a**.

The scalar product of two state vectors is represented by the scalar product of the corresponding column vectors:

$$\langle\phi|\chi\rangle = \sum_{m,n} (\langle\psi_m|b_m\rangle(a_n|\psi_n\rangle)$$

$$= \sum_{m,n} b_m{}^* a_n \langle\psi_m|\psi_n\rangle$$

$$= \sum_{m,n} b_m{}^* a_n \delta_{mn} = \sum_n b_n{}^* a_n \tag{40}$$

The last expression is the usual scalar product $\mathbf{b}^* \cdot \mathbf{a}$ of two column vectors.

Since state vectors are represented by column vectors, linear operators will have to be represented by matrices. Suppose that an operator Q transforms $|\chi\rangle$ into $|\phi\rangle$:

$$|\phi\rangle = Q|\chi\rangle \tag{41}$$

If $|\chi\rangle$ is represented by $\mathbf{a}$ and $|\phi\rangle$ by $\mathbf{b}$, then we have

$$b_n = \langle\psi_n|\phi\rangle = \langle\psi_n|Q|\chi\rangle$$

$$= \langle\psi_n|Q \left(\sum_k a_k|\psi_k\rangle\right)$$

$$= \sum_k \langle\psi_n|Q|\psi_k\rangle a_k \tag{42}$$

This shows that the column vector $\mathbf{b}$ is obtained from $\mathbf{a}$ by multiplication by a matrix. We designate the components of this matrix by Q_{nk}, with

$$Q_{nk} = \langle\psi_n|Q|\psi_k\rangle \tag{43}$$

and we can then write

$$b_n = \sum_k Q_{nk} a_k \tag{44}$$

In vector notation this reads

$$\mathbf{b} = Q\mathbf{a} \tag{45}$$

where Q is the matrix with components Q_{nk}.

Not only is the multiplication of a state vector by an operator represented by matrix multiplication, but the multiplication of two operators by each other is represented by the corresponding multiplication of two matrices. Thus, suppose we want to know what matrix represents QR. Obviously, this matrix is

$$(QR)_{nk} = \langle \psi_n | QR | \psi_k \rangle \tag{46}$$

But if we insert the completeness relation $\mathbf{1} = \sum |\psi_l\rangle \langle \psi_l|$ between the operators Q and R, we find that the left side of Eq. (46) equals

$$\langle \psi_n | Q \mathbf{1} R | \psi_k \rangle = \sum_l \langle \psi_n | Q | \psi_l \rangle \langle \psi_l | R | \psi_k \rangle$$

$$= \sum_l Q_{nl} R_{lk} \tag{47}$$

The last expression is precisely the matrix product of the matrices $\mathbf{Q}$ and $\mathbf{R}$.

Exercise 2. Show that a hermitian operator is represented by a hermitian matrix, that is, a matrix such that

$$Q_{mn} = Q_{nm}{}^*$$

Exercise 3. Show that if the basis vectors $|\psi_1\rangle$, $|\psi_2\rangle$, $|\psi_3\rangle$, . . . of the representation are eigenvectors of an operator R, then the matrix representing R is diagonal.

Let us now ask what happens if we change the basis vectors. Suppose that we have a second complete set of orthonormal vectors: $|\psi_1'\rangle$, $|\psi_2'\rangle$, $|\psi_3'\rangle$, ... With respect to this new basis, the state vector $|\chi\rangle$ of Eq. (37) is

$$|\chi\rangle = \sum_n a_n' |\psi_n'\rangle \tag{48}$$

The state vector now has components $a_n' = \langle \psi_n' | \chi \rangle$, which can be written as a column vector a'. The column vectors a and a' represent the same state vector $|\chi\rangle$, but they are different. In the language of the familiar displacement vectors in three-dimensional space, we would say that the components of a vector change when we change the coordinate system. How are the components related?

We can regard the change of basis as an operation. We define an operator $U^\dagger$ that transforms the old basis vectors into the new ones:[5]

---

[5] There is no good reason for the use of $U^\dagger$ (rather than U) in Eq. (49); it is a convention.

$$|\psi_n{}'\rangle = U^\dagger|\psi_n\rangle \tag{49}$$

We will show that the operator U has the property

$$UU^\dagger = 1 \qquad \text{and} \qquad U^\dagger U = 1 \tag{50}$$

This means that $U^\dagger$ is the inverse of U,

$$U^\dagger = U^{-1}. \tag{51}$$

An operator having the property (50) [or, equivalently, the property (51)] is said to be *unitary*.

To show that Eq. (50) holds, consider

$$\langle\psi_m{}'|\psi_n{}'\rangle = (\langle\psi_m|U^\dagger)\,(U^\dagger|\psi_n\rangle) = \langle\psi_m|UU^\dagger|\psi_n\rangle \tag{52}$$

Since $\langle\psi_m{}'|\psi_n{}'\rangle = \delta_{mn}$, we have

$$\langle\psi_m|UU^\dagger|\psi_n\rangle = \delta_{mn} \tag{53}$$

This can be true for all m and n only if $UU^\dagger = 1$.

Exercise 4. Show that if $UU^\dagger|\psi_n\rangle \neq |\psi_n\rangle$ for some n, then Eq. (53) fails for some m.

Exercise 5. Starting with $|\psi_n\rangle = U|\psi_n{}'\rangle$, show that $U^\dagger U = 1$.

The column vectors **a** and **a**$'$ are related in the following way:

$$\begin{aligned}
a_n{}' &= \langle\psi_n{}'|\psi\rangle = (\langle\psi_n|U^\dagger)\,|\psi\rangle \\
&= \langle\psi_n|U|\psi\rangle \\
&= \sum_m \langle\psi_n|U|\psi_m\rangle\,\langle\psi_m|\psi\rangle \\
&= \sum_m U_{nm}a_m
\end{aligned} \tag{54}$$

This can also be written as

$$\mathbf{a}' = U\mathbf{a} \tag{55}$$

that is, the column vectors in the two different representations are related by a matrix transformation using the unitary matrix U.

Exercise 6. Show that

$$U_{nm} = \langle\psi_n{}'|\psi_m\rangle \tag{56}$$

The scalar product of two state vectors should have a value independent of which representation we use to evaluate it. We can check that Eq. (55) does satisfy this requirement:

$$\sum_n b_n'^* a_n' = \sum_n \left(\sum_k U_{nk} b_k \right)^* \left(\sum_l U_{nl} a_l \right)$$

$$= \sum_{n,k,l} b_k^* U_{nk}^* U_{nl} a_l$$

$$= \sum_{n,k,l} b_k^* (U^\dagger)_{kn} U_{nl} a_l$$

$$= \sum_{k,l} b_k^* \delta_{kl} a_l = \sum_k b_k^* a_k \tag{57}$$

In particular, the norm of any vector is preserved under a change of representation.

Finally, we want to establish the relation between the matrices representing an operator in different representations. If we have an operator Q, it is represented by a matrix Q_{mn} in one representation and by Q_{mn}' in the other, and these matrices are related as follows:

$$Q_{mn}' = \langle \psi_m' | Q | \psi_n' \rangle = (\langle \psi_m | U^\dagger) Q (U^\dagger | \psi_n \rangle)$$

$$= \langle \psi_m | U Q U^\dagger | \psi_n \rangle$$

Inserting the completeness relation once between U and Q and once more between Q and $U^\dagger$, we obtain

$$Q'_{mn} = \sum_{k,l} \langle \psi_m | U | \psi_k \rangle \langle \psi_k | Q | \psi_l \rangle \langle \psi_l | U^\dagger | \psi_n \rangle$$

$$= \sum_{k,l} U_{mk} Q_{kl} U_{ln}^\dagger \tag{58}$$

In matrix notation this is simply

$$Q' = UQU^\dagger \tag{59}$$

Such a transformation of a matrix (left multiplication by U and right multiplication by $U^\dagger$) is called a *similarity transformation* of the matrix Q by the unitary matrix U. This transformation leaves algebraic relations among several matrices invariant. For example, suppose that we have three matrices A, B, C, and we know that

$$C = AB \tag{60}$$

Then

$$C' = UCU^\dagger = UABU^\dagger$$

$$= UAU^\dagger UBU^\dagger$$

$$= A'B' \tag{61}$$

which has the same *form* as Eq. (60).

Our earlier result that the commutator of the operators x_{op}, p_{op} has the same form in the position, momentum, and energy representations is a special case of the rule that similarity transformations leave the algebra of matrices invariant. There is only a slight difficulty in that the position and the momentum representations do not use a discrete set of basis vectors, but rather a continuous set. In dealing with a representation involving a continuum of basis vectors, summations over indices must be replaced by integrations. For instance, in the position representation, we use as basis vectors the position vectors $|x\rangle$, where x is any real number. As we know from our earlier discussion of this representation, states are represented by $\langle x|\psi\rangle$, usually called the wavefunction $\psi(x)$. The product of two state vectors is represented by an integral:

$$\langle \phi|\psi\rangle = \int dx\, \langle \phi|x\rangle\, \langle x|\psi\rangle = \int dx\, \phi^*(x)\, \psi(x) \tag{62}$$

This is essentially Eq. (40), with $\phi^*(x)$ and $\psi(x)$ taking the place of b_n^* and a_n. The continuous variable x has taken the place of the discrete index n.

Operators are represented by matrix elements with two continuous indices taking the place of the two discrete indices. By analogy with Eq. (43), the matrix elements in the position representation are $\langle x'|Q|x\rangle$, and the action of an operator on a state vector is obtained by the usual rules of matrix multiplication. Thus, in the position representation, Eq. (44) becomes

$$\phi(x) = \int dx'\, \langle x|Q|x'\rangle\, \psi(x') \tag{63}$$

As a specific example, consider $Q = x_{op}$. In this case

$$\langle x|Q|x'\rangle = \langle x|x_{op}|x'\rangle = \langle x|x'|x'\rangle = x'\langle x|x'\rangle = x'\delta(x' - x)$$

and if we multiply the "vector" $\psi(x')$ by the "matrix" $\langle x|x_{op}|x'\rangle$, we obtain

$$\int dx'\, \langle x|x_{op}|x'\rangle\, \psi(x') = \int dx'\, \delta(x - x')\, x'\, \psi(x')$$

$$= x\psi(x) \tag{64}$$

Thus, in this simple example, the result is ordinary multiplication of the function $\psi(x)$ by x. In the general case, Eq. (63) shows that operators in the x representation are *integral operators*.

Let us now reexamine the calculations in Section 5.1 that led us to the expressions for the Schrödinger equation in the position and in the momentum representations. From the point of view of the general theory of representations, the integral appearing on the right side of Eq. (5) is the product of the "vector" $\psi(x', t)$ by the "matrix" $\langle x|H|x'\rangle$. In the evaluation of this product, we found that the term arising from the potential [that is, the term $\langle x|V(x_{\mathrm{op}})|x'\rangle$ in Eq. (7)] was easy to calculate. As in Eq. (64), the product of the vector $\psi(x', t)$ by the matrix $\langle x|V(x_{\mathrm{op}})|x'\rangle$ is simply the ordinary product $V(x)\psi(x, t)$. However, we found that the term arising from the kinetic energy (that is, the term $\langle x|\ p_{\mathrm{op}}^2/2m\ |x'\rangle$) was more difficult to handle, and we evaluated it by a devious comparison between the forms this term takes in the position and in the momentum representations. For a more straightforward evaluation of this term, we can proceed as follows. As a first step, insert the completeness relation $\mathbf{1} = \int |p\rangle\langle p|\ dp$ on each side of the kinetic-energy operator:

$$\langle x|\frac{p_{\mathrm{op}}^2}{2m}\ |x'\rangle = \iint dp\ dp'\ \langle x|p\rangle\langle p|\frac{p_{\mathrm{op}}^2}{2m}|p'\rangle\langle p'|x'\rangle \tag{65}$$

This expresses the matrix components $\langle x|\ p_{\mathrm{op}}^2/2m\ |x'\rangle$ of the position representation in terms of the matrix components $\langle p|\ p_{\mathrm{op}}^2/2m\ |p'\rangle$ of the momentum representation. Note that Eq. (65) is essentially Eq. (58) written with continuous indices and (integrals) rather than discrete indices (and sums). Since $|p'\rangle$ is an eigenstate of p_{op}, the left side of Eq. (65) equals

$$\iint dp\ dp'\ \langle x|p\rangle\langle p|p'\rangle\frac{p'^2}{2m}\langle p'|x'\rangle$$

$$= \iint dp\ dp'\ \langle x|p\rangle\ \delta(p - p')\frac{p'^2}{2m}\langle p'|x'\rangle$$

$$= \int dp\ \langle x|p\rangle\frac{p^2}{2m}\langle p|x'\rangle \tag{66}$$

If we use the explicit expressions for the wavefunctions $\langle x|p\rangle$ and $\langle p'|x'\rangle$, we obtain

$$\int dp\ \frac{e^{ipx/\hbar}}{\sqrt{2\pi\hbar}}\frac{p^2}{2m}\frac{e^{-ipx'/\hbar}}{\sqrt{2\pi\hbar}} = \int dp\left(-\frac{\hbar^2}{2m}\right)\frac{\partial^2}{\partial x^2}\frac{e^{ip(x - x')/\hbar}}{2\pi\hbar} \tag{67}$$

$$= -\frac{\hbar^2}{2m}\delta''(x - x') \tag{68}$$

The product of this matrix and the vector $\psi(x', t)$ is then

$$\int dx' \, \langle x| \frac{p_{op}^2}{2m} |x'\rangle \, \psi(x', t) = \int dx' \left[-\frac{\hbar^2}{2m} \delta''(x - x') \right] \psi(x', t)$$

$$= -\frac{\hbar^2}{2m} \frac{\partial^2}{\partial x^2} \psi(x, t) \tag{69}$$

which is the same result as we obtained by the shortcut in Section 5.1.

We can perform a similar straightforward evaluation of the potential-energy term $\langle p|V(x_{op})|p'\rangle$ that appears in the momentum representation. A calculation similar to Eqs. (65)–(68) leads to

$$\langle p|V(x_{op})|p'\rangle = \int dx \, V(x) \frac{e^{i(p' - p)x/\hbar}}{2\pi\hbar} \tag{70}$$

If $V(x)$ is a polynomial, then the factor $V(x)$ in the integrand can be replaced by $V(i\hbar \, \partial/\partial p)$, and the result for the product of the matrix $\langle p|V(x_{op})|p'\rangle$ and the vector $\phi(p', t)$ is as given in Eq. (14).

Exercise 7. Perform the calculations leading to Eq. (70), and verify that the final result is Eq. (14).

However, if $V(x)$ is not a polynomial, then we cannot replace x by $i\hbar \, \partial/\partial p$, and we can do nothing further with Eq. (70), unless we are given an explicit expression for the function $V(x)$. With an explicit expression, we can integrate Eq. (70). The integral in Eq. (70) is simply the Fourier transform of $V(x)$. Designating this Fourier transform by $v(p - p')$, we can write the product of the matrix $\langle p|V(x_{op})|p'\rangle$ and the vector $\phi(p', t)$ as

$$\int dp' \, v(p - p')\phi(p', t) \tag{71}$$

The Schrödinger equation in the momentum representation then becomes

$$i\hbar \frac{\partial}{\partial t} \phi(p, t) = \frac{p^2}{2m} \phi(p, t) + \int dp' \, v(p - p') \, \phi(p', t) \tag{72}$$

Thus, the Schrödinger equation in this representation is a rather complicated differential-integral equation.

PROBLEMS   1. Verify that

$$|\psi(t)\rangle = e^{-iHt/\hbar} |\psi(0)\rangle$$

is a formal solution of the differential equation (1). Here, the operator $e^{-iHt/\hbar}$ must be interpreted as a Taylor-series expansion.

2. Show that if $V(x_{op})$ is a polynomial in x_{op}, then $V(x_{op})|x\rangle = V(x)|x\rangle$.

3. Consider a Gaussian packet representing a particle at rest, that is, a packet of zero expectation value of the momentum. From Eq. (2.70), with $p_0 = 0$, evaluate the uncertainty ΔE of the energy (as rms deviation from the mean). Compare ΔE with the characteristic spreading time t given by Eq. (2.78). Is the energy–time uncertainty relation satisfied?

4. Show that if $|\psi(t)\rangle$ is any state of a free particle, then the mean-square deviation of the momentum is constant, that is,

$$\frac{d}{dt} (\Delta p)^2 = 0$$

where $(\Delta p)^2 = \langle \psi(t)|p_{op}^2|\psi(t)\rangle - (\langle \psi(t)|p_{op}|\psi(t)\rangle)^2$

5. Show that for any stationary state of a particle moving in a potential in one dimension the expectation value of $x_{op}p_{op} + p_{op}x_{op}$ is zero,

$$\langle \psi(t)| (x_{op}p_{op} + p_{op}x_{op}) |\psi(t)\rangle = 0$$

(Hint: Consider the expectation value of $[H, x_{op}^2]$.)

6. Prove the quantum-mechanical virial theorem,

$$\langle \psi(t)| x_{op} \frac{\partial V}{\partial x_{op}} |\psi(t)\rangle = 2\langle \psi(t)| K |\psi(t)\rangle$$

where V is the potential-energy operator and K is the kinetic-energy operator.

7. (a) Suppose that a particle in an infinite square well is in an energy eigenstate $|\psi(t)\rangle$, with one of the eigenfunctions given by Eq. (3.32). Evaluate the expectation values

$$\langle \psi(t)|x_{op}|\psi(t)\rangle \quad \text{and} \quad \langle \psi(t)|p_{op}|\psi(t)\rangle$$

for such a state. Evaluate the time derivatives

$$\frac{d}{dt} \langle \psi(t)|x_{op}|\psi(t)\rangle \quad \text{and} \quad \frac{d}{dt} \langle \psi(t)|p_{op}|\psi(t)\rangle$$

Are the results consistent with the Ehrenfest equations (22) and (23)?

(b) Suppose that the particle is in a superposition of the ground state and the first excited state,

$$|\psi(t)\rangle = \frac{1}{\sqrt{2}} |\psi_1(t)\rangle + \frac{1}{\sqrt{2}} |\psi_2(t)\rangle$$

Evaluate the expectation values $\langle\psi(t)|x_{op}|\psi(t)\rangle$ and $\langle\psi(t)|p_{op}|\psi(t)\rangle$ for this state. Evaluate the time derivatives $d/dt\ \langle\psi(t)|x_{op}|\psi(t)\rangle$ and $d/dt\ \langle\psi(t)|p_{op}|\psi(t)\rangle$. Are the results consistent with the Ehrenfest equations (22) and (23)?

8. Show that the matrix elements x_{kn} and p_{kn} for the stationary states of a particle bound in any kind of potential are related by

$$\frac{1}{m} p_{kn} = \frac{E_n - E_k}{i\hbar} x_{kn}$$

Verify that this relation is satisfied by the matrix elements of the infinite square well given in Eqs. (3.64) and (3.72).

9. Consider a particle in an infinite square well, with energy eigenfunctions given by Eq. (3.32). Suppose the particle is in the state

$$|\psi(t)\rangle = \frac{1}{2} |\psi_1(t)\rangle + \frac{\sqrt{3}}{2} |\psi_2(t)\rangle$$

(a) What is the value of ΔE for this state?

(b) What are the values of Δx and of $d/dt\ \langle\psi|x_{op}|\psi\rangle$ at time t? What is the maximum value of $d/dt\ \langle\psi|x_{op}|\psi\rangle$?

(c) If we take $\delta t = \Delta x/(d/dt\ \langle\psi|x_{op}|\psi\rangle)$, calculated with the maximum value of $d/dt\ \langle\psi|x_{op}|\psi\rangle$ and the corresponding value of Δx, as an estimate for the characteristic time for the evolution of the system, is the energy-time uncertainty relation satisfied?

(d) Suggest some other way of defining a characteristic time for the evolution of the system, by means of some observable operator other than x_{op}.

10. Show that if c is an eigenvalue of a unitary operator, then $cc^* = 1$.

11. Consider the operators $x_{op}p_{op}$ and $\frac{1}{2}(x_{op}p_{op} + p_{op}x_{op})$. What are the explicit forms of these operators in the position representation? In the momentum representation? Which of the two operators $x_{op}p_{op}$ and $\frac{1}{2}(x_{op}p_{op} + p_{op}x_{op})$ can be regarded as a suitable quantum-mechanical equivalent of the classical observable [position] × [momentum]?

12. Consider the classical observable $x^2 p^2$. Several possible quantum-mechanical operators that can be regarded as equivalent to this classical observable are $x_{op} p_{op}^2 x_{op}$, $p_{op} x_{op}^2 p_{op}$, $\frac{1}{2}(x_{op}^2 p_{op}^2 + p_{op}^2 x_{op}^2)$, and $\frac{1}{2}(x_{op} p_{op} x_{op} p_{op} + p_{op} x_{op} p_{op} x_{op})$

(a) Show that all these quantum-mechanical operators are hermitian.

(b) Show that first two operators are equal, but that the last two differ from the first two, and evaluate the difference.

6

The Factorization Method

If we want to determine the stationary states of a system, we have to solve the energy eigenvalue equation (5.17). In the position representation, this is a differential equation, the time-independent Schrödinger equation, and we are faced with the problem of finding those values of the energy E for which there exists a solution of this differential equation with the "right" boundary conditions. This is a familiar problem in mathematical physics, where it appears in the theory of normal modes of classical systems such as vibrating strings, vibrating membranes, electromagnetic resonant cavities, and so on. The traditional method of solution of the differential equation describing such a system is by an infinite series with unknown coefficients.

However, we will *not* use this traditional method here, because it would confine us to the position representation, and we want to avoid excessive emphasis on this representation. Instead of solving the eigenvalue equation in the position representation, we will solve it in a way that is independent of the choice of representation. We will deal with the eigenvalue problem by means of a general and elegant operator method, called the *factorization method.* This method outflanks the traditional power-series method for the solution of differential equations; it proceeds to a direct stepwise construction of the eigenvalues and eigenvectors by manipulation of the quantum-mechanical operators and their commutation relations. In the last section of this chapter, we will compare this method with the traditional method of solving the Schrödinger equation, and we will examine the advantages and the disadvantages of each method.

6.1 *The Harmonic Oscillator*

We begin with the solution of the eigenvalue equation for a particle in a harmonic-oscillator potential, $V = \frac{1}{2}kx^2$. This solution is of considerable interest, because, with a suitable choice of the "spring" constant k, the harmonic-oscillator potential provides an approximate description of the forces that act on diverse physical systems vibrating about their equilibrium points. For instance, the oscillation of the atoms in a diatomic molecule can be described by a harmonic-oscillator potential. Furthermore, the equations for the normal modes of vibration of complicated physical systems are often mathematically equivalent to harmonic-oscillator equations. Thus, the vibrations of a crystal lattice and the vibrations of the electromagnetic field are described by harmonic-oscillator equations; the energy quanta associated with these harmonic oscillators of the crystal lattice and of the electromagnetic field are, respectively, phonons and photons. Besides, the harmonic-oscillator eigenfunctions form a very convenient complete set of functions, which is sometimes used even in problems where there is no harmonic-oscillator potential.

For a particle of mass m moving in a harmonic-oscillator potential, the Hamiltonian operator is

$$H = \frac{p_{\mathrm{op}}^2}{2m} + \frac{1}{2}\, m\omega^2 x^2 \tag{1}$$

where $\omega^2 \equiv k/m$. This Hamiltonian operator has been obtained from the classical Hamiltonian by the usual prescription involving the replacements $x \to x_{\mathrm{op}}$, $p \to p_{\mathrm{op}}$. We will look for stationary states; that is, we will look for state vectors of the form

$$e^{-iE_n t/\hbar}|E_n\rangle \tag{2}$$

where $|E_n\rangle$ is a time-independent energy eigenvector. Such an energy eigenvector must satisfy the eigenvalue equation $H|E_n\rangle = E_n|E_n\rangle$, or

$$\left(\frac{p_{\mathrm{op}}^2}{2m} + \frac{1}{2}\, m\omega^2 x_{\mathrm{op}}^2\right)|E_n\rangle = E_n|E_n\rangle \tag{3}$$

To construct the solution of this eigenvalue equation, define operators a and $a^\dagger$ as follows:

$$a = \frac{1}{\sqrt{2}}\left(\frac{1}{x_0}\, x_{\mathrm{op}} + \frac{ix_0}{\hbar}\, p_{\mathrm{op}}\right) \tag{4}$$

$$a^\dagger = \frac{1}{\sqrt{2}} \left(\frac{1}{x_0} x_{op} - \frac{ix_0}{\hbar} p_{op} \right) \tag{5}$$

where x_0 is a constant, $x_0 = \sqrt{\hbar/m\omega}$. We will hereafter omit the "op" labels on x and p, and simply write

$$a = \frac{1}{\sqrt{2}} \left(\frac{1}{x_0} x + \frac{ix_0}{\hbar} p \right) \tag{6}$$

$$a^\dagger = \frac{1}{\sqrt{2}} \left(\frac{1}{x_0} x - \frac{ix_0}{\hbar} p \right) \tag{7}$$

The operators x and p can be expressed in terms of a and $a^\dagger$ as follows:

$$x = \frac{x_0}{\sqrt{2}} (a + a^\dagger)$$

$$p = \frac{\hbar}{i\sqrt{2}x_0} (a - a^\dagger) \tag{8}$$

Exercise 1. Show that the expressions (8) follow from Eq. (7).

The operators $a^\dagger$ and a are called, respectively, *creation* and *destruction* operators, or, alternatively, *raising* and *lowering* operators; the reason for these names will become clear later. From the commutation relation $[x, p] = i\hbar$, it is easy to find the commutation relation for a and $a^\dagger$:

$$[a, a^\dagger] = 1 \tag{9}$$

Exercise 2. Verify this commutation relation.

By means of Eqs. (6) and (7), we can rewrite the Hamiltonian in terms of a and $a^\dagger$:

$$H = \frac{1}{2m} p^2 + \frac{1}{2} m\omega^2 x^2$$

$$= -\frac{\hbar^2}{4mx_0^2} (a - a^\dagger)(a - a^\dagger) + \frac{m\omega^2 x_0^2}{4} (a + a^\dagger)(a + a^\dagger)$$

$$= \frac{\hbar\omega}{2} (aa^\dagger + a^\dagger a)$$

In view of the commutation relation (9), this equals

$$H = \hbar\omega \left(a^\dagger a + \frac{1}{2} \right) \tag{10}$$

With this expression for H, we can readily show that

$$Ha = aH - a\hbar\omega$$
$$Ha^\dagger = a^\dagger H + a^\dagger \hbar\omega \tag{11}$$

These two equations will prove useful for the calculations in the next paragraph.

Exercise 3. Use the commutation relation for a and $a^\dagger$ to obtain the two equations (11).

Suppose now that $|E_n\rangle$ is an energy eigenvector, that is, $H|E_n\rangle = E_n|E_n\rangle$. Then, in consequence of Eq. (11), we find

$$H(a^\dagger|E_n\rangle) = (a^\dagger H + a^\dagger \hbar\omega)|E_n\rangle = a^\dagger E_n|E_n\rangle + a^\dagger \hbar\omega|E_n\rangle$$
$$= (E_n + \hbar\omega)(a^\dagger|E_n\rangle) \tag{12}$$

This means that $a^\dagger|E_n\rangle$ is an eigenvector of H with eigenvalue $E_n + \hbar\omega$. Similarly, we find

$$H(a|E_n\rangle) = (E_n - \hbar\omega)(a|E_n\rangle) \tag{13}$$

which means that $a|E_n\rangle$ is an eigenvector of H with eigenvalue $E_n - \hbar\omega$. This is why the operators $a^\dagger$ and a are called raising and lowering operators—when they act on an energy eigenvector, they produce an eigenvector of increased and decreased eigenvalue, respectively.

Exercise 4. Verify Eq. (13).

Let us designate the energy eigenvalue of the ground state, or the state of lowest energy, by E_0:

$$H|E_0\rangle = E_0|E_0\rangle \tag{14}$$

But then

$$H(a|E_0\rangle) = (E_0 - \hbar\omega)a|E_0\rangle \tag{15}$$

This is impossible, since it says that $a|E_0\rangle$ is an eigenvector with eigenvalue *below* E_0. The only way to avoid this contradiction is to assume that when a acts on $|E_0\rangle$, the result is no vector; that is, the result is zero:

$$a|E_0\rangle = 0 \tag{16}$$

With this, we find that the expectation value of the energy in the ground state is

$$\langle E_0|H|E_0\rangle = \hbar\omega \langle E_0| \left(a^\dagger a + \frac{1}{2} \right) |E_0\rangle$$

$$= \hbar\omega \langle E_0| a^\dagger a |E_0\rangle + \frac{\hbar\omega}{2} \langle E_0|E_0\rangle$$

$$= 0 + \frac{\hbar\omega}{2} \tag{17}$$

Since the expectation value of the energy in an eigenstate equals the eigenvalue ($\langle E_0|H|E_0\rangle = E_0$), Eq. (17) implies that

$$E_0 = \tfrac{1}{2}\hbar\omega \tag{18}$$

We are now ready to construct all the energy eigenvalues and eigenvectors. According to Eq. (12), each time we multiply an eigenvector by $a^\dagger$, we obtain an eigenvector of an eigenvalue larger by $\hbar\omega$. Thus, if we multiply $|E_0\rangle$ repeatedly by $a^\dagger$, we obtain the eigenvectors

$$a^\dagger|E_0\rangle, \quad a^\dagger a^\dagger|E_0\rangle, \quad a^\dagger a^\dagger a^\dagger|E_0\rangle, \quad \ldots$$

with eigenvalues

$$\tfrac{1}{2}\hbar\omega + \hbar\omega, \quad \tfrac{1}{2}\hbar\omega + 2\hbar\omega, \quad \tfrac{1}{2}\hbar\omega + 3\hbar\omega, \quad \ldots.$$

In general, if we multiply the $|E_0\rangle$ by $(a^\dagger)^n$, we obtain an eigenvector of eigenvalue $\tfrac{1}{2}\hbar + n\hbar\omega$:

$$|E_n\rangle \propto (a^\dagger)^n|E_0\rangle \tag{19}$$

with

$$E_n = n\hbar\omega + \tfrac{1}{2}\hbar\omega \tag{20}$$

Since repeated multiplication of $|E_0\rangle$ by $a^\dagger$ yields a step-by-step construction of the eigenvectors, the raising operator $a^\dagger$ is sometimes called a "ladder" operator.

Figure 6.1 summarizes these eigenvalues and eigenvectors of the harmonic oscillator.

We have written Eq. (19) as a proportionality rather than an equality, because the right side is not correctly normalized. To adjust the normalization, let us examine what happens when we multiply $|E_n\rangle$ by $a^\dagger$. We know that the result must be proportional to $|E_{n+1}\rangle$, that is,

$$|E_{n+1}\rangle = ca^\dagger|E_n\rangle \tag{21}$$

where c is a constant of proportionality. What is the value of this constant c? We have

$$1 = \langle E_{n+1}|E_{n+1}\rangle = |c|^2(\langle E_n|a^\dagger)(a^\dagger|E_n\rangle) = |c|^2\langle E_n|aa^\dagger|E_n\rangle$$

In the last equation, it is understood that $aa^\dagger$ acts to the right. We now use

$$aa^\dagger = a^\dagger a + 1 = \left(a^\dagger a + \frac{1}{2}\right) + \frac{1}{2} = \frac{H}{\hbar\omega} + \frac{1}{2}$$

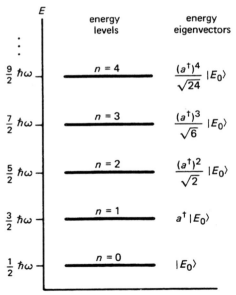

	energy levels	energy eigenvectors
$\frac{9}{2}\hbar\omega$	$n = 4$	$\dfrac{(a^\dagger)^4}{\sqrt{24}}\,|E_0\rangle$
$\frac{7}{2}\hbar\omega$	$n = 3$	$\dfrac{(a^\dagger)^3}{\sqrt{6}}\,|E_0\rangle$
$\frac{5}{2}\hbar\omega$	$n = 2$	$\dfrac{(a^\dagger)^2}{\sqrt{2}}\,|E_0\rangle$
$\frac{3}{2}\hbar\omega$	$n = 1$	$a^\dagger\,|E_0\rangle$
$\frac{1}{2}\hbar\omega$	$n = 0$	$|E_0\rangle$

Fig. 6.1 Energy eigenvalues and eigenvectors of the harmonic oscillator (a normalization factor has been included in the eigenvectors).

to obtain

$$1 = |c|^2 \left(\langle E_n | \frac{H}{\hbar\omega} | E_n \rangle + \frac{1}{2} \langle E_n | E_n \rangle \right)$$

$$= |c|^2 \left(n + \frac{1}{2} + \frac{1}{2} \right) = |c|^2 (n + 1) \qquad (22)$$

Hence,

$$|c| = \frac{1}{\sqrt{n+1}}$$

Since neither the eigenvalue equation nor the normalization condition determines the phase of the eigenvector, we are free to make any convenient choice of phase for c; for instance, we can choose c real and positive:

$$c = \frac{1}{\sqrt{n+1}}$$

We can then write Eq. (21) as

$$|E_{n+1}\rangle = \frac{1}{\sqrt{n+1}} \, a^\dagger |E_n\rangle \qquad (24)$$

Thus, $a^\dagger/\sqrt{n+1}$ can be regarded as a normalized raising operator, which preserves the normalization when it acts on the eigenvector $|E_n\rangle$.

If we now apply Eq. (24) successively with $n = 0, 1, 2, \ldots$, we see that the normalized eigenvector $|E_n\rangle$ is

$$|E_n\rangle = \frac{(a^\dagger)^n}{\sqrt{n!}} |E_0\rangle \qquad (25)$$

Exercise 5. Show that $a/\sqrt{n}$ is a normalized lowering operator, that is,

$$|E_{n-1}\rangle = \frac{1}{\sqrt{n}} \, a|E_n\rangle \qquad (26)$$

The ladder construction (25) for the excited states leaves open the question of whether we have found *all* the energy eigenvectors and eigenvalues. Could there be an energy eigenvalue *between* the values E_n given by Eq. (20), say, an energy eigenvalue between $\frac{3}{2}\hbar\omega$ and $\frac{5}{2}\hbar\omega$? If there were an eigenvector of such an intermediate eigenvalue, we could multiply it repeatedly by the lowering operator a. Each multiplication would yield an eigenvector of an eigenvalue lower by $\hbar\omega$ [see Eq. (13)]. This sequence of eigenvectors cannot end with an eigenvalue lower than E_0, since E_0 is supposed to be the lowest eigenvalue. Thus, the sequence of eigenvectors must end with some eigenvector $|E_0'\rangle$ for which $a|E_0'\rangle = 0$. But according to Eq. (17), if $a|E_0'\rangle = 0$, then the energy eigenvalue is necessarily $E_0' = \hbar\omega/2$, and thus the eigenvector $|E_0'\rangle$ coincides with the usual ground state $|E_0\rangle$. This means that there can be no eigenvectors besides those given by the ladder construction (25).

Thus, we have found all the eigenvectors and eigenvalues for the harmonic oscillator. It is to be emphasized that Eqs. (24) and (26) provide us with all the physically relevant information about the states of the harmonic oscillator—the expectation value of any physical observable can be evaluated by means of these equations. For instance, suppose that we want to know the expectation value of the potential energy in the state $|E_n\rangle$:

$$
\begin{aligned}
\langle E_n| \tfrac{1}{2}m\omega^2 x^2 |E_n\rangle &= \langle E_n| \tfrac{1}{4}m\omega^2 x_0^2(a + a^\dagger)(a + a^\dagger) |E_n\rangle \\
&= \tfrac{1}{4}m\omega^2 x_0^2 \, \langle E_n| \, (aa + aa^\dagger + a^\dagger a + a^\dagger a^\dagger) \, |E_n\rangle \\
&= \tfrac{1}{4}m\omega^2 x_0^2 \, \langle E_n| \, (a\sqrt{n} \, |E_{n-1}\rangle + a\sqrt{n+1} \, |E_{n+1}\rangle \\
&\quad + a^\dagger \sqrt{n} \, |E_{n-1}\rangle + a^\dagger \sqrt{n+1} \, |E_{n+1}\rangle) \\
&= \tfrac{1}{4}m\omega^2 x_0^2 \, \langle E_n| \, (\sqrt{n} \sqrt{n-1} \, |E_{n-2}\rangle \\
&\quad + \sqrt{n+1} \sqrt{n+1} \, |E_n\rangle + \sqrt{n} \sqrt{n} \, |E_n\rangle \\
&\quad + \sqrt{n+1} \sqrt{n+2} \, |E_{n+2}\rangle) \\
&= \tfrac{1}{4}m\omega^2 x_0^2 \, (0 + \sqrt{n+1} \sqrt{n+1} + \sqrt{n} \sqrt{n} + 0) \\
&= \tfrac{1}{4}\hbar\omega \, (2n + 1) \\
&= \tfrac{1}{2}E_n \qquad\qquad\qquad\qquad\qquad\qquad (27)
\end{aligned}
$$

This says that the expectation value of the potential energy is exactly one-half the energy of the state. This result corresponds to the classical virial theorem, according to which the *average* value of the potential energy of the harmonic oscillator is one-half the total energy.

Exercise 6. By a similar, explicit calculation show that the expectation value of the kinetic energy is also $\tfrac{1}{2}E_n$:

$$
\langle E_n| \frac{p^2}{2m} |E_n\rangle = \frac{1}{2} E_n
$$

Note that in the above calculation we had no need for the wavefunctions in the position representation. As a matter of fact, the calculation would have been much harder to do if we had used the wavefunctions. In almost all cases, expectation values are more easily calculated by operator methods than by explicit integration with the explicit wavefunctions in the position representation.

The general state of the harmonic oscillator will be a superposition of the stationary states $e^{-iE_n t/\hbar} |E_n\rangle$, as follows:

$$|\psi(t)\rangle = \sum_{n=0}^{\infty} c_n e^{-iE_n t/\hbar} |E_n\rangle$$

$$= \sum_{n=0}^{\infty} c_n e^{-i(n + 1/2)\omega t} |E_n\rangle \qquad (28)$$

This superposition has an interesting property: Suppose that we increase the time by an amount $\Delta t = 2\pi/\omega$, which in the classical theory would simply be the period of the oscillator. We then obtain

$$|\psi(t + \Delta t)\rangle = \sum_{n=0}^{\infty} e^{-i(n + 1/2)(t + 2\pi/\omega)\omega} |E_n\rangle$$

$$= e^{-i\pi} \sum_{n=0}^{\infty} e^{-i(n + 1/2)\omega t} |E_n\rangle$$

$$= - |\psi(t)\rangle$$

This says that after this time, only the sign of the state vector has changed. In the position representation, it means that the wavefunction differs only in sign, that is, the probability distributions $|\psi(x, t)|^2$ and $|\psi(x, t + \Delta t)|^2$ are the same. Thus, the probability distribution will return to its original shape after a classical period. This phenomenon is a peculiarity of the harmonic oscillator, and it occurs because the energy levels are evenly spaced.

We will next determine the explicit wavefunctions in the position representation. We begin with the ground state. In the position representation, the operator x is represented by ordinary multiplication, whereas the operator p is represented by $(\hbar/i)(d/dx)$. Hence the operators a and $a^{\dagger}$ must be

$$a = \frac{1}{\sqrt{2}} \left(\frac{x}{x_0} + x_0 \frac{d}{dx} \right)$$

$$a^\dagger = \frac{1}{\sqrt{2}} \left(\frac{x}{x_0} - x_0 \frac{d}{dx} \right) \tag{29}$$

The equation

$$a|E_0\rangle = 0 \tag{30}$$

can then be written in the position representation as

$$\frac{1}{\sqrt{2}} \left(\frac{x}{x_0} + x_0 \frac{d}{dx} \right) \psi_{E_0}(x) = 0 \tag{31}$$

This differential equation has the solution

$$\psi_{E_0}(x) \propto e^{-\frac{1}{2}x^2/x_0^2} \tag{32}$$

The normalization of this solution is fixed by requiring that

$$\int_{-\infty}^{\infty} |\psi_{E_0}|^2 \, dx = 1$$

From this, we readily find that the normalized ground-state wave-function is

$$\psi_{E_0}(x) = \frac{1}{\sqrt{x_0 \sqrt{\pi}}} e^{-\frac{1}{2}x^2/x_0^2} \tag{33}$$

Exercise 7. Verify that $\psi_{E_0}(x)$ is a solution of the differential equation (32), and verify that it satisfies the normalization condition.

All the other (normalized) eigenfunctions can now be found by making use of the version of Eq. (25) in the position representation:

$$\psi_{E_n}(x) = \frac{1}{\sqrt{n!}} \left[\frac{1}{\sqrt{2}} \left(\frac{x}{x_0} - x_0 \frac{d}{dx} \right) \right]^n \frac{1}{\sqrt{x_0 \sqrt{\pi}}} e^{-\frac{1}{2}x^2/x_0^2} \tag{34}$$

If we introduce the notation $\xi \equiv x/x_0$, then this becomes

$$\psi_{E_n}(x) = \frac{1}{\sqrt{n!}} \frac{1}{2^{n/2}} \frac{1}{\sqrt{x_0 \sqrt{\pi}}} \left(\xi - \frac{d}{d\xi} \right)^n e^{-\frac{1}{2}\xi^2} \tag{35}$$

Exercise 8. By induction, show that

$$\left(\xi - \frac{d}{d\xi}\right)^n e^{-\frac{1}{2}\xi^2} = (-1)^n e^{\frac{1}{2}\xi^2} \frac{d^n}{d\xi^n} e^{-\xi^2} \tag{36}$$

Equation (36) enables us to write $\psi_{E_n}(x)$ as follows:

$$\psi_{E_n}(x) = \frac{1}{2^{n/2}} \frac{1}{\sqrt{n!}} \frac{1}{\sqrt{x_0 \sqrt{\pi}}} e^{-\frac{1}{2}\xi^2} H_n(\xi) \tag{37}$$

where

$$H_n(\xi) \equiv (-1)^n e^{\xi^2} \frac{d^n}{d\xi^n} e^{-\xi^2} \tag{38}$$

From this formula, we readily see that $H_n(\xi)$ is a polynomial of order n. The polynomials $H_n(\xi)$ are called *Hermite polynomials*. The first few of these are

$$H_0 = 1$$

$$H_1 = 2\xi$$

$$H_2 = 4\xi^2 - 2$$

$$H_3 = 8\xi^3 - 12\xi \tag{39}$$

Exercise 9. From Eq. (38), show that H_n is a polynomial of order n, and obtain the explicit expressions (39) for H_0, H_1, H_2, and H_3.

Figure 6.2 gives plots of the first few eigenfunctions $\psi_{E_n}(x)$. Note that the wavefunction for the ground state is an even function of x, that is,

$$\psi_{E_0}(-x) = \psi_{E_0}(x)$$

The wavefunction for the first excited state is odd, that is,

$$\psi_{E_1}(-x) = -\psi_{E_1}(x)$$

The wavefunction for the second excited state is even, and so on.

The reversal of the coordinate x, from x to $-x$, is called the *parity operation*. Formally, it is represented by an operator P such that

$$P\psi(x) = \psi(-x) \tag{40}$$

When this operator acts on the energy eigenfunctions $\psi_{E_n}(x)$, the result is

$$P\psi_{E_n}(x) = \psi_{E_n}(-x) = (-1)^n\psi_{E_n}(x) \tag{41}$$

Thus, $\psi_{E_n}(x)$ is an eigenfunction of P with eigenvalue $(-1)^n$.

The harmonic-oscillator wavefunctions $\psi_{E_n}(x)$ form a complete orthonormal set. The completeness relation for these wavefunctions is

$$\delta(x - x') = \sum_n \psi_{E_n}(x)\psi_{E_n}(x')$$

$$= \sum_n \frac{1}{2^n} \frac{1}{n!} \frac{1}{x_0\sqrt{\pi}} \, e^{-\frac{1}{2}x^2/x_0^2} H_n(x/x_0) e^{-\frac{1}{2}x'^2/x_0^2} H_n(x'/x_0) \tag{42}$$

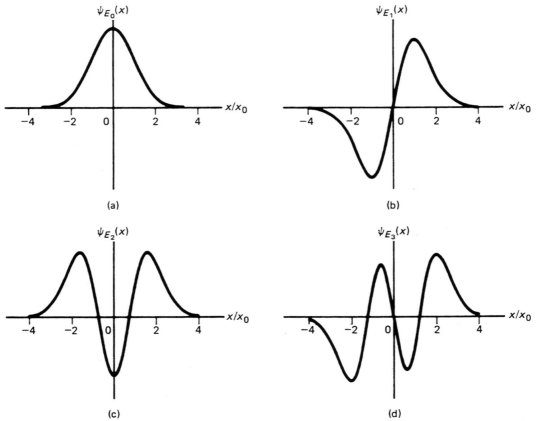

Fig. 6.2 The energy eigenfunctions ψ_{E_0}, ψ_{E_1}, ψ_{E_2}, and ψ_{E_3} for the harmonic oscillator.

The harmonic-oscillator wavefunctions can of course also be obtained by a brute-force solution of the time-independent Schrödinger equation. With the potential $V(x) = \frac{1}{2}m\omega^2 x^2$, the time-independent Schrödinger equation in the position representation takes the form

$$-\frac{\hbar^2}{2m}\frac{d^2}{dx^2}\psi(x) + \frac{1}{2}m\omega^2 x^2\,\psi(x) = E\psi(x) \tag{43}$$

This kind of differential equation can be solved by the power-series method. We will deal with this method in Section 6.4.

6.2 *Eigenvectors and Eigenvalues of a General Hamiltonian*

Suppose that we have some Hamiltonian operator H and we want to find its eigenvectors and eigenvalues. There exists a general operator method for doing this, called the "factorization method." As in the case of the harmonic oscillator, this factorization method hinges on expressing the Hamiltonian as a product of two operators, and these operators are then used for a ladder construction of the eigenvectors. However, in contrast to the simple case of the harmonic oscillator, where one single ladder operator was sufficient to construct all the eigenvectors, the general case requires a new ladder operator for each eigenvector.

The factorization method was first proposed by Schrödinger,[1] and it was further developed by Infeld and Hull[2] and by Green.[3] Our discussion of this method is patterned after that given by Green.

The construction of the required ladder operators proceeds as follows. We begin by finding operators η_1, η_2, η_3, ... and real constants E_1, E_2, E_3, ... such that

$$\eta_1{}^\dagger\eta_1 + E_1 = H \tag{44}$$

$$\eta_2{}^\dagger\eta_2 + E_2 = \eta_1\eta_1{}^\dagger + E_1$$

$$\eta_3{}^\dagger\eta_3 + E_3 = \eta_2\eta_2{}^\dagger + E_2, \dots$$

---

[1] E. Schrödinger, Proc. Roy. Irish Acad., **A46**, 9 (1940); Proc. Roy. Irish Acad., **A46**, 183 (1941); Proc. Roy. Irish Acad., **A47**, 53 (1941).

[2] L. Infeld and T. E. Hull, Rev. Mod. Phys. **23**, 21 (1951) and references cited therein.

[3] H. S. Green, *Matrix Methods in Quantum Mechanics* (Barnes & Noble, New York, 1968).

and, in general,

$$\eta_{j+1}{}^\dagger \eta_{j+1} + E_{j+1} = \eta_j \eta_j{}^\dagger + E_j \qquad j = 1, 2, 3, \ldots \qquad (45)$$

These equations are recursion relations for the operators η_j and the constants E_j. In general there will exist several solutions of these equations. But suppose that the η_j are chosen in such a way that all the E_j are as large as possible; then the solution of Eqs. (44) and (45) is unique.[4] We will show that the constants E_j chosen in this way are the eigenvalues. Furthermore, the operators η_j are ladder operators for the construction of the eigenvectors. The precise result is contained in the following theorem:

Suppose that the operators $\eta_1, \eta_2, \eta_3, \ldots$ and the real constants $E_1, E_2, E_3, \ldots$ satisfy Eqs. (44) and (45). Suppose further that each η_j has an eigenvector $|\zeta_j\rangle$ of eigenvalue zero

$$\eta_j |\zeta_j\rangle = 0 \qquad (46)$$

Then:

(i) *The constant E_j is the jth eigenvalue of H (arranged in ascending order).[5]*

(ii) *The corresponding eigenvector is (except for normalization)*

$$|E_j\rangle = \eta_1{}^\dagger \eta_2{}^\dagger \ldots \eta_{j-1}{}^\dagger |\zeta_j\rangle \qquad (47)$$

Before we give the proof, a few remarks will be helpful. Item (i) not only says that E_j is an eigenvalue, but also implies that there is no eigenvalue *between* E_{j-1} and E_j. In other words, if E is an eigenvalue of H, then either E equals one of the E_j, or else E is larger than all of the E_j. If we introduce the notation E_{max} for the least upper bound of the sequence $E_1, E_2, E_3, \ldots$, then the theorem gives us all the eigenvalues below E_{max}. In some cases (for instance, the harmonic oscillator), $E_{max} = \infty$, and then the ladder construction (47) gives us *all* the eigenvalues. In other cases (for instance, the hydrogen atom), $E_{max} = 0$, and then the ladder construction gives us all the negative eigenvalues. This means that the factorization method yields the discrete set of energy eigenvalues, but not the continuum.

---

[4] Except for an unimportant ambiguity in the phase of η_j.

[5] This means that in the notation of the theorem, E_1 is the energy of the ground state, E_2 the energy of the first excited state, and so on.

The proof of the theorem is as follows. For convenience, we define operators Λ_j by

$$\Lambda_j \equiv \eta_j^\dagger \eta_j + E_j \tag{48}$$

In view of Eq. (45), we have

$$\Lambda_{j+1} = \eta_j \eta_j^\dagger + E_j \tag{49}$$

It is then easy to check that

$$\Lambda_{j+1} \eta_j = \eta_j \Lambda_j \tag{50}$$

and that

$$\Lambda_j \eta_j^\dagger = \eta_j^\dagger \Lambda_{j+1} \tag{51}$$

Exercise 10. Check these equations.

We can now examine what the operator H does to the vector $|E_j\rangle$ defined by Eq. (47). Since $H = \Lambda_1$, we have

$$
\begin{aligned}
H|E_j\rangle &= \Lambda_1 \eta_1^\dagger \eta_2^\dagger \cdots \eta_{j-1}^\dagger |\zeta_j\rangle \\
&= \eta_1^\dagger \Lambda_2 \eta_2^\dagger \cdots \eta_{j-1}^\dagger |\zeta_j\rangle \\
&= \eta_1^\dagger \eta_2^\dagger \Lambda_3 \cdots \eta_{j-1}^\dagger |\zeta_j\rangle \\
&= \eta_1^\dagger \eta_2^\dagger \cdots \eta_{j-1}^\dagger \Lambda_j |\zeta_j\rangle
\end{aligned} \tag{52}
$$

To finish this calculation, we need the result that $|\zeta_j\rangle$ is an eigenvector of Λ_j. This follows from Eq. (46):

$$\Lambda_j |\zeta_j\rangle = (\eta_j^\dagger \eta_j + E_j)|\zeta_j\rangle = E_j |\zeta_j\rangle \tag{53}$$

Inserting this into Eq. (52), we obtain

$$
\begin{aligned}
H|E_j\rangle &= E_j \eta_1^\dagger \eta_2^\dagger \cdots \eta_{j-1}^\dagger |\zeta_j\rangle \\
&= E_j |E_j\rangle
\end{aligned} \tag{54}
$$

Hence the number E_j and the vector $|E_j\rangle$ are, respectively, an eigenvalue and eigenvector of H.

To complete the proof of our theorem, we need to show that the constants $E_1, E_2, E_3, \ldots$ form an ascending sequence, and that if E is an eigenvalue of H, then E cannot lie *between* E_j and E_{j+1}. Consider $E_{j+1} - E_j$. Assuming that the vector $|\zeta_{j+1}\rangle$ is normalized, we have

$$E_{j+1} - E_j = \langle \zeta_{j+1} | (E_{j+1} - E_j) | \zeta_{j+1} \rangle \tag{55}$$

According to Eq. (45), this equals

$$E_{j+1} - E_j = \langle \zeta_{j+1} | (\eta_j \eta_j^\dagger - \eta_{j+1}^\dagger \eta_{j+1}) | \zeta_{j+1} \rangle$$

But $\eta_{j+1} | \zeta_{j+1} \rangle = 0$, and hence the right side reduces to $\langle \zeta_{j+1} | \eta_j \eta_j^\dagger | \zeta_{j+1} \rangle$, which is the magnitude of the vector $\eta_j^\dagger | \zeta_{j+1} \rangle$, and therefore cannot be negative. This establishes that

$$E_{j+1} - E_j \geq 0$$

and that

$$E_1 \leq E_2 \leq E_3 \leq \cdots \tag{56}$$

Consider, next, an eigenvector $|E\rangle$ of H, and define a set of auxiliary vectors $|\xi_n\rangle$ by

$$|\xi_n\rangle = \eta_n \eta_{n-1} \cdots \eta_2 \eta_1 |E\rangle \tag{57}$$

Then

$$0 \leq \langle \xi_1 | \xi_1 \rangle = (\langle E | \eta_1 \rangle (\eta_1 | E \rangle)) = \langle E | \eta_1^\dagger \eta_1 | E \rangle$$
$$= \langle E | (\Lambda_1 - E_1) | E \rangle = E - E_1 \tag{58}$$

This says that the eigenvalue satisfies $E \geq E_1$. Next,

$$0 \leq \langle \xi_2 | \xi_2 \rangle = \langle E | \eta_1^\dagger \eta_2^\dagger \eta_2 \eta_1 | E \rangle$$
$$= \langle E | \eta_1^\dagger (\Lambda_2 - E_2) \eta_1 | E \rangle$$
$$= \langle E | \eta_1^\dagger \eta_1 (\Lambda_1 - E_2) | E \rangle$$
$$= (E - E_2) \langle E | \eta_1^\dagger \eta_1 | E \rangle = (E - E_2)(E - E_1) \tag{59}$$

This implies that either $E \geq E_2$ or else $E = E_1$. In general, we obtain

$$0 \leq \langle \xi_n | \xi_n \rangle = (E - E_n)(E - E_{n-1}) \cdots (E - E_2)(E - E_1) \tag{60}$$

Exercise 11. Derive this inequality.

From Eq. (60), we deduce that either E is larger than all of the E_n, or else E equals one of them.

This concludes the proof of the theorem. Note that the theorem does not invoke the condition, mentioned on p. 163, that the η_j must be chosen so as to make the constants E_j as large as possible.

This maximum requirement for the constants E_j is actually implied by the assumptions of the theorem. Consider a change in η_j and in E_j, with η_{j-1} and E_{j-1} held fixed:

$$\delta E_j = \langle \zeta_j | \delta E_j | \zeta_j \rangle = \langle \zeta_j | (-\delta \eta_j^{\dagger} \eta_j - \eta_j^{\dagger} \delta \eta_j) | \zeta_j \rangle$$

$$= - \langle \zeta_j | \delta \eta_j^{\dagger} \eta_j | \zeta_j \rangle - (\langle \zeta_j | \eta_j) \delta \eta_j | \zeta_j \rangle$$

This is zero, since $\eta_j | \zeta_j \rangle = 0$. Thus, the maximum condition for E_j is implied by Eq. (46).

The theorem does not tell us how to go about finding the operators η_j that satisfy the recursion relations (44) and (45). Fortunately, the Hamiltonians preferred by physicists (and by nature) are often fairly simple, and it is usually not hard to guess operators η_j that satisfy these recursion relations. If the Hamiltonian describes the motion of a particle in some potential, then the operators η_j must be some functions of the operators p and x, since these are the only relevant operators. The crucial step is the factorization of the Hamiltonian, as required by the first of the recursion relations:

$$H = \eta_1^{\dagger} \eta_1 + E_1$$

Roughly, this expresses H as the "square" of an operator η_1 plus a numerical "remainder." Once the operator η_1 has been found, the other operators η_j are usually easy to construct, because their dependence on p and x is usually similar to that of η_1, with only some differences in multiplicative and additive constants. When adjustable constants occur in the construction of the operators η_j, they must be chosen so as to make the "remainders" E_j as large as possible, in accordance with the maximum requirement.

Once we have constructed some operators η_j satisfying the recursion relations, the next step is to test whether Eq. (46) has a solution. We will call Eq. (46) the *subsidiary condition*. The test of whether this subsidiary condition is satisfied can be conveniently performed in the position representation, where Eq. (46) is a first-order differential equation. The ladder construction in Eq. (47) will then give us the eigenvectors. If no solution of the subsidiary condition exists, then our construction of the operators η_j is faulty, and it must be improved.

As a simple illustration, we will apply the factorization method to the solution of the eigenvector equation for the harmonic oscillator. We must then find operators η_j that satisfy the recursion relations (44) and (45), with $H = p^2/2m + m\omega^2 x^2/2$. We will make the following *Ansatz* for η_j:

$$\eta_j = \frac{1}{\sqrt{2m}}(p + if_j(x)) \tag{61}$$

where $f_j(x)$ is a real function of the operator x. This *Ansatz* is suggested by the form of H. Since $\eta_1^\dagger \eta_1$ equals H to within a constant, the operator η_1 ought to contain a term $\propto p$ to match the kinetic energy term $\propto p^2$ in H. According to the recursion relations (45), each of the operators η_j then also ought to contain a term $\propto p$. The functions f_j in Eq. (61) will have to be adjusted to match the potential energy term in H. Actually, with suitable choices for these functions f_j, the *Ansatz* (61) is valid not only for the harmonic-oscillator Hamiltonian, but also for other Hamiltonians of the form $H = p^2/2m + V(x)$, with a variety of potentials $V(x)$.

With the general commutator $[f(x), p] = i\hbar \, df/dx$, Eq. (61) leads to the results

$$\eta_j^\dagger \eta_j = \frac{1}{2m}p^2 + \frac{1}{2m}f_j^2 + \frac{\hbar}{2m}\frac{df_j}{dx} \tag{62}$$

$$\eta_j \eta_j^\dagger = \frac{1}{2m}p^2 + \frac{1}{2m}f_j^2 - \frac{\hbar}{2m}\frac{df_j}{dx} \tag{63}$$

Exercise 12. Obtain these results.

Equation (44) therefore becomes

$$\frac{1}{2m}p^2 + \frac{1}{2m}f_1^2 + \frac{\hbar}{2m}\frac{df_1}{dx} + E_1 = \frac{1}{2m}p^2 + \frac{1}{2}m\omega^2 x^2$$

or

$$\frac{1}{2m}f_1^2 + \frac{\hbar}{2m}\frac{df_1}{dx} + E_1 = \frac{1}{2}m\omega^2 x^2 \tag{64}$$

An obvious solution of this equation is

$$f_1 = \pm m\omega x \qquad E_1 = \mp\tfrac{1}{2}\hbar\omega \tag{65}$$

Note that we do not have to find the most general solution of Eq. (64). All we need is a particular solution (however, the correct particular solution for η_1 must satisfy the subsidiary condition $\eta_1|\zeta_1\rangle = 0$, that is, η_1 must have an eigenvector of eigenvalue zero; we will check later that η_1 indeed has such an eigenvector). To proceed further, we must make a choice of sign in Eq. (65). Since the E_j are subject to a maximum condition, it is obvious that we must always make the choice that yields the largest value for E_j. Hence

$$f_1 = -m\omega x \qquad E_1 = \tfrac{1}{2}\hbar\omega \tag{66}$$

We next have to find f_2. Inserting (66) into $\eta_2{}^\dagger\eta_2 + E_2 = \eta_1\eta_1{}^\dagger + E_1$, we obtain

$$\frac{1}{2m} f_2{}^2 + \frac{\hbar}{2m}\frac{df_2}{dx} + E_2 = \frac{1}{2} m\omega^2 x^2 + \hbar\omega \tag{67}$$

A solution of this equation is

$$f_2 = -m\omega x \qquad E_2 = \tfrac{3}{2}\hbar\omega \tag{68}$$

where the correct choice of signs has already been made. Likewise, we can find all the other functions f_j from their recursion relations. The general result is

$$f_j = -m\omega x \qquad E_j = (j - \tfrac{1}{2})\hbar\omega \tag{69}$$

This means that all the operators η_j are exactly the same:

$$\eta_j = \frac{1}{\sqrt{2m}}(p - im\omega x) \tag{70}$$

Equation (47) then gives us the eigenvectors:

$$|E_j\rangle \propto (p + im\omega x)^{j-1}|\zeta\rangle \tag{71}$$

where $|\zeta\rangle$ is a solution of the subsidiary condition

$$(p - im\omega x)|\zeta\rangle = 0 \tag{72}$$

Equation (71) is the same as Eq. (19), except for a change of notation. In the present notation, the ground-state energy is designated by E_1, whereas in Section 6.1, this energy was designated by E_0. This change of notation must be kept in mind when comparing the results.

To complete the solution of the harmonic-oscillator problem by the factorization method, we must check that $|\zeta\rangle$ exists. This is conveniently done by writing Eq. (72) as a differential equation in the position representation. That a solution of this equation exists is shown by Eqs. (31)–(32). The eigenfunctions can then be constructed as before [see Eqs. (34)–(37)].

We will next apply the factorization method to a more difficult problem: the particle in a one-dimensional infinite square well. This problem does not lend itself very well to solution by the factorization method, because the potential is not defined by an explicit, analytic function of x, and enters only implicitly through the boundary conditions at the walls of the well. Hence the Hamiltonian has no explicit dependence on x, but we must nevertheless introduce a dependence on x in the factorization. The mathematical arguments required to adapt the factorization method to this problem are rather tricky and complicated, which led Schrödinger to remark that it is "like shooting at sparrows with heavy artillery." Nevertheless, Schrödinger felt the factorization for the infinite square well was a worthwhile exercise, because it illustrates the worst that can happen with this method.

We suppose that, as in Chapter 3, the well extends from $x = 0$ to $x = L$. The Hamiltonian for a particle in this well is

$$H = \frac{p^2}{2m} \tag{73}$$

We again use the *Ansatz* (61) for η_j. This leads to the following equation for f_1:

$$\frac{1}{2m} f_1{}^2 + \frac{\hbar}{2m} \frac{df_1}{dx} + E_1 = 0 \tag{74}$$

This equation can be integrated directly, with the result

$$f_1 = \sqrt{2mE_1} \cot \left[\frac{\sqrt{2mE_1}}{\hbar} (x - a) \right] \tag{75}$$

where a is the constant of integration. The possible choices for the values of E_1 and a are restricted by the behavior of the cotangent function. In the position representation, x is a number within the interval $0 < x < L$, and we must insist that the cotangent remain finite in this interval. Since the singularities of the cotangent are separated by π radians, we attain the largest possible value of E_1 if one of the singularities of the cotangent lies at $x = 0$, and the next at $x = L$. (At the endpoints $x = 0$ and $x = L$ it is permissible for the cotangent to become infinite since there the potential, and hence H, are infinite.) This means that $a = 0$ and that $\sqrt{2mE_1}\, L/\hbar = \pi$. Thus,

$$E_1 = \frac{\pi^2 \hbar^2}{2mL^2} \tag{76}$$

$$f_1 = \frac{\pi \hbar}{L} \cot \frac{\pi x}{L} \tag{77}$$

and

$$\eta_1 = \frac{1}{\sqrt{2m}} \left(p + \frac{i\pi\hbar}{L} \cot \frac{\pi x}{L} \right) \tag{78}$$

Next we can write down the equations for $f_2, f_3, \ldots$. We will not carry out this calculation step by step, because we can save ourselves much labor by taking as our starting point the following clever *Ansatz* for $\eta_2, \eta_3, \ldots$:

$$\eta_j = \frac{1}{\sqrt{2m}} \left(p + i c_j \cot b_j x \right) \tag{79}$$

where c_j and b_j are real constants which must be chosen so as to give the correct recursion relations among the operators η_j. This *Ansatz* is based on the guess that all the operators η_j are similar to Eq. (78), but with different constants c_j and b_j. The value of b_j is restricted by $0 \le b_j \le \pi/L$. To see that this restriction is necessary, we note that in the position representation, x is a number in the interval $0 < x < L$, and the condition on b_j is needed in order to keep the cotangent finite.

By a straightforward computation, we find that

$$\eta_j{}^\dagger \eta_j = \frac{1}{2m} \left(p^2 - c_j b_j \hbar + c_j(c_j - b_j \hbar) \cot^2 b_j x \right) \tag{80}$$

and that

$$\eta_j \eta_j{}^\dagger = \frac{1}{2m} \left(p^2 + c_j b_j \hbar + c_j(c_j + b_j \hbar) \cot^2 b_j x \right) \tag{81}$$

Exercise 13. Check these equations.

The recursion relations (45) then become

$$\frac{1}{2m} \left(p^2 - c_{j+1} b_{j+1} \hbar + c_{j+1}(c_{j+1} - b_{j+1} \hbar) \cot^2 b_{j+1} x \right) + E_{j+1}$$

$$= \frac{1}{2m} \left(p^2 + c_j b_j \hbar + c_j(c_j + b_j \hbar) \cot^2 b_j x \right) + E_j \tag{82}$$

The coefficients of different powers of x on both sides of this equation must be equal. This requires that the arguments of the cotangents agree:

$$b_{j+1} = b_j \tag{83}$$

Furthermore, the coefficients multiplying the cotangents must agree:

$$c_{j+1}(c_{j+1} - b_{j+1}\hbar) = c_j(c_j + b_j\hbar) \tag{84}$$

and the additive constants on the two sides of the equation must agree:

$$2mE_{j+1} - c_{j+1}b_{j+1}\hbar = 2mE_j + c_j b_j\hbar \tag{85}$$

If we subtract Eq. (84) from Eq. (85), we obtain

$$2mE_{j+1} - (c_{j+1})^2 = 2mE_j - (c_j)^2 \tag{86}$$

This recursion relation implies that

$$2mE_j - (c_j)^2 = 2mE_1 - (c_1)^2 = 0 \tag{87}$$

and hence

$$E_j = \frac{(c_j)^2}{2m} \tag{88}$$

If we use Eq. (83), we obtain $b_j = b_1 = \pi/L$ and, by Eq. (84),

$$c_{j+1}\left(c_{j+1} - \frac{\pi\hbar}{L}\right) = c_j\left(c_j + \frac{\pi\hbar}{L}\right) \tag{89}$$

There are two solutions of this equation: $c_{j+1} = -c_j$ and

$$c_{j+1} = c_j + \frac{\pi\hbar}{L} \tag{90}$$

The solution (90) is the one that gives the largest value for E_{j+1}. From this we obtain $c_j = j\pi\hbar/L$ and hence

$$E_j = \frac{j^2\pi^2\hbar^2}{2mL^2} \tag{91}$$

This gives us all the eigenvalues provided we can show that there exists a solution of the subsidiary condition $\eta_j |\zeta_j\rangle = 0$. In the position representation, this subsidiary condition is a differential equation:

$$\left(\frac{\hbar}{i}\frac{d}{dx} + i\frac{j\pi\hbar}{L}\cot\frac{\pi x}{L}\right)\zeta_j(x) = 0 \tag{92}$$

where $\zeta_j(x) \equiv \langle x|\zeta_j\rangle$ is the wavefunction that corresponds to the state vector $|\zeta_j\rangle$. Equation (92) is easy to solve:

$$\zeta_j(x) = \left(\sin\frac{\pi x}{L}\right)^j \qquad (93)$$

The eigenfunctions can now be constructed as follows:

$$\psi_1(x) = \zeta_1(x) = \sin\frac{\pi x}{L}$$

$$\psi_2(x) = \eta_1{}^\dagger\zeta_2(x) = \left(\frac{\hbar}{i}\frac{d}{dx} - \frac{i\pi\hbar}{L}\cot\frac{\pi x}{L}\right)\sin^2\frac{\pi x}{L} \qquad (94)$$

$$\psi_3(x) = \eta_1{}^\dagger\eta_2{}^\dagger\zeta_3(x)$$

$$= \left(\frac{\hbar}{i}\frac{d}{dx} - \frac{i\pi\hbar}{L}\cot\frac{\pi x}{L}\right)\left(\frac{\hbar}{i}\frac{d}{dx} - \frac{2i\pi\hbar}{L}\cot\frac{\pi x}{L}\right)\sin^3\frac{\pi x}{L}$$

If we simplify these expressions, we find that they reduce to the familiar formulas for the wavefunctions of the infinite square well.

Exercise 14. Show that

$$\psi_2(x) \propto \sin\frac{2\pi x}{L} \qquad \text{and} \qquad \psi_3(x) \propto \sin\frac{3\pi x}{L}$$

6.3 *Normalization of the Eigenvectors*

Equation (47) gives us the eigenvectors except for normalization. In general we must write

$$|E_j\rangle = c_j\eta_1{}^\dagger\eta_2{}^\dagger \cdots \eta_{j-1}{}^\dagger |\zeta_j\rangle \qquad (95)$$

where c_j is some constant to be determined from the normalization requirement, $\langle E_j|E_j\rangle = 1$. With Eq. (95), this becomes

$$1 = |c_j|^2 \langle\zeta_j|\, \eta_{j-1}\eta_{j-2} \cdots \eta_2\eta_1\eta_1{}^\dagger\eta_2{}^\dagger \cdots \eta_{j-1}{}^\dagger |\zeta_j\rangle \qquad (96)$$

Assume that $|\zeta_j\rangle$ is normalized. For the case $j = 1$, Eq. (96) will be satisfied provided $|c_1| = 1$. For the case $j = 2$ we obtain, by means of Eq. (53),

$$1 = |c_2|^2 \langle\zeta_2|\, \eta_1\eta_1{}^\dagger |\zeta_2\rangle = |c_2|^2 \langle\zeta_2|\, (\Lambda_2 - E_1) |\zeta_2\rangle$$

$$= |c_2|^2 \langle\zeta_2|\, (E_2 - E_1) |\zeta_2\rangle = |c_2|^2 (E_2 - E_1) \qquad (97)$$

Hence

$$|c_2| = (E_2 - E_1)^{-1/2} \tag{98}$$

For $j = 3$,

$$1 = |c_3|^2 \langle \zeta_3| \, \eta_2 \eta_1 \eta_1^\dagger \eta_2^\dagger \, |\zeta_3\rangle = |c_3|^2 \langle \zeta_3| \, \eta_2 (\Lambda_2 - E_1) \eta_2^\dagger \, |\zeta_3\rangle \tag{99}$$

But $\Lambda_2 \eta_2^\dagger = \eta_2^\dagger \Lambda_3$, and so

$$1 = |c_3|^2 \langle \zeta_3| \, \eta_2 \eta_2^\dagger \, (\Lambda_3 - E_1) \, |\zeta_3\rangle = |c_3|^2 (E_3 - E_1) \langle \zeta_3| \, \eta_2 \eta_2^\dagger \, |\zeta_3\rangle$$

$$= |c_3|^2 (E_3 - E_1) \langle \zeta_3| \, (\Lambda_3 - E_2) \, |\zeta_3\rangle = |c_3|^2 (E_3 - E_1)(E_3 - E_2) \tag{100}$$

From this we see that

$$|c_3| = [(E_3 - E_2)(E_3 - E_1)]^{-1/2} \tag{101}$$

The general result is

$$|c_j| = [(E_j - E_{j-1})(E_j - E_{j-2}) \cdots (E_j - E_1)]^{-1/2} \tag{102}$$

With this value of the constant $|c_j|$, Eq. (95) yields a normalized vector.

Exercise 15. Derive the general result (102).

6.4 *Power-Series Solution of the Schrödinger Equation*

The most common method employed for finding the eigenvectors and the eigenvalues for a particle moving in some given potential is the solution of the time-independent Schrödinger wave equation in the position representation,

$$\left[-\frac{\hbar^2}{2m} \frac{d^2}{dx^2} + V(x) \right] \psi(x) = E\psi(x)$$

This is a second-order linear differential equation. Such an equation can be solved by substituting a power series with arbitrary coefficients:

$$\psi(x) = b_0 + b_1 x + b_2 x^2 + b_3 x^3 + \cdots \tag{103}$$

The differential equation then implies a recursion relation among the coefficients b_j, which permits us to obtain the values of these coefficients, starting with some assumed values for b_0 or b_1.

We will illustrate the application of this traditional power-series method for the case of the harmonic oscillator. The Schrödinger wave equation is

$$\left[-\frac{\hbar^2}{2m}\frac{d^2}{dx^2} + \frac{1}{2}m\omega^2 x^2 \right]\psi(x) = E\psi(x) \tag{104}$$

With the notation $\xi = x/x_0$, this becomes

$$\left[\frac{d^2}{d\xi^2} + \frac{2E}{\hbar\omega} - \xi^2 \right]\psi(\xi) = 0 \tag{105}$$

Instead of direct substitution of a power series for $\psi(\xi)$, it proves convenient to express $\psi(\xi)$ as the product of two factors:

$$\psi(\xi) = H(\xi)e^{-\frac{1}{2}\xi^2} \tag{106}$$

where $H(\xi)$ is a power series. This form of the function $\psi(\xi)$ is suggested by the asymptotic behavior of the differential equation (105) in the limit of large ξ. Obviously, if ξ is very large, so $2E/\hbar\omega$ can be neglected compared with ξ^2, the differential equation has the asymptotic solution $\psi(\xi) \propto e^{-\frac{1}{2}\xi^2}$. In Eq. (106), this asymptotic solution has been separated out, and thus the remaining factor $H(\xi)$ represents the deviation from the asymptotic behavior. Substituting Eq. (106) into Eq. (105), we obtain a differential equation for $H(\xi)$:

$$\left[\frac{d^2}{d\xi^2} - 2\xi\frac{d}{d\xi} + \left(\frac{2E}{\hbar\omega} - 1\right) \right]H(\xi) = 0 \tag{107}$$

Exercise 16. Perform the calculations leading from Eq. (105) to Eq. (107).

We now try a power series for $H(\xi)$:

$$H(\xi) = b_0 + b_1\xi + b_2\xi^2 + b_3\xi^3 + \cdots \tag{108}$$

With this, the differential equation (107) becomes

$$\sum_{j=0}^{\infty}\left[j(j-1)b_j\xi^{j-2} - \left(2j - \frac{2E}{\hbar\omega} + 1\right)b_j\xi^j \right] = 0 \tag{109}$$

This equation requires that the coefficient of each power of ξ be zero. Thus, if we examine the coefficient of ξ^j, we find

$$(j+2)(j+1)b_{j+2} - \left(2j - \frac{2E}{\hbar\omega} + 1\right)b_j = 0$$

which yields

$$b_{j+2} = \frac{2j - 2E/\hbar\omega + 1}{(j + 2)(j + 1)} b_j \tag{110}$$

This equation determines all the coefficients b_j, starting with some given values for b_0 and b_1. Note that Eq. (110) does not mix even and odd coefficients. Thus, the coefficients b_0, b_2, b_4, . . . are completely independent of the coefficients b_1, b_3, b_5, . . .; the former are determined by the value assigned to b_0, and the latter are determined by the value assigned to b_1. This means that the differential equation has two independent kinds of solutions, with even and with odd powers of ξ (even and odd parity).

For each of these two kinds of solutions, the ratio of consecutive terms in the power series is

$$\frac{b_{j+2}\xi^{j+2}}{b_j\xi^j} = \frac{2j - 2E/\hbar\omega + 1}{(j + 2)(j + 1)} \xi^2 \tag{111}$$

which approaches $2\xi^2/j$ when j is large, that is, when $2E/\hbar\omega - 1$ can be neglected compared with $2j$. But this ratio of consecutive terms is the same as that for the series

$$e^{\xi^2} = \sum_{j=0}^{\infty} \frac{(\xi^2)^j}{j!} = \sum_{j=0,2,4,...}^{\infty} \frac{\xi^j}{(j/2)!} \tag{112}$$

Hence the wavefunction $H(\xi)e^{-\frac{1}{2}\xi^2}$ would asymptotically behave as $e^{\frac{1}{2}\xi^2}$, that is, it would diverge as $\xi \to \infty$, and it would not be normalizable. To avoid this unacceptable behavior, we must assume that the power series (108) ends after a finite number of terms, and therefore that the function $H(\xi)$ is a polynomial rather than an infinite series. For the coefficients b_j, this requires that $b_j = 0$ for some value of j. According to Eq. (110), this requirement is

$$2j - \frac{2E}{\hbar\omega} + 1 = 0$$

or

$$E_j = \hbar\omega \left(j + \frac{1}{2}\right) \tag{113}$$

These are the usual energy eigenvalues for the harmonic oscillator.

For a given energy E_j, all the coefficients beyond b_j are all zero; thus, $H(\xi)$ is a polynomial of order j. It easy to verify that with suitable choices of b_0 and b_1, the functions $H(\xi)$ coincide with the Hermite polynomials defined in Section 6.1.

To end this chapter, let us consider the advantages and disadvantages of this power-series solution of the Schrödinger equation as compared with the factorization method. The power-series solution of the Schrödinger equation does not demand as much mathematical sophistication, but it also does not yield as much. The factorization method of Section 6.1 not only gave us the energy eigenfunctions and eigenvalues, but also provided us with a simple way to normalize the eigenfunctions and a simple way to calculate the expectation value of any physical operator involving x and p. The normalization of the eigenfunctions and the calculation of expectation values is quite messy if we attempt a brute-force integration of the Hermite polynomials; in fact, the most practical way to perform the required integrations is by exploiting the identity (38); this identity emerges very naturally from the operator formalism, but it is much harder to derive from the recursion relations (110) for the coefficients of the Hermite polynomials. Thus, for the harmonic oscillator, the factorization method is clearly superior.

However, for the infinite square well, the factorization method becomes quite messy, and loses its advantages. The square well, and other piecewise constant potentials, are not well suited to the factorization method, because the potential is not defined by an explicit, analytic function of x, but by step functions. This kind of problem is much easier to handle by the direct solution of the Schrödinger wave equation.

Which method is best depends on the details of the potential. The factorization method is feasible for all the potentials for which the Schrödinger differential equation is known to admit of an exact, rigorous solution, that is, a solution that does not rely on perturbation methods or numerical approximation methods. Thus the two methods are equally feasible, and we must make the decision between the two methods according to convenience and according to what we want to calculate (Only eigenvalues? Or eigenvalues, and normalization factors, and expectation values?).

In any case, the factorization method provides us with valuable experience in the manipulation of operators. As we will see in the next chapter, operators analogous to the raising and lowering operators used in the construction of the energy eigenvectors also prove useful for the construction of the eigenvectors of angular momentum.

Lastly, it is of some interest to note that the factorization method is not only a method for the solution of eigenvalue problems in quantum mechanics, but it is also a general method for the solution of differential equations. A list of all the differential equations for which there is a known factorization has been assembled by Infeld and Hull (see footnote 2).

PROBLEMS   1. An operator is said to be *Fredholm* if the nonexistence of its inverse implies the existence of an eigenvector of eigenvalue zero of this operator. Assume that the operators η_j are Fredholm. Show that in this case the condition that E_j is maximum guarantees the existence of an eigenvector of eigenvalue zero, that is, it guarantees the subsidiary condition. (Hint: Consider what happens to E_j if the operator η_j is changed by an amount $\delta\eta_j^\dagger = -\varepsilon\eta_j^{-1}$, where $\varepsilon > 0$.)

2. Equation (72) is a first-order differential equation in the position representation, but it is also a first-order differential equation in the momentum representation (where $x_{op} = i\hbar\, d/dp$). Solve this differential equation in the momentum representation, and normalize the solution. Then check that the Fourier transform of your solution agrees with Eq. (33).

3. The frequency of vibration of the HCl molecule is 6.3×10^{13} Hz.

   (a) Suppose that such a molecule is in its ground vibrational state. What is the rms deviation of the position of the hydrogen atom from its (classical) equilibrium position? Assume that the chlorine atom remains at rest, since it is much heavier than the hydrogen atom.

   (b) Repeat the calculation if the molecule is in its third excited vibrational state.

4. Evaluate the uncertainties Δx and Δp for the ground state of the harmonic oscillator and compare your results with the uncertainty relation.

5. At $t = 0$, a particle in a harmonic-oscillator potential is in the initial state

$$|\psi(0)\rangle = \frac{1}{\sqrt{5}}|E_1\rangle + \frac{2}{\sqrt{5}}|E_2\rangle$$

   where $|E_1\rangle$ is the first excited state and $|E_2\rangle$ the second.

   (a) What is the expectation value of the energy in the state $|\psi(0)\rangle$?

   (b) Find $|\psi(t)\rangle$. Is this a stationary state?

   (c) Evaluate the expection value $\langle\psi(t)|x|\psi(t)\rangle$. What is the frequency of oscillation of this expectation value?

6. At $t = 0$, a particle in a harmonic-oscillator potential is in the initial state

$$|\psi(0)\rangle = \frac{1}{2}|E_0\rangle + \frac{\sqrt{3}}{2}|E_2\rangle$$

Find $|\psi(t)\rangle$ and evaluate the expectation value $\langle\psi(t)|x^2|\psi(t)\rangle$. What is the frequency of oscillation of this expectation value?

7. Evaluate the expectation values $\langle E_1|x^2|E_1\rangle$, $\langle E_1|x^3|E_1\rangle$, and $\langle E_1|x^4|E_1\rangle$ for the first excited state of the harmonic oscillator.

8. Show that the expectation value of x^{2k} (where k is an integer) for the ground state of the harmonic oscillator is

$$\langle E_0| x^{2k} |E_0\rangle = \frac{(2k)!}{2^{2k}k!} x_0^{2k}$$

9. Show that if $|E_n\rangle$ is the nth excited state of the harmonic oscillator, then

$$\langle E_n| x |E_{n+1}\rangle = \sqrt{\frac{n+1}{2}}\, x_0$$

$$\langle E_n| x^2 |E_{n+1}\rangle = 0$$

$$\langle E_n| x |E_{n+2}\rangle = 0$$

$$\langle E_n| x^2 |E_{n+2}\rangle = \tfrac{1}{2}\sqrt{n+2}\,\sqrt{n+1}\, x_0^2$$

10. Using the results of the preceding problem, find all the matrix elements of x_{op} in the energy representation and write them as a matrix.

11. Show that for the nth excited state of the harmonic oscillator, the expectation value of x^4 is

$$\langle E_n|x^4|E_n\rangle = \frac{3}{2}\left(n^2 + n + \frac{1}{2}\right) x_0^4$$

12. Show that for the harmonic oscillator

$$\langle E_0| x^n |E_n\rangle = \frac{\sqrt{n!}}{2^{n/2}} x_0^n$$

13. Suppose that the initial state of a particle in a harmonic-oscillator potential is

$$|\psi(0)\rangle = \frac{1}{\sqrt{3}}|E_0\rangle + \sqrt{\frac{2}{3}}|E_1\rangle$$

Find the uncertainties Δx and Δp as a function of time. Plot $\Delta x \, \Delta p$ vs. time. Is the uncertainty principle always satisfied?

14. (a) Show that the eigenstate of position can be expressed as

$$|x\rangle = \sum_n |E_n\rangle \langle E_n|x\rangle$$

$$= \frac{1}{\sqrt{x_0\sqrt{\pi}}} \sum_n \frac{1}{2^{n/2}\,n!} \, e^{-\frac{1}{2}x^2/x_0^2} \, H_n(x/x_0)(a^\dagger)^n \, |E_0\rangle$$

   (b) If a particle is in this eigenstate of position, what is probability that a measurement of its energy yields E_0? E_1?

15. Show that the wavefunction $\psi_{E_n}(x)$ of the nth excited state of the harmonic oscillator has n nodes (not counting $x = \pm\infty$).

16. Suppose that a particle in a harmonic-oscillator potential initially has the wavefunction

$$\psi(x,\, 0) = \frac{1}{\sqrt{a\sqrt{\pi}}} \, e^{-\frac{1}{2}x^2/a^2}$$

   where a is a positive constant.

   (a) Express this wavefunction as a superposition of the energy eigenfunctions. [Hint: Use the completeness relation (42).]

   (b) What is the probability for the value E_0 of the energy? For E_1? For E_2?

17. At the initial time $t = 0$, a particle in a harmonic-oscillator potential has the wavefunction

$$\psi(x,\, 0) = \frac{1}{\sqrt{a\sqrt{\pi}}} \, e^{-\frac{1}{2}(x-b)^2/a^2}$$

   where a and b are constants.

   (a) What will be the wavefunction at the time $t = 2\pi/\omega$?

   (b) What is the uncertainty Δx at time $t = 0$? At time $t = 2\pi/\omega$? Can you guess whether the uncertainty at an intermediate time will be larger or smaller?

18. In the abstract state-vector notation, the effect of the parity operator on the harmonic-oscillator eigenvectors is

$$P|E_n\rangle = (-1)^n \, |E_n\rangle$$

We can obtain this result directly from the ladder construction of the eigenvectors by examining the effect of the parity operator P on the raising operator $a^\dagger$.

(a) Given that the effect of the parity operator P on the operators x_{op} and p_{op} is

$$P x_{op} = -x_{op} P \quad \text{and} \quad P p_{op} = -p_{op} P$$

show that $P a^\dagger = -a^\dagger P$

(b) Given that $P|E_0\rangle = |E_0\rangle$, show that

$$P(a^\dagger)^n |E_0\rangle = (-1)^n (a^\dagger)^n |E_0\rangle$$

19. Show that

$$f_1 = \pm m\omega x + \frac{\hbar}{x}, \qquad E_1 = \mp \frac{3}{2} \hbar\omega$$

also solves Eq. (64). Why is the resulting operator η_1 unsuitable for the harmonic oscillator?

20. Check that Eq. (102) leads to the normalization (25) for the harmonic oscillator.

21. Show that the coefficients b_j determined from Eq. (110) agree with the coefficients in the Hermite polynomials H_0, H_1, H_2, and H_3 given by Eq. (39).

22. Find the expectation value $\langle E_2 | x^2 | E_2 \rangle$ for the second excited state of the harmonic oscillator by explicit evaluation of the integral

$$\int_{-\infty}^{\infty} x^2 (\psi_{E_2}(x))^2 \, dx$$

7

The Particle in Three Dimensions
and Angular Momentum

The position and momentum operators for a particle moving in three dimensions are straightforward generalizations of the one-dimensional operators. What is radically new in three dimensions is that the particle is endowed with orbital angular momentum. In this chapter we will examine the eigenvectors and the eigenvalues of the orbital angular-momentum operator, and we will see how the quantization of the orbital angular momentum emerges as a direct consequence of the commutation relations of the position and momentum operators. Thus, the orbital angular momentum is always quantized, regardless of whether the particle is free or bound.

We know from classical physics that the angular momentum is especially useful in the solution of problems involving a central force, where the angular momentum is a constant of the motion. We will see that this is also true in quantum physics. For a particle moving under the influence of a central force, the angular-momentum operator commutes with the Hamiltonian, and the expectation value of the angular momentum is constant. Furthermore, we can construct simultaneous eigenvectors of the energy and the angular momentum; these simultaneous eigenvectors form a convenient complete set of basis vectors for the description of the states of a particle moving under the influence of a central force.

7.1   *The Free Particle*

The Hamiltonian operator for a free particle moving in three dimensions is

$$H = \frac{1}{2m}\,\mathbf{p}_{\text{op}}^2 = \frac{1}{2m}\,(p_{x,\text{op}}^2 + p_{y,\text{op}}^2 + p_{z,\text{op}}^2) \tag{1}$$

where $p_{x,\text{op}}$, $p_{y,\text{op}}$, and $p_{z,\text{op}}$ are the components of the momentum operator. It is convenient to write x_1, x_2, x_3 and p_1, p_2, p_3 for x, y, z and p_x, p_y, p_z, respectively. The commutation relations between all the operators for position and momenta are then

$$[x_{j,\text{op}}, p_{k,\text{op}}] = i\hbar\delta_{jk} \tag{2}$$

$$[p_{j,\text{op}}, p_{k,\text{op}}] = 0 \tag{3}$$

$$[x_{j,\text{op}}, x_{k,\text{op}}] = 0 \tag{4}$$

that is, everything commutes except the position operator with its *corresponding* momentum operator.

Since the momentum operators $p_{1,\text{op}}$, $p_{2,\text{op}}$, and $p_{3,\text{op}}$ commute with each other, they have simultaneous eigenvectors. We designate a simultaneous eigenvector of these momentum operators by $|p_1 p_2 p_3\rangle$. This eigenvector describes a particle of definite momentum, with components p_1, p_2, p_3:

$$p_{1,\text{op}}\,|p_1 p_2 p_3\rangle = p_1\,|p_1 p_2 p_3\rangle$$

$$p_{2,\text{op}}\,|p_1 p_2 p_3\rangle = p_2\,|p_1 p_2 p_3\rangle$$

$$p_{3,\text{op}}\,|p_1 p_2 p_3\rangle = p_3\,|p_1 p_2 p_3\rangle$$

In concise notation, these three equations can be expressed as

$$\mathbf{p}_{\text{op}}\,|\mathbf{p}\rangle = \mathbf{p}\,|\mathbf{p}\rangle \tag{5}$$

where $\mathbf{p}_{\text{op}}$ stands for the three operators $p_{1,\text{op}}$, $p_{2,\text{op}}$, $p_{3,\text{op}}$, and $\mathbf{p}$ stands for the three numbers p_1, p_2, p_3. Note that $\mathbf{p}_{\text{op}}$ is a vector in three-dimensional space, and concurrently, an operator in Hilbert space. The eigenvectors of momentum are not normalizable, but it is convenient to choose the multiplicative factor so the completeness relation takes the familiar form [compare Eq. (4.46)]

$$I = \int dp_1\, dp_2\, dp_3\, |p_1 p_2 p_3\rangle\, \langle p_1 p_2 p_3|$$

With $d^3p \equiv dp_1\, dp_2\, dp_3$ we can write this as

$$I = \int d^3p\, |\mathbf{p}\rangle\, \langle\mathbf{p}| \tag{6}$$

Similarly, the position operators or $x_{1,op}$, $x_{2,op}$, and $x_{3,op}$ commute with each other, and they have simultaneous eigenvectors designated by $|x_1 x_2 x_3\rangle$:

$$x_{1,op} |x_1 x_2 x_3\rangle = x_1 |x_1 x_2 x_3\rangle$$

$$x_{2,op} |x_1 x_2 x_3\rangle = x_2 |x_1 x_2 x_3\rangle$$

$$x_{3,op} |x_1 x_2 x_3\rangle = x_3 |x_1 x_2 x_3\rangle$$

In concise notation, these equations become

$$\mathbf{r}_{op} |\mathbf{r}\rangle = \mathbf{r} |\mathbf{r}\rangle \tag{7}$$

where $\mathbf{r}_{op}$ stands for the three operators $x_{1,op}$, $x_{2,op}$, $x_{3,op}$, and $\mathbf{r}$ stands for the three numbers x_1, x_2, x_3. The completeness relation for the eigenvectors of position is

$$1 = \int d^3 r \, |\mathbf{r}\rangle \langle \mathbf{r}| \tag{8}$$

where $d^3 r = dx_1 \, dx_2 \, dx_3 = dx \, dy \, dz$.

The wavefunction $\psi_{\mathbf{p}}(\mathbf{r})$ corresponding to momentum $\mathbf{p}$ (the time dependence is omitted) is given by

$$\psi_{\mathbf{p}}(\mathbf{r}) = \langle \mathbf{r}|\mathbf{p}\rangle = \frac{1}{(2\pi\hbar)^{3/2}} e^{i\mathbf{p}\cdot\mathbf{r}/\hbar} \tag{9}$$

Note that this three-dimensional wavefunction is the product of three separate one-dimensional wavefunctions [compare Eq. (4.48)]:

$$\psi_{\mathbf{p}}(\mathbf{r}) = \left(\frac{1}{\sqrt{2\pi\hbar}} e^{ip_1 x_1/\hbar}\right)\left(\frac{1}{\sqrt{2\pi\hbar}} e^{ip_2 x_2/\hbar}\right)\left(\frac{1}{\sqrt{2\pi\hbar}} e^{ip_3 x_3/\hbar}\right) \tag{10}$$

In the position representation, the completeness relation (6) becomes

$$\delta(\mathbf{r} - \mathbf{r}') = \int d^3 p \, \psi_{\mathbf{p}}^*(\mathbf{r}') \, \psi_{\mathbf{p}}(\mathbf{r}) \tag{11}$$

The object appearing on the left is the three-dimensional δ function, which is the product of three one-dimensional delta functions:

$$\delta(\mathbf{r} - \mathbf{r}') \equiv \frac{1}{(2\pi\hbar)^3} \int e^{i\mathbf{p}\cdot(\mathbf{r}-\mathbf{r}')/\hbar} d^3 p$$

$$= \left[\frac{1}{2\pi\hbar} \int e^{ip_1(x_1 - x_1')/\hbar} \, dp_1\right]\left[\frac{1}{2\pi\hbar} \int e^{ip_2(x_2 - x_2')/\hbar} \, dp_2\right]\left[\frac{1}{2\pi\hbar} \int e^{ip_3(x_3 - x_3')/\hbar} \, dp_3\right]$$

$$= \delta(x_1 - x_1')\delta(x_2 - x_2')\delta(x_3 - x_3') \tag{12}$$

The notation $\delta^3(\mathbf{r} - \mathbf{r}')$ is sometimes used for the three-dimensional δ function. Of course, the three-dimensional delta function has the fundamental property that, for any function $f(\mathbf{r})$,

$$f(0) = \int d^3r' \, \delta(\mathbf{r}') \, f(\mathbf{r}') \tag{13}$$

Finally, we write down the Fourier theorem in three dimensions. This is obtained from Eq. (11) by multiplying both sides by an arbitrary function $f(\mathbf{r}')$ and integrating over $\mathbf{r}'$, with the result

$$f(\mathbf{r}) = \int d^3p \int d^3r' \, \frac{1}{(2\pi\hbar)^{3/2}} \, e^{-i\mathbf{p}\cdot\mathbf{r}'/\hbar} \, f(\mathbf{r}') \, \frac{1}{(2\pi\hbar)^{3/2}} \, e^{i\mathbf{p}\cdot\mathbf{r}/\hbar}$$

$$= \frac{1}{(2\pi\hbar)^{3/2}} \int d^3p \, e^{i\mathbf{p}\cdot\mathbf{r}/\hbar} \, \phi(\mathbf{p}) \tag{14}$$

where

$$\phi(\mathbf{p}) = \frac{1}{(2\pi\hbar)^{3/2}} \int d^3r' \, e^{-i\mathbf{p}\cdot\mathbf{r}'/\hbar} \, f(\mathbf{r}') \tag{15}$$

This provides us with the connection between the wavefunctions in the coordinate representation and in the momentum representation.

The Schrödinger equation for a free particle in three dimensions is

$$-\frac{\hbar^2}{2m} \left(\frac{\partial^2}{\partial x^2} + \frac{\partial^2}{\partial y^2} + \frac{\partial^2}{\partial z^2} \right) \psi = i\hbar \frac{\partial \psi}{\partial t} \tag{16}$$

As in the one-dimensional case, we obtain solutions of this Schrödinger equation for a free particle if we multiply the momentum eigenfunctions in Eq. (9) by $e^{-iEt/\hbar}$:

$$\psi_{\mathbf{p}}(\mathbf{r}, t) = e^{-iEt/\hbar} \psi_{\mathbf{p}}(\mathbf{r})$$

$$= e^{-ip^2t/2m\hbar} \psi_{\mathbf{p}}(\mathbf{r}) \tag{17}$$

Exercise 1. Check that Eq. (17) is a solution of the Schrödinger equation.

7.2 Angular Momentum

The crucial difference between one and three dimensions is that in the latter case a particle can have orbital angular momentum. Like every physical observable, the angular momentum in quantum mechanics is represented by an operator. We will see that for a particle moving in the potential of a central force, the angular-momentum operator commutes with the Hamiltonian. Thus, the energy and the angular momentum form a set of commuting observables which can be used for the specification of the possible states of the particle.

We need a hermitian operator to represent the angular momentum. By analogy with classical physics, we take

$$\mathbf{L}_{op} \equiv \mathbf{r}_{op} \times \mathbf{p}_{op} \tag{18}$$

or, in component form,

$$L_x = yp_z - zp_y$$

$$L_y = zp_x - xp_z \tag{19}$$

$$L_z = xp_y - yp_x$$

For the sake of brevity, we have omitted the label $_{op}$ in these last equations. We will hereafter omit the label op whenever it is obvious from the context that an operator is meant. Note that we can equally well write Eq. (18) as $\mathbf{L} = -\mathbf{p} \times \mathbf{r}$; this is obvious from the component form of this equation, since on the right side of Eq. (19) the factors in each term commute (for instance, $yp_z = p_z y$)

Exercise 2. Show that the hermiticity of $\mathbf{p}$ and $\mathbf{r}$ implies that of $\mathbf{L}$.

Using the commutation relations (2)–(4) for x_j and p_k, we can easily obtain the commutation relations of the components of angular momentum:

$$[L_x, L_y] = i\hbar L_z$$

$$[L_y, L_z] = i\hbar L_x \tag{20}$$

$$[L_z, L_x] = i\hbar L_y$$

Exercise 3. Derive these equations.

Let us introduce the operator $\mathbf{L}^2 = L_x{}^2 + L_y{}^2 + L_z{}^2$, called the *angular momentum squared*. This operator commutes with the angular momentum operator:

$$[\mathbf{L}, \mathbf{L}^2] = 0 \tag{21}$$

that is,

$$[L_x, \mathbf{L}^2] = 0 \quad [L_y, \mathbf{L}^2] = 0 \quad \text{and} \quad [L_z, \mathbf{L}^2] = 0$$

Exercise 4. Show this.

Exercise 5. Show that $\mathbf{L}^2$ is hermitian.

For a free particle, the operators L_x, L_y, L_z, and $\mathbf{L}^2$ commute with the Hamiltonian. For instance, to evaluate the commutator of L_x with $p^2/2m$, we note that the commutator of x_j with any function of p_k is

$$[f(p_k), x_j] = \frac{\hbar}{i} \frac{\partial f}{\partial p_j} \tag{22}$$

This formula is an obvious generalization of Eq. (2.103) to three dimensions. With this, we immediately obtain

$$[p^2, L_x] = [p^2, yp_z - zp_y] = [p^2, yp_z] - [p^2, zp_y]$$

$$= \frac{\hbar}{i} \frac{\partial p^2}{\partial p_y} p_z - \frac{\hbar}{i} \frac{\partial p^2}{\partial p_z} p_y = \frac{\hbar}{i} 2p_y p_z - \frac{\hbar}{i} 2p_z p_y = 0 \tag{23}$$

Thus, L_x commutes p^2, and the same is true for the other components of the angular momentum. Since the components of the angular momentum commute with the Hamiltonian, their expectation values are constant in time, according to Eq. (5.15). This result corresponds to the classical theorem of the conservation of angular momentum for a free particle.

Classical physics leads us to conjecture that the expectation values of the components of the angular momentum should also be constant for a particle moving in a central potential, that is, a potential of the form $V(r)$, where $r^2 = x^2 + y^2 + z^2$. In classical physics, such a potential corresponds to a central force of magnitude $-\partial V/\partial r$ directed toward the origin. For this potential, the Hamiltonian is

$$H = \frac{p^2}{2m} + V(r) \tag{24}$$

Since we already know that the angular momentum commutes with the term $p^2/2m$ in this Hamiltonian, it only remains to check that the angular momentum commutes with the potential $V(r)$. For this purpose, it is convenient to regard V as a function of r^2, and it then suffices to check that the angular momentum commutes with r^2. For instance, to evaluate the commutator of L_x with r^2, we note that the commutator of p_j and any function of position is

$$[g(x_k), p_j] = i\hbar \frac{\partial g}{\partial x_j} \qquad (25)$$

For $g(x_k) = r^2$, this becomes

$$[r^2, p_j] = 2i\hbar x_j$$

With this, we can carry out a calculation similar to that in Eq. (23), and we find the result

$$[r^2, L_x] = 0.$$

Exercise 6. Carry out this calculation.

Thus, L_x commutes with r^2, and so do the other components of the angular momentum. As in the case of the free particle, the commutativity of the components of the angular momentum with the Hamiltonian implies that their expectation values are constant in time.

For a particle or a system of particles interacting with a central potential, the angular-momentum operators are very useful for specifying the possible states of the system. Suppose that we want to find a complete set of commuting observables. This means that we seek simultaneous eigenfunctions of a set of commuting operators, chosen so that specification of the eigenvalues determines the state completely. As one of the operators in this set, we will usually take the Hamiltonian, and therefore all the others must commute with the Hamiltonian (and their expectation values are conserved). We saw above that, in a central potential, the components of the angular momentum, and therefore also the magnitude squared $\mathbf{L}^2$ of the angular momentum, commute with the Hamiltonian. So we may take $\mathbf{L}^2$ as another operator in the set of commuting observables. Next, we take L_z as another operator in the set. Since L_z commutes with $\mathbf{L}^2$, simultaneous eigenfunctions of H, $\mathbf{L}^2$, and L_z exist. Next, we might want to take L_x as another operator in the set. But this is impossible, because L_x and L_z do not commute, and if the system is in an eigenstate of L_z it will *not* be in an eigenstate of L_x or L_y. There is one trivial exception: if the angular momentum is zero, then the system is in a simultaneous eigenstate of all three components of $\mathbf{L}$. Apart from this special case, if a state is an eigenstate of any one component of $\mathbf{L}$, it cannot be an eigenstate of any other. Only one component at a time can have a sharp value.

Exercise 7. Show that if a system is in a simultaneous eigenstate of L_z and L_x, then the state is an eigenstate of $\mathbf{L}$ with eigenvalue zero. [Hint: Use Eq. (20).]

We will now construct the simultaneous eigenstates of L^2 and L_z. Of course, instead of selecting L_z for special treatment, we could equally well select L_x or L_y. What is relevant is that the component of the angular momentum operator in some chosen direction has simultaneous eigenvectors with $\mathbf{L}^2$. We can call this component L_z, since we can always reorient our coordinate system so the chosen direction is the z direction.

Suppose, then, that $|\lambda, m_l\rangle$ is a simultaneous eigenvector of L^2 and L_z with eigenvalues $\hbar^2\lambda$ and $\hbar m$, respectively:

$$\mathbf{L}^2 |\lambda, m_l\rangle = \hbar^2\lambda |\lambda, m_l\rangle \tag{26}$$

$$L_z |\lambda, m_l\rangle = \hbar m_l |\lambda, m_l\rangle \tag{27}$$

It is easy to prove that the expectation value of the square of any hermitian operator is never negative.

Exercise 8. Prove this theorem.

Hence, with the assumption that $|\lambda, m_l\rangle$ is properly normalized,

$$0 \le \langle\lambda, m_l| (L_x^2 + L_y^2) |\lambda, m_l\rangle = \langle\lambda, m_l| (\mathbf{L}^2 - L_z^2) |\lambda, m_l\rangle$$

$$= (\lambda\hbar^2 - m_l^2\hbar^2) \langle\lambda, m_l|\lambda, m_l\rangle$$

$$= \lambda\hbar^2 - m_l^2\hbar^2 \tag{28}$$

which implies

$$\lambda \ge m_l^2 \tag{29}$$

We now define two operators

$$L_+ = L_x + iL_y \quad \text{and} \quad L_- = L_x - iL_y \tag{30}$$

These are called, respectively, the angular momentum *raising* and *lowering operators*. Since L_x and L_y commute with $\mathbf{L}^2$, we obviously have

$$[\mathbf{L}^2, L_+] = 0 \quad [\mathbf{L}^2, L_-] = 0 \tag{31}$$

We then note that $L_{\pm}|\lambda,m_l\rangle$ is an eigenstate of $\mathbf{L}^2$:

$$\mathbf{L}^2(L_{\pm}|\lambda,m_l\rangle) = L_{\pm}\mathbf{L}^2|\lambda,m_l\rangle = L_{\pm}\hbar^2\lambda|\lambda,m_l\rangle$$
$$= \hbar^2\lambda(L_{\pm}|\lambda,m_l\rangle) \tag{32}$$

Thus, $L_{\pm}|\lambda,m_l\rangle$ is an eigenstate of $\mathbf{L}^2$ with the same eigenvalue $\hbar^2\lambda$ as that corresponding to $|\lambda,m_l\rangle$. Next, we note that it is also an eigenstate of L_z:

$$L_z(L_+|\lambda,m_l\rangle) = L_z(L_x + iL_y)|\lambda,m_l\rangle = (L_zL_x + iL_zL_y)|\lambda,m_l\rangle$$
$$= [(L_xL_z + i\hbar L_y) + i(L_yL_z - i\hbar L_x)]|\lambda,m_l\rangle$$
$$= (L_x\hbar m_l + i\hbar L_y + iL_y\hbar m_l + \hbar L_x)|\lambda,m_l\rangle$$
$$= (\hbar m_l + \hbar)(L_x + iL_y)|\lambda,m_l\rangle = \hbar(m_l + 1)(L_+|\lambda,m_l\rangle) \tag{33}$$

Thus, $L_+|\lambda,m_l\rangle$ is an eigenstate of L_z with eigenvalue $\hbar(m_l + 1)$. Combining both results we have that

$$L_+|\lambda,m_l\rangle = c_{\lambda,m_l+1}|\lambda,m_l+1\rangle \tag{34}$$

where c_{λ,m_l+1} is some constant of proportionality that remains to be evaluated. Likewise, we can show that

$$L_-|\lambda,m_l\rangle = c'_{\lambda,m_l-1}|\lambda,m_l-1\rangle \tag{35}$$

Exercise 9. Show this.

Exercise 10. Show that

$$\langle\lambda,m_l|\, L_x\,|\lambda,m_l\rangle = 0 \quad\text{and}\quad \langle\lambda,m_l|\, L_y\,|\lambda,m_l\rangle = 0$$

Equations (34) and (35) mean that L_+ and L_- are ladder operators for the eigenvectors of L_z. The operator L_+ raises the eigenvalue m_l to $m_l + 1$, and the operator L_- lowers the eigenvalue m_l to $m_l - 1$, while leaving the eigenvalue $\hbar^2\lambda$ fixed.

With these raising and lowering operators we are now ready to determine the allowed values of λ and m_l and to construct the eigenvectors. The procedure here is quite similar to that we used in Chapter 6 in the construction of the energy eigenvectors of the harmonic oscillator. Consider some fixed value of λ. Then any possible value of m_l must be such that $m_l^2 \leq \lambda$. There will be some maximum and some minimum value of m_l consistent with this inequality; we designate these values of m_{max} and m_{min}, respectively. Clearly, $L_+|\lambda,m_{max}\rangle$ must be zero since if it were not, it would be a state with a value of m_l larger than the maximum. Likewise, $L_-|\lambda,m_{min}\rangle$ must be zero:

$$L_+|\lambda,m_{max}\rangle = 0$$
$$L_-|\lambda,m_{min}\rangle = 0$$

(36)

Next, we calculate $L^2|\lambda,m_{max}\rangle$ using the identity

$$\mathbf{L}^2 = L_-L_+ + L_z^2 + \hbar L_z$$

(37)

Exercise 11. Verify this identity.

We obtain

$$\mathbf{L}^2|\lambda,m_{max}\rangle = (L_-L_+ + L_z^2 + \hbar L_z)|\lambda,m_{max}\rangle$$
$$= (0 + \hbar^2 m_{max}^2 + \hbar^2 m_{max})|\lambda,m_{max}\rangle$$
$$= \hbar^2 m_{max}(m_{max} + 1)|\lambda,m_{max}\rangle$$

(38)

Likewise, we can show that

$$\mathbf{L}^2|\lambda,m_{min}\rangle = \hbar^2 m_{min}(m_{min} - 1)|\lambda,m_{min}\rangle$$

(39)

Exercise 12. Show this

Now, let us compare Eqs. (38) and (39). We know that both $|\lambda,m_{min}\rangle$ and $|\lambda,m_{max}\rangle$ are eigenstates of $\mathbf{L}^2$ with the same eigenvalue $\hbar^2\lambda$. It follows that

$$m_{min}(m_{min} - 1) = m_{max}(m_{max} + 1)$$

(40)

This equation implies that either $m_{min} = m_{max} + 1$ (which is absurd) or else

$$m_{max} = -m_{min}$$

(41)

So we can write $m_{max} = |m_{min}| = l$, where l is some number. The possible values for m_l are then

$$m_l = l, l - 1, l - 2, \ldots, -l + 1, -l$$

(42)

This equation says that there are $2l + 1$ possible values of m_l, and this, in turn, implies that $2l + 1$ is a positive integer. Consequently,

$$l = 0, \tfrac{1}{2}, 1, \tfrac{3}{2}, 2, \tfrac{5}{2}, \ldots$$

(43)

We will see later that the half-integer values for l are not possible if we are dealing (as we are here) with *orbital* angular momentum. These values are possible only if there is *spin* angular momentum.

To summarize: Our eigenvalue equation is

$$\mathbf{L}^2|l(l+1),m_l\rangle = \hbar^2 l(l + 1)|l(l+1),m_l\rangle \qquad (44)$$

where we have inserted for λ its value $l(l + 1)$. The angular-momentum eigenvectors are usually written in the concise notation $|l,m_l\rangle$, and therefore

$$\mathbf{L}^2|l,m_l\rangle = \hbar^2 l(l + 1)|l,m_l\rangle \qquad l = 0, 1, 2, 3, \ldots \qquad (45)$$

$$L_z|l,m_l\rangle = \hbar m_l|l,m_l\rangle \qquad m_l = -l, -l + 1, \ldots, l - 1, l \qquad (46)$$

These equations show that the magnitudes of the angular momentum and of the z component of the angular momentum are quantized. Note that in contrast to the quantization of the energy, the quantization of the angular momentum does not require that the particle be bound. A particle in a state of positive energy is subject to the same quantization of angular momentum as a bound particle, and so is a free particle moving in the absence of any central force. Such a quantization of the angular momentum of a free particle seems puzzling—we know that the free particle can be described by eigenstates of momentum (plane waves), and these, obviously, are not eigenstates of angular momentum. The answer to this puzzle is that the momentum eigenstates can be expressed as superpositions of the angular-momentum eigenstates, and conversely. Both kinds of states are complete sets of states, and whether we adopt one set or the other for the description of a free particle is a matter of choice. However, the construction of a momentum eigenstate by the superposition of angular-momentum eigenstates, or conversely, is a fairly complicated exercise. We will perform this construction in the next chapter.

Note that the possible magnitudes of the angular momentum are $\sqrt{l(l + 1)}\hbar$, that is, 0, $\sqrt{2}\hbar$, $\sqrt{6}\hbar$, $\sqrt{12}\hbar$, ... If we want to draw a picture of the angular-momentum vector, we must select one of these magnitudes, and we must select its angular orientation with the z axis in accord with Eq. (46). For instance, if the magnitude of the angular momentum is $\sqrt{6}\hbar$, then the possible orientations of the angular-momentum vector are as shown in Fig. 7.1a. However, since the x and y components of the angular-momentum vector are not well defined, we cannot think of the vector as having a definite direction in space. Instead, we must think of

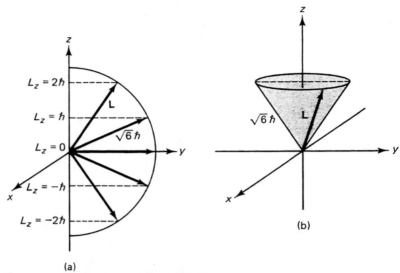

(a)

Fig. 7.1   (a) Possible orientations of the angular-momentum vector for $l = 2$ and $m_l = +2, +1, 0, -1, -2$. (b) The vector model for the state $l = 2$, $m_l = +2$.

the vector as distributed over all directions in the x-y plane at once; thus, the vector is distributed over a cone around the z axis (see Fig. 7.1b). This graphical representation of the angular momentum is called the *vector model*.

Let us now return to Eq. (34) and evaluate the constant of proportionality. First we rewrite this equation in our new notation:

$$L_+|l,m_l\rangle = c_{l,m_l+1}|l,m_l+1\rangle$$
$$L_-|l,m_l\rangle = c_{l,m_l-1}|l,m_l-1\rangle \tag{47}$$

The normalization condition is

$$|c_{l,m_l+1}|^2 \langle l,m_l+1|l,m_l+1\rangle = \langle l,m_l| L_+^\dagger L_+ |l,m_l\rangle$$
$$= \langle l,m_l| L_- L_+ |l,m_l\rangle \tag{48}$$

Using the identity (37), we find that

$$\langle l,m_l| L_- L_+ |l,m_l\rangle = \langle l,m_l| (\mathbf{L}^2 - L_z^2 - \hbar L_z) |l,m_l\rangle$$
$$= \langle l,m_l| [\hbar^2 l(l+1) - m_l^2\hbar^2 - \hbar^2 m_l] |l,m_l\rangle \tag{49}$$

Hence

$$c_{l,m_l+1} = \sqrt{\hbar^2[l(l+1) - m_l(m_l+1)]} \tag{50}$$

and

$$L_+|\,l,m_l\rangle = \hbar\sqrt{l(l+1) - m_l(m_l+1)}\,|l,m_l+1\rangle \tag{51}$$

where we have made the simplest possible choice for the arbitrary phase in the constant c_{l,m_l+1}.

Exercise 13. Likewise, show that

$$L_-|l,m_l\rangle = \hbar\sqrt{l(l+1) - m_l(m_l-1)}\,|l,m_l-1\rangle \tag{52}$$

With the raising operator L_+ we can construct the eigenvectors $|l,m_l\rangle$, starting with $|l,-l\rangle$, the eigenvector of the lowest m_l. We have

$$|l,-l+1\rangle = \frac{1}{c_{l,-l+1}}\,L_+|l,-l\rangle = \frac{1}{\hbar\sqrt{2l}}\,L_+|l,-l\rangle$$

$$|l,-l+2\rangle = \frac{1}{c_{l,-l+2}}\,L_+|l,-l+1\rangle = \frac{1}{\hbar^2}\,\frac{1}{\sqrt{4l-2}}\,\frac{1}{\sqrt{2l}}\,L_+^2|l,-l\rangle \tag{53}$$

The general result is

$$|l,m_l\rangle = \frac{1}{\hbar^{l+m_l}}\,\sqrt{\frac{(l-m_l)!}{(2l)!\,(l+m_l)!}}\,(L_+)^{l+m_l}|l,-l\rangle \tag{54}$$

This general result may be obtained as follows. With

$$l(l+1) - m_l(m_l-1) = (l-m_l+1)(l+m_l)$$

we can rewrite Eq. (51) as

$$L_+|l,m_l-1\rangle = \hbar\sqrt{(l-m_l+1)(l+m_l)}\,|l,m_l\rangle \tag{55}$$

Repeated application of Eq. (55) yields

$$(L_+)^k\,|l,m_l-k\rangle = \hbar^k\{(l-m_l+1)(l+m_l)(l-m_l+2)(l+m_l-1)\cdots$$
$$(l-m_l+k)(l+m_l-k+1)\}^{1/2}\,|l,m_l\rangle \tag{56}$$

For $k = l + m_l$ this becomes

$$(L_+)^{l+m_l}\,|l,-l\rangle = \hbar^{l+m_l}\{(l-m_l+1)(l+m_l)(l-m_l+2)(l+m_l-1)$$
$$\cdots (2l)(1)\}^{1/2}\,|l,m_l\rangle \tag{57}$$

With some rearrangement, we recognize that this is the same as Eq. (54).

Equation (54) expresses the state with arbitrary m_l in terms of the state with lowest m_l (for given l). We will now find the wavefunction corresponding to $|l,-l\rangle$ in the position representation. This will permit us to construct the wavefunctions corresponding to $|l,m_l\rangle$.

The equation that determines $|l,-l\rangle$ is

$$L_-|l,-l\rangle = 0 \tag{58}$$

In the position representation, we designate the eigenfunctions by $Y_l^m(\mathbf{r})$, that is,

$$Y_l^{m_l}(\mathbf{r}) = \langle \mathbf{r}|l,m_l\rangle$$

Equation (58) then reads

$$(L_x - iL_y)Y_l^{-l} = 0 \tag{59}$$

Here, L_x and L_y are the differential operators

$$L_x = \frac{\hbar}{i}\left(y\frac{\partial}{\partial z} - z\frac{\partial}{\partial y}\right)$$

$$L_y = \frac{\hbar}{i}\left(z\frac{\partial}{\partial x} - x\frac{\partial}{\partial z}\right) \tag{60}$$

The differential operators L_x, L_y, and L_z can be expressed in spherical coordinates r, θ, ϕ (see Fig. 7.2), with the result

$$L_x = -\frac{\hbar}{i}\left(\sin\phi\frac{\partial}{\partial\theta} + \cot\theta\cos\phi\frac{\partial}{\partial\phi}\right)$$

$$L_y = \frac{\hbar}{i}\left(\cos\phi\frac{\partial}{\partial\theta} - \cot\theta\sin\phi\frac{\partial}{\partial\phi}\right)$$

$$L_z = \frac{\hbar}{i}\frac{\partial}{\partial\phi} \tag{61}$$

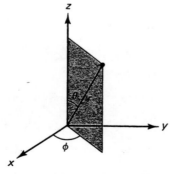

Fig. 7.2   The spherical coordinates r, θ, and ϕ.

Exercise 14. Perform the transformation of L_x, L_y, and L_z into spherical coordinates.

Exercise 15. Show that

$$\mathbf{L}^2 = -\hbar^2 \left(\frac{1}{\sin\theta} \frac{\partial}{\partial\theta} \sin\theta \frac{\partial}{\partial\theta} + \frac{1}{\sin^2\theta} \frac{\partial^2}{\partial\phi^2} \right) \tag{62}$$

The raising and lowering operators L_+ and L_- in spherical coordinates are then

$$L_+ = \hbar e^{i\phi} \left(\frac{\partial}{\partial\theta} + i \cot\theta \frac{\partial}{\partial\phi} \right)$$

$$L_- = -\hbar e^{-i\phi} \left(\frac{\partial}{\partial\theta} - i \cot\theta \frac{\partial}{\partial\phi} \right) \tag{63}$$

and Eq. (59) becomes

$$-\hbar e^{-i\phi} \left(\frac{\partial}{\partial\theta} - i \cot\theta \frac{\partial}{\partial\phi} \right) Y_l^{-l}(\theta, \phi) = 0 \tag{64}$$

Here we have written Y_l^{-l} as a function of the variables θ and ϕ, since these are the only variables that appear in the differential operator. The dependence on r is left undetermined by the differential equation. We may as well assume that there is no radial dependence in the eigenfunctions $Y_l^{m_l}$. Since Y_l^{-l} is an eigenfunction of L_z with eigenvalue $-\hbar l$, it must also satisfy the equation

$$\frac{\hbar}{i} \frac{\partial}{\partial\phi} Y_l^{-l}(\theta, \phi) = -\hbar l \, Y_l^{-l}(\theta, \phi) \tag{65}$$

This equation has the solution

$$Y_l^{-l}(\theta, \phi) = e^{-il\phi} f(\theta) \tag{66}$$

where $f(\theta)$ is some function of θ. If we insert this into Eq. (64), we obtain

$$-\hbar e^{-i\phi} \left(\frac{\partial}{\partial\theta} - l \cot\theta \right) e^{-il\phi} f(\theta) = 0 \tag{67}$$

This equation has the solution

$$f(\theta) = c(\sin\theta)^l \tag{68}$$

where c is a constant. Therefore,

$$Y_l^{-l}(\theta, \phi) = ce^{-il\phi} (\sin \theta)^l \tag{69}$$

where c is to be regarded as a normalization constant.

Exercise 16. Using $L_+ Y_l^l(\theta, \phi) = 0$, show that

$$Y_l^l = c'e^{il\phi} (\sin \theta)^l \tag{70}$$

For convenience, we will chose the constant in Eq. (69) so the integral of $|Y_l^{-l}|^2$ over all angles is unity:

$$1 = \int (Y_l^{-l})^* \, Y_l^{-l} \, d\Omega = \iint |c|^2 (\sin \theta)^{2l} \sin \theta \, d\theta \, d\phi \tag{71}$$

where $d\Omega = \sin \theta \, d\theta \, d\phi$ is the element of solid angle.

Exercise 17. Show that, with $\mu \equiv \cos \theta$,

$$\int_0^{2\pi} \int_0^{\pi} (\sin \theta)^{2l} \sin \theta \, d\theta \, d\phi = 2\pi \int_{-1}^{1} (1 - \mu^2)^l \, d\mu$$

$$= (2^l l!)^2 \frac{4\pi}{(2l + 1)!} \tag{72}$$

and that (with a choice of phase)

$$c = \frac{1}{2^l l!} \sqrt{\frac{(2l + 1)!}{4\pi}} \tag{73}$$

The normalized eigenfunction is then

$$Y_l^{-l}(\theta, \phi) = \frac{1}{2^l l!} \sqrt{\frac{(2l + 1)!}{4\pi}} \, e^{-il\phi} (\sin \theta)^l \tag{74}$$

Before we go on, let us see what happens if l is a "half-integer," say $l = \frac{1}{2}$. Then Eq. (69) yields

$$Y_{1/2}^{-1/2} \propto e^{-i\phi/2} \sqrt{\sin \theta} \tag{75}$$

It is often said that the function (75) is not acceptable because when ϕ is increased by 2π, $e^{-i\phi/2}$ becomes $-e^{-i\phi/2}$, and thus $Y_{1/2}^{-1/2}$ is not single valued. But this is no reason for discarding the function (75), because the only things we ever measure are probabilities and expectation values computed with these probabilities. Hence we need to demand only that the absolute value $|Y_{1/2}{}^{-1/2}|^2$ be single valued, and it is. Wavefunctions that change sign upon a rotation by 2π are actually very important in physics, since electrons (and other particles) are described by such wavefunctions when spin is taken into account.

The real trouble with the function (75) is not its behavior under an increase of ϕ by 2π, but its behavior under the action of the raising operator L_+. If we apply the raising operator L_+ to $Y_{1/2}{}^{-1/2}$, we should get $Y_{1/2}{}^{1/2}$:

$$Y_{1/2}{}^{1/2} \propto L_+ Y_{1/2}{}^{-1/2}$$

$$\propto \hbar e^{i\phi} \left(\frac{\partial}{\partial\theta} + i \cot\theta \, \frac{\partial}{\partial\phi} \right) e^{-i\phi/2} \sqrt{\sin\theta}$$

$$\propto \hbar e^{i\phi/2} \frac{\cos\theta}{\sqrt{\sin\theta}} \tag{76}$$

But this is wrong, since, according to Eq. (70),

$$Y_{1/2}{}^{1/2} \propto e^{i\phi/2} \sqrt{\sin\theta}$$

The contradiction between these two expressions for $Y_{1/2}{}^{1/2}$ can be resolved only by excluding $l = \frac{1}{2}$. It is easy to show that the same trouble occurs whenever l is a half-integer, and therefore all these half-integer values of l must be discarded. Thus, there exist no wavefunctions for half-integer angular momentum, and we can conclude that the orbital angular momentum is restricted to integer values of l. This argument for integer values of the orbital angular momentum is, in essence, due to Pauli.[1] Note that the argument does not rule out half-integer values for the spin, since the existence of a wavefunction (that is, a function of the coordinates in the position representation) is not required for the spin [the angular momentum operators for spin obey the commutation relations (20), but they do not have a representation as differential operators]. From now on we will always assume that l is an integer.

---

[1] An alternative argument may be found in H. S. Green, *Matrix Methods in Quantum Mechanics*.

The wavefunction $Y_l^{m_l}$ is obtained from Y_l^{-l} by making use of Eq. (54):

$$Y_l^{m_l}(\theta, \phi) = \sqrt{\frac{(l - m_l)!}{(2l)! \, (l + m_l)!}} \left(\frac{L_+}{\hbar}\right)^{l+m_l} Y_l^{-l}(\theta, \phi)$$

$$= \sqrt{\frac{(l - m_l)!}{(2l)! \, (l + m_l)!}} \left[e^{i\phi} \left(\frac{\partial}{\partial\theta} + i \cot\theta \, \frac{\partial}{\partial\phi}\right)\right]^{l+m_l}$$

$$\frac{1}{2^l l!} \sqrt{\frac{(2l + 1)!}{4\pi}} \, e^{-il\phi}(\sin\theta)^l \qquad (77)$$

We can put this into a more convenient form by some manipulation. If we provisionally ignore the normalization factors, we can write Eq. (77) as

$$Y_l^{m_l} \propto \left[e^{i\phi} \left(\frac{\partial}{\partial\theta} + i \cot\theta \, \frac{\partial}{\partial\phi}\right)\right]^{l+m_l} e^{-il\phi}(\sin\theta)^l \qquad (78)$$

Let us consider the effect of the raising operator $e^{i\phi} (\partial/\partial\theta + i \cot\theta \, \partial/\partial\phi)$ on $Y_l^{m_l-1}$:

$$Y_l^{m_l} \propto e^{i\phi} \left(\frac{\partial}{\partial\theta} + i \cot\theta \, \frac{\partial}{\partial\phi}\right) Y_l^{m_l-1}$$

$$= e^{i\phi} \left(\frac{\partial}{\partial\theta} - (m_l - 1) \cot\theta\right) Y_l^{m_l-1}$$

$$= \left(e^{i\phi}(\sin\theta)^{(m_l-1)} \frac{\partial}{\partial\theta} (\sin\theta)^{-(m_l-1)}\right) Y_l^{m_l-1}$$

$$= \left(-e^{i\phi}(\sin\theta)^{m_l} \frac{\partial}{\partial(\cos\theta)} (\sin\theta)^{-(m_l-1)}\right) Y_l^{m_l-1} \qquad (79)$$

With the abbreviation $\mu \equiv \cos\theta$, Eq. (79) becomes

$$Y_l^{m_l} \propto \left(-e^{i\phi}(1 - \mu^2)^{\frac{1}{2}m_l} \frac{\partial}{\partial\mu} (1 - \mu^2)^{-\frac{1}{2}(m_l-1)}\right) Y_l^{m_l-1} \qquad (80)$$

It is easy to see that this recursion relation is satisfied by the following expression for $Y_l^{m_l}$:

$$Y_l^{m_l} \propto (-1)^{m_l}(1 - \mu^2)^{\frac{1}{2}m_l} \frac{\partial^{l+m_l}}{\partial\mu^{l+m_l}} (\mu^2 - 1)^l e^{im_l\phi} \qquad (81)$$

Exercise 18. Check that Eq. (81) satisfies the recursion relation (80), and that it also satisfies the condition $Y_l^{-l} \propto e^{-il\phi}(1 - \mu^2)^{\frac{1}{2}l}$.

In view of these results, Eq. (77) takes the form

$$Y_l^{m_l}(\theta, \phi) = \sqrt{\frac{2l + 1}{4\pi} \frac{(l - m_l)!}{(l + m_l)!}} \frac{(-1)^{m_l}}{2^l l!} (1 - \mu^2)^{\frac{1}{2}m_l} \frac{\partial^{l+m_l}}{\partial \mu^{l+m_l}} (\mu^2 - 1)^l e^{im_l\phi} \quad (82)$$

The functions $Y_l^{m_l}(\theta, \phi)$ are called *spherical harmonics.* They form a complete set for the expansion of any function of θ and ϕ. Their orthonormality condition is

$$\int Y_l^{m_{l'}}{}^* Y_l^{m_l} \, d\Omega = \delta_{ll'}\delta_{m_l m_{l'}} \quad (83)$$

where $d\Omega \equiv \sin \theta \, d\theta \, d\phi$, and the integration goes over all values of θ and ϕ. The spherical harmonics of positive and negative values of m_l are related by complex conjugation; from Eq. (82) it is easy to see that

$$Y_l^{-m_l} = (-1)^{m_l}(Y_l^{m_l})^* \quad (84)$$

Table 7.1 lists the first few spherical harmonics $Y_l^{m_l}$, and Fig. 7.3 displays plots of some of these spherical harmonics for $\phi = 0$.

The spherical harmonics are often written as

$$Y_l^{m_l} = \sqrt{\frac{2l + 1}{4\pi} \frac{(l - m_l)!}{(l + m_l)!}} \, e^{im_l\phi} P_l^{m_l}(\mu) \quad (85)$$

TABLE 7.1 Spherical Harmonics

$$Y_0^0 = \sqrt{\frac{1}{4\pi}}$$

$$Y_1^0 = \sqrt{\frac{3}{4\pi}} \cos \theta$$

$$Y_1^1 = -\sqrt{\frac{3}{8\pi}} \sin \theta \, e^{i\phi}$$

$$Y_2^0 = \sqrt{\frac{5}{16\pi}} (3 \cos^2 \theta - 1)$$

$$Y_2^1 = -\sqrt{\frac{15}{8\pi}} \sin \theta \cos \theta \, e^{i\phi}$$

$$Y_2^2 = \sqrt{\frac{15}{32\pi}} \sin^2 \theta \, e^{2i\phi}$$

$$Y_3^0 = \sqrt{\frac{7}{16\pi}} (5 \cos^3 \theta - 3 \cos \theta)$$

$$Y_3^1 = -\sqrt{\frac{21}{64\pi}} (5 \cos^2 \theta - 1) \sin \theta \, e^{i\phi}$$

$$Y_3^2 = \sqrt{\frac{105}{32\pi}} \sin^2 \theta \cos \theta \, e^{2i\phi}$$

$$Y_3^3 = -\sqrt{\frac{35}{64\pi}} \sin^3 \theta \, e^{3i\phi}$$

where

$$P_l^{m_l}(\mu) = \frac{1}{2^l l!} (-1)^{m_l}(1 - \mu^2)^{m_l/2} \frac{d^{l+m_l}}{d\mu^{l+m_l}} (\mu^2 - 1)^l \qquad (86)$$

are the *associated Legendre functions*.

Exercise 19. Show that $P_l^0(\mu)$ is a polynomial of order l. Show that

$$
\begin{aligned}
P_0^0(\mu) &= 1 \\
P_1^0(\mu) &= \mu \\
P_2^0(\mu) &= \tfrac{1}{2}(3\mu^2 - 1) \\
P_3^0(\mu) &= \tfrac{1}{2}(5\mu^3 - 3\mu)
\end{aligned}
\qquad (87)
$$

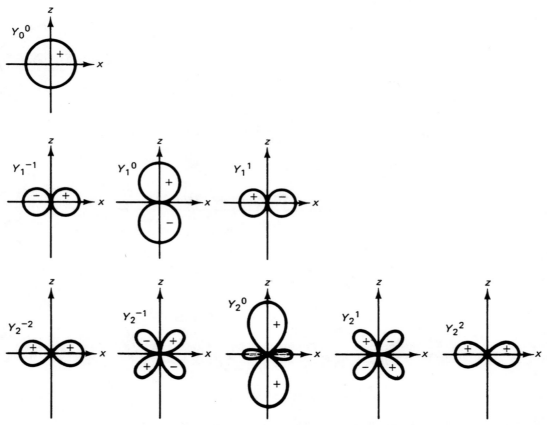

Fig. 7.3 Polar plots of the spherical harmonics $Y_l^{m_l}(\theta, 0)$ as a function of θ. These plots may be regarded as displaying $Y_l^{m_l}$ in the z-x plane, where $\phi = 0$. Positive and negative values of $Y_l^{m_l}$ are indicated by + and −.

Exercise 20. Show that

$$\mu P_l^{m_l} = \frac{l + 1 - m_l}{2l + 1} P_{l+1}^{m_l} + \frac{l + m_l}{2l + 1} P_{l-1}^{m_l} \tag{88}$$

Exercise 21. Show that

$$(1 - \mu^2) \frac{dP_l^{m_l}}{d\mu} = (l + 1)\mu P_l^{m_l} - (l + 1 - m_l)P_{l+1}^{m_l} \tag{89}$$

7.3 *Parity*

In classical mechanics, there exists symmetry between right and left, in the sense that for any given motion of an isolated system of interacting classical particles the mirror-reflected motion is also possible.[2] For example, all the planets move around the Sun in a counterclockwise direction (as seen from the Pole star). But they could just as well move in a clockwise direction—the preference of the Solar System for counterclockwise rotation is not due to the equation of motion, but rather to the initial conditions.

If we consider a mirror in the x-y plane (see Fig. 7.4), then the operation of mirror reflection changes the position of a particle from (x, y, z) to $(x, y, -z)$. Mathematically, the right–left symmetry of the system means that the equations of motion are invariant (that is, unchanged) by a mirror reflection of the position coordinates of all the particles in the system.

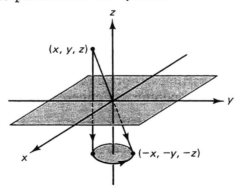

Fig. 7.4 Reflection through the origin and reflection in the x-y plane followed by a rotation of 180° around the z axis. These two operations are equivalent.

---

[2] In this context, *isolated* means that there are no external forces acting on the particles, that is, anything the particles interact with is included in the system.

Instead of mirror reflection in a plane, we will find it more convenient to deal with reflection through the origin. By the latter is meant a change in the position of the particle from (x, y, z) to $(-x, -y, -z)$. Obviously, reflection through the origin is equivalent to a mirror reflection in the x-y plane followed by a rotation of the particle's position by 180° about the z axis (see Fig. 7.4). Since a rotation of any isolated system of particles does not change the equation of motion, right–left symmetry is mathematically equivalent to symmetry with respect to reflection through the origin. If the position and the velocity of a particle are $\mathbf{r}$ and $\mathbf{v}$, then reflection through the origin gives $-\mathbf{r}$ and $-\mathbf{v}$. The right–left symmetry of classical mechanics then amounts to this: The equations of motion, or, equivalently, the Hamiltonian, remain invariant under the transformations $\mathbf{r} \to -\mathbf{r}$ and $\mathbf{v} \to -\mathbf{v}$. The operation of replacing quantities by the corresponding quantities reflected through the origin is called the (classical) parity operation.

In quantum mechanics, we define the parity operator P as follows: Consider a state that describes a particle at a definite position $\mathbf{r}$. The parity operator P is defined by what it does to this state:[3]

$$P|\mathbf{r}\rangle = |-\mathbf{r}\rangle \tag{90}$$

This means that the parity operator changes the position of the particle from $\mathbf{r}$ to $-\mathbf{r}$, that is, it reflects the position though the origin. Since the states $|\mathbf{r}\rangle$ form a complete set, the operator P is unambiguously defined by its effect on the members of this complete set. Its effect on any arbitrary state can be determined by expressing such a state as a superposition of the states $|\mathbf{r}\rangle$.

A word of caution is necessary. The definition (90) applies to electrons, protons, neutrons, and many other particles; but it does not apply to all particles. For pions, positrons, and some other particles, the parity operator is defined with an extra minus sign:

$$P|\mathbf{r}\rangle = -|-\mathbf{r}\rangle \tag{91}$$

Particles to which this alternative definition applies are said to have odd (or negative) *intrinsic parity*. By contrast, particles to which the original definition (90) applies are said to have even (or

---

[3] Alternatively, we could define P by what it does to momentum eigenstates or to wavefunctions. For nonrelativistic quantum mechanics, the definition (90) is the most obvious.

positive) intrinsic parity. We will hereafter consider only particles of even intrinsic parity.

Parity is a hermitian operator. We can establish this by using $\langle \mathbf{r}'|-\mathbf{r}\rangle = \langle -\mathbf{r}'|\mathbf{r}\rangle$, which is true since both of these expressions equal $\delta(\mathbf{r} + \mathbf{r}')$. We then have

$$\langle \mathbf{r}'|P|\mathbf{r}\rangle = \langle \mathbf{r}'|-\mathbf{r}\rangle = \langle -\mathbf{r}'|\mathbf{r}\rangle = (\langle \mathbf{r}'|P)|\mathbf{r}\rangle$$

It is obvious from the definition (90) that when the operator P acts twice on any given state vector, it leaves the state vector unchanged; thus,

$$P^2 = 1 \tag{92}$$

From this, we see that if $|\psi\rangle$ is an eigenstate of parity, with $P|\psi\rangle = a|\psi\rangle$, then $a^2 = 1$, and hence $a = \pm 1$. Thus, the only possible eigenvalues of the parity operator are $+1$ and -1.

To see what the parity operator does in the position representation, we consider an arbitrary state and express it as a superposition of position eigenvectors:

$$|\psi\rangle = \int d^3r \, |\mathbf{r}\rangle \, \langle \mathbf{r}|\psi\rangle \tag{93}$$

If we operate with P on this state,

$$P|\psi\rangle = \int d^3r \, P(|\mathbf{r}\rangle) \, \langle \mathbf{r}|\psi\rangle$$

$$= \int d^3r \, |-\mathbf{r}\rangle \, \langle \mathbf{r}|\psi\rangle$$

$$= \int d^3r \, |\mathbf{r}\rangle \, \langle -\mathbf{r}|\psi\rangle \tag{94}$$

where the last step in Eq. (94) is a simple change of the variable of integration from $\mathbf{r}$ to $-\mathbf{r}$. Comparing Eq. (94) with Eq. (93), we see that under the parity operation the wavefunction $\psi(\mathbf{r}) = \langle \mathbf{r}|\psi\rangle$ is changed into $\psi(-\mathbf{r})$. Thus, in the position representation,

$$P\psi(\mathbf{r}) = \psi(-\mathbf{r}) \tag{95}$$

This result is as expected; in fact, we could have adopted Eq. (95) as the definition of the parity operator.

Since the effect of the parity operator in the position representation is so simple, it is often most convenient to deal with parity in this representation. The eigenfunctions of parity with eigenvalue $+1$ are the *even* functions of $\mathbf{r}$:

$$\psi_{\text{even}}(\mathbf{r}) = \psi_{\text{even}}(-\mathbf{r})$$

and the eigenfunctions of eigenvalue -1 are the *odd* functions of $\mathbf{r}$:

$$\psi_{odd}(\mathbf{r}) = -\psi_{odd}(-\mathbf{r})$$

These two kinds of functions form a complete set, since if ψ is an arbitrary function, then

$$\psi(\mathbf{r}) = \tfrac{1}{2}[\psi(\mathbf{r}) + \psi(-\mathbf{r})] + \tfrac{1}{2}[\psi(\mathbf{r}) - \psi(-\mathbf{r})] \qquad (96)$$

where the term in the first bracket is even, and the term in the second bracket is odd. This expresses ψ as a superposition of an even and an odd function.

The commutation relation between $\mathbf{r}_{op}$ and P can be discovered by considering the matrix element of $P\mathbf{r}_{op} + \mathbf{r}_{op}P$:

$$\langle \mathbf{r}'|(P\mathbf{r}_{op} + \mathbf{r}_{op}P)|\mathbf{r}\rangle = \mathbf{r}\langle \mathbf{r}'|P|\mathbf{r}\rangle + \langle \mathbf{r}'|\mathbf{r}_{op}|-\mathbf{r}\rangle$$

$$= \mathbf{r}\langle \mathbf{r}'|-\mathbf{r}\rangle - \mathbf{r}\langle \mathbf{r}'|-\mathbf{r}\rangle = 0 \qquad (97)$$

Since this holds for all possible states $|\mathbf{r}\rangle$ and $|\mathbf{r}'\rangle$, we must have

$$P\mathbf{r}_{op} = -\mathbf{r}_{op}P \qquad (98)$$

This means that the operators P and $\mathbf{r}_{op}$ anticommute.

The effect of the parity operator on momentum eigenstates and on the momentum operator can be deduced similarly,[4]

$$P|\mathbf{p}\rangle = |-\mathbf{p}\rangle \qquad (99)$$

$$P\mathbf{p}_{op} = -\mathbf{p}_{op}P \qquad (100)$$

Exercise 22. Derive Eqs. (99) and (100).

We can now evaluate the commutator of the Hamiltonian with P:

$$[H, P] = H(\mathbf{r}_{op}, \mathbf{p}_{op})\, P - PH(\mathbf{r}_{op}, \mathbf{p}_{op})$$

$$= H(\mathbf{r}_{op}, \mathbf{p}_{op})\, P - H(-\mathbf{r}_{op}, -\mathbf{p}_{op})P \qquad (101)$$

If $H = \mathbf{p}_{op}^2/2m + V(\mathbf{r}_{op})$, this becomes

$$[H, P] = (V(\mathbf{r}_{op}) - V(-\mathbf{r}_{op}))P \qquad (102)$$

This shows that H and P commute, provided that the potential $V(\mathbf{r})$ is an even function of $\mathbf{r}$. This will clearly be the case for any

---

[4] Note that P stands for parity and $\mathbf{p}$ for momentum.

central potential, since then the potential is a function of only the radial distance from the origin, that is, a function of r. For any such potential, H and P commute, and the expectation values of the parity is conserved:

$$\frac{d}{dt}\langle\psi|P|\psi\rangle = \frac{i}{\hbar}\langle\psi|[H, P]|\psi\rangle = 0 \tag{103}$$

Parity is conserved in almost all interactions between particles. However, there is an exception: the weak interaction, which produces beta decay and some other decays, involves a Hamiltonian that does not commute with the parity operator. In this case, parity need not be conserved—the parity after a beta-decay reaction can be different from that before. Beta decay violates right–left symmetry.

Finally, let us examine the parity of the eigenfunctions of angular momentum. The angular momentum operator commutes with the parity operator:

$$PL = LP \tag{104}$$

Exercise 23. Deduce this from the effect of the parity operator on $\mathbf{r}_{op}$ and on $\mathbf{p}_{op}$.

This commutativity implies the existence of simultaneous eigenfunctions of angular momentum and parity. In fact, the angular-momentum eigenfunctions we have constructed in the preceding section are eigenfunctions of parity. To establish this, note that in spherical coordinates the parity operation (95) is (see Fig. 7.5)

$$P\psi(r, \theta, \phi) = \psi(r, \pi - \theta, \phi + \pi) \tag{105}$$

If we perform this operation on the spherical harmonics, we readily find

$$PY_l^{m_l}(\theta, \phi) = Y_l^{m_l}(\pi - \theta, \phi + \pi)$$
$$= (-1)^l Y_l^{m_l}(\theta, \phi) \tag{106}$$

Thus, the spherical harmonic $Y_l^{m_l}(\theta, \phi)$ is an eigenfunction of parity with the eigenvalue $(-1)^l$.

Exercise 24. Derive Eq. (106).

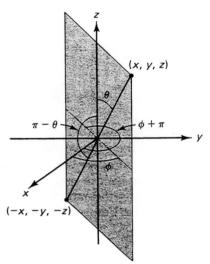

Fig. 7.5   The parity operation in spherical coordinates.

7.4   *The Kinetic Energy*

For a particle moving in three dimension, the kinetic energy is

$$K = \frac{1}{2m}(p_x{}^2 + p_y{}^2 + p_z{}^2)$$

or

$$K = \frac{1}{2m}\mathbf{p}^2 \tag{107}$$

From classical mechanics, we know that in spherical coordinates this kinetic energy can be expressed as a sum of radial part (involving the radial component of the momentum) and an angular part (involving the angular momentum):

$$K = \frac{1}{2m}p_r{}^2 + \frac{1}{2mr^2}\mathbf{L}^2 \tag{108}$$

This expression for the kinetic energy is very useful for the study of the motion of a particle in central potential, that is, a potential $V(r)$ that depends only on the radial distance r from the origin. For such a potential, the angular momentum is a constant of the motion, and the kinetic energy therefore depends only on the radial momentum and the radial distance. This means that the motion can be treated as one-dimensional motion in the radial direction.

We will show that with a suitable definition of p_r, the kinetic energy also can be expressed in the form (108) in quantum mechanics. And in the next chapter, we will apply this expression in the solution of eigenvalue problems with central potentials.

The starting point of our derivation of the expression (108) is the following vector identity, valid for ordinary, commuting, vectors **A, B, C, D**:

$$(\mathbf{A} \times \mathbf{B}) \cdot (\mathbf{C} \times \mathbf{D}) = \mathbf{A} \cdot \mathbf{C}\, \mathbf{B} \cdot \mathbf{D} - \mathbf{A} \cdot \mathbf{D}\, \mathbf{B} \cdot \mathbf{C} \tag{109}$$

Exercise 25. Verify this identity. [Hint: $\mathbf{B} \times (\mathbf{C} \times \mathbf{D}) = \mathbf{B} \cdot \mathbf{D}\, \mathbf{C} - \mathbf{B} \cdot \mathbf{C}\, \mathbf{D}$.]

If the vectors **A, B, C, D** are objects that do not commute, then Eq. (109) still holds, provided we rewrite it so all the vectors on the right and on the left sides appear in the same order, that is,

$$(\mathbf{A} \times \mathbf{B}) \cdot (\mathbf{C} \times \mathbf{D}) = \sum_{j,l} A_j B_l C_j D_l - \sum_{j,l} A_j B_l C_l D_j \tag{110}$$

With this identity, we find

$$\mathbf{L}^2 = (\mathbf{r} \times \mathbf{p}) \cdot (\mathbf{r} \times \mathbf{p}) = \sum_{j,l} x_j p_l x_j p_l - \sum_{j,l} x_j p_l x_l p_j$$

$$= \sum_{j,l} (x_j x_j p_l p_l - i\hbar \delta_{lj} x_j p_l) - \sum_{j,l} (p_l x_j x_l p_j + i\hbar \delta_{lj} x_l p_j)$$

$$= \sum_{j,l} x_j x_j p_l p_l - i\hbar \sum_j x_j p_j - \sum_{j,l} p_l x_j x_l p_j - i\hbar \sum_j x_j p_j$$

$$= r^2 \mathbf{p}^2 - 2i\hbar\, \mathbf{r} \cdot \mathbf{p} - \sum_{j,l} (x_l p_l x_j p_j - i\hbar \delta_{ll} x_j p_j)$$

$$= r^2 \mathbf{p}^2 - 2i\hbar\, \mathbf{r} \cdot \mathbf{p} - (\mathbf{r} \cdot \mathbf{p})^2 + 3i\hbar\, \mathbf{r} \cdot \mathbf{p}$$

$$= r^2 \mathbf{p}^2 - (\mathbf{r} \cdot \mathbf{p})^2 + i\hbar\, \mathbf{r} \cdot \mathbf{p} \tag{111}$$

If we solve this for $\mathbf{p}^2$, we find

$$\mathbf{p}^2 = \frac{1}{r^2} (\mathbf{r} \cdot \mathbf{p})^2 - \frac{i\hbar}{r^2} \mathbf{r} \cdot \mathbf{p} + \frac{\mathbf{L}^2}{r^2} \tag{112}$$

We can simplify this some more by introducing a new operator p_r which represents the radial component of the momentum:

$$p_r = \frac{1}{2} \left(\frac{1}{r} \mathbf{r} \cdot \mathbf{p} + \mathbf{p} \cdot \mathbf{r} \frac{1}{r} \right) \tag{113}$$

This can also be written in the equivalent form

$$p_r = \frac{1}{2} (\hat{\mathbf{r}} \cdot \mathbf{p} + \mathbf{p} \cdot \hat{\mathbf{r}}) \tag{114}$$

where $\hat{\mathbf{r}} = \mathbf{r}/r$ is the radial unit vector. If we compare Eq. (114) with the familiar classical equation $p_r = \hat{\mathbf{r}} \cdot \mathbf{p}$ for the radial component of the momentum, we see that the quantum-mechanical equation is obtained from the classical equation by symmetrization of the order of the factors. This symmetrization is required to make p_r hermitian.

Exercise 26. Show that $\frac{1}{2}(\hat{\mathbf{r}} \cdot \mathbf{p} + \mathbf{p} \cdot \hat{\mathbf{r}})$ is hermitian, but that $\mathbf{p} \cdot \hat{\mathbf{r}}$ is not.

The commutator of r and p_r has the standard form of the commutator of a coordinate and the corresponding momentum:

$$[r, p_r] = i\hbar \tag{115}$$

Exercise 27. Show that

$$[r, \mathbf{p}] = i\hbar \frac{\mathbf{r}}{r} \tag{116}$$

and deduce Eq. (115) from this.

We want to replace $\mathbf{r} \cdot \mathbf{p}$ in Eq. (112) by p_r; but we must first eliminate the term $\mathbf{p} \cdot \mathbf{r}$ from our equation for p_r. We can do this by using the commutator $[p_k, f] = (\hbar/i) \, \partial f/\partial x_k$, or, in vector notation, $[\mathbf{p}, f] = (\hbar/i) \nabla f$. With this, Eq. (113) becomes[5]

$$p_r = \frac{1}{2} \left(\frac{1}{r} \mathbf{r} \cdot \mathbf{p} + \frac{1}{r} \mathbf{r} \cdot \mathbf{p} + \frac{\hbar}{i} \nabla \cdot \frac{\mathbf{r}}{r} \right)$$

$$= \frac{1}{r} \mathbf{r} \cdot \mathbf{p} - \frac{i\hbar}{r} \tag{117}$$

Consequently,

$$\mathbf{r} \cdot \mathbf{p} = r p_r + i\hbar \tag{118}$$

---

[5] Note that here ∇ acts on $\mathbf{r}/r$ only; it does not act on whatever state vector we place to the right of the operator p_r.

With this, the first two terms on the right side of Eq. (112) become

$$\frac{1}{r^2}(\mathbf{r}\cdot\mathbf{p})^2 - \frac{i\hbar}{r^2}\mathbf{r}\cdot\mathbf{p} = \frac{1}{r^2}(\mathbf{r}\cdot\mathbf{p} - i\hbar)(\mathbf{r}\cdot\mathbf{p})$$

$$= \frac{1}{r}p_r(rp_r + i\hbar)$$

$$= \frac{1}{r}(rp_rp_r - i\hbar p_r + i\hbar p_r)$$

$$= p_r^2 \tag{119}$$

Equation (112) therefore reduces to

$$\mathbf{p}^2 = p_r^2 + \frac{\mathbf{L}^2}{r^2} \tag{120}$$

and the kinetic energy is

$$K = \frac{1}{2m}p_r^2 + \frac{1}{2mr^2}\mathbf{L}^2 \tag{121}$$

In the position representation, Eq. (121) is a differential operator involving the Laplacian in spherical coordinates. We will not be needing this representation in the next chapter, but for reference purposes we state the formulas for this differential operator in the following exercises.

Exercise 28. Use Eq. (117) to show that, in the position representation, p_r is the differential operator

$$p_r = \frac{\hbar}{i}\left(\frac{\partial}{\partial r} + \frac{1}{r}\right) \tag{122}$$

or, equivalently,

$$p_r = \frac{\hbar}{i}\frac{1}{r}\frac{\partial}{\partial r}r \tag{123}$$

Exercise 29. Show that p_r^2 is represented by the following equivalent differential operators

$$p_r^2 = -\hbar^2\frac{1}{r}\frac{\partial^2}{\partial r^2}r \tag{124}$$

$$p_r^2 = -\frac{\hbar^2}{r^2}\frac{\partial}{\partial r}r^2\frac{\partial}{\partial r} \tag{125}$$

$$p_r^2 = -\hbar^2 \left(\frac{\partial^2}{\partial r^2} + \frac{2}{r}\frac{\partial}{\partial r} \right) \tag{126}$$

Exercise 30. Show that, by Eqs. (124) and (62),

$$p_r^2 + \frac{\mathbf{L}^2}{r^2} = -\hbar^2 \nabla^2$$

where

$$\nabla^2 = \frac{1}{r}\frac{\partial^2}{\partial r^2} r + \frac{1}{r^2 \sin\theta}\frac{\partial}{\partial\theta} \sin\theta \frac{\partial}{\partial\theta} + \frac{1}{r^2 \sin^2\theta}\frac{\partial^2}{\partial\phi^2} \tag{127}$$

The differential operator (127) is the Laplacian in spherical coordinates.

PROBLEMS

1. Consider an infinite potential well extending from $x = 0$ to $x = a$, $y = 0$ to $y = b$, and $z = 0$ to $z = c$. Find the energy eigenvalues and the eigenfunctions for a particle of mass m in this well. What are the three lowest eigenvalues if $a = \frac{3}{2}b = \frac{5}{2}c$?

2. Prove that $\mathbf{L} \cdot \mathbf{p} = 0$ and $\mathbf{L} \cdot \mathbf{r} = 0$.

3. Verify that the commutation relations for the angular momentum imply $\mathbf{L} \times \mathbf{L} = i\hbar\mathbf{L}$.

4. Prove that

$$[L_x^2, L_y^2] = [L_y^2, L_z^2] = [L_z^2, L_x^2]$$

   (Hint: Start with $[\mathbf{L}^2, L_y^2] = 0$.)

5. Show that

$$[\mathbf{L}^2, \mathbf{r}] = -2i\hbar\mathbf{L} \times \mathbf{r} - 2\hbar^2\mathbf{r}$$

   and that

$$[\mathbf{L}^2, \mathbf{p}] = -2i\hbar\mathbf{L} \times \mathbf{p} - 2\hbar^2\mathbf{p}$$

6. Show that $[\mathbf{L}, \mathbf{L} \times \mathbf{L}] = 0$.

7. Show that

   (a) $[L_z, x] = i\hbar y$, $\quad [L_z, y] = -i\hbar x$, $\quad [L_z, z] = 0$

   (b) $[L_z, p_x] = i\hbar p_y$, $\quad [L_z, p_y] = -i\hbar p_x$, $\quad [L_z, p_z] = 0$

8. (a) Show that if $|\psi\rangle$ is an eigenstate of L_z, then $\langle\psi|p_y|\psi\rangle = 0$ and $\langle\psi|p_x|\psi\rangle = 0$. (Hint: Use the results stated in the preceding problem.)

   (b) Show that if $|\psi\rangle$ is an eigenstate of L_z, then $\langle\psi|y|\psi\rangle = 0$ and $\langle\psi|x|\psi\rangle = 0$.

9. Show that for the eigenstate $|l,m_l\rangle$ of $\mathbf{L}^2$ and of L_z, the expectation values of L_x^2 and of L_y^2 are

$$\langle l,m_l| \, L_x^2 \, |l,m_l\rangle = \langle l,m_l| \, L_y^2 \, |l,m_l\rangle = \tfrac{1}{2}(l(l+1) - m_l^2)\,\hbar^2$$

and show that the rms uncertainties are

$$\Delta L_x = \Delta L_y = \sqrt{\tfrac{1}{2}(l(l+1) - m_l^2)}\;\hbar$$

10. Consider the three states $|1,1\rangle$, $|1,0\rangle$, and $|1,-1\rangle$ with angular-momentum quantum numbers $l = 1$, $m_l = 1, 0, -1$. Suppose that we adopt a matrix representation and represent these three states by the three column vectors

$$\begin{pmatrix}1\\0\\0\end{pmatrix}, \quad \begin{pmatrix}0\\1\\0\end{pmatrix}, \quad \text{and} \quad \begin{pmatrix}0\\0\\1\end{pmatrix}$$

respectively. The operators L_x, L_y, L_z, L_+, L_-, and $\mathbf{L}^2$ will then be represented by 3×3 matrices. Construct all these matrices. By explicit multiplication of your matrices, verify that $[L_x, L_y] = i\hbar L_z$.

11. Express $|1,1\rangle$ as a superposition of eigenstates of L_x. (Hint: Use the results obtained in the preceding problem.)

12. Suppose we rotate our coordinate system so the new z' axis lies in the old z-x plane and makes an angle α with the old z axis (see Fig. 7.6). Consider the eigenstate $|1,-1\rangle$ of $\mathbf{L}^2$ and L_z. Is this an eigenstate of $L_{z'}$? What is the expectation value of $L_{z'}$ for this state?

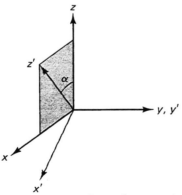

Fig. 7.6   Rotated coordinates, x', y', z'.

13. (a) Show that the commutator $[\phi, L_z]$ equals $i\hbar$.

    (b) Show that it is *not* possible to deduce a meaningful uncertainty relation of the form $\Delta\phi\,\Delta L_z \geq \hbar/2$ from this commutator. (Hint:

The average values of ϕ and of ϕ^2 are not well defined, since these values depend on what range of integration we adopt. For example, if the probability distribution is uniform over all angles, then the average value of ϕ is π if we adopt a range of integration from 0 to 2π, but the average value is 0 if we adopt a range of integration from $-\pi$ to $+\pi$. Give some other examples of such ambiguities in the average values of ϕ and ϕ^2.)

14. (a) Evaluate the commutator $[\sin \phi, L_z]$.

   (b) Show that $\sin \phi$ and L_z obey the uncertainty relation

$$\Delta(\sin \phi) \, \Delta L_z \geq \frac{\hbar}{2} \, |\langle \cos \phi \rangle|$$

   where $\langle \cos \phi \rangle$ is the expectation value of $\cos \phi$ in the quantum state under consideration.

   (c) Verify this uncertainty relation explicity for the eigenfunction $Y_0{}^0$ and for the eigenfunction $Y_1{}^1$.

   (d) Show that if $f(\sin \phi)$ is any function of $\sin \phi$, then the uncertainty relation is

$$\Delta f \, \Delta L_z \geq \frac{\hbar}{2} \, |\langle f' \cos \phi \rangle|$$

15. Verify Eq. (84).

16. Verify that $(L_+)^3 Y_{3/2}{}^{-3/2}$ does not yield a function proportional to $Y_{3/2}{}^{3/2}$.

17. Suppose that a particle has the wavefunction

$$\psi(r, \theta, \phi) = \frac{1}{4} \sqrt{\frac{5}{\pi}} \sin^2 \theta \, (1 + \sqrt{14} \cos \theta) \cos 2\phi \, f(r)$$

   where $f(r)$ is a (normalized) radial wavefunction.

   (a) Express the angular part of this wavefunction as a superposition of the eigenfunctions $Y_l{}^{m_l}(\theta, \phi)$. What are the probabilities for the different possible outcomes of a measurement of L^2 and L_z?

   (b) What are the expectation values of L^2 and L_z?

   (c) What are the rms uncertainties in L^2 and in L_z?

18. Consider the operator $e^{i\alpha L_z/\hbar}$. Show that when this operator acts on a function $f(\phi)$, the result is $f(\phi + \alpha)$, that is,

$$e^{i\alpha L_z/\hbar} f(\phi) = f(\phi + \alpha)$$

19. According to Eq. (122), in the position representation the operator p_r is

$$p_r = \frac{\hbar}{i} \frac{\partial}{\partial r} - \frac{i\hbar}{r}$$

Show explicitly that this operator is hermitian; that is, show that if χ and ψ are two wavefunctions, then

$$\int_0^\infty \chi^*(p_r\psi)r^2 \, dr = \int_0^\infty (p_r\chi)^*\psi r^2 \, dr$$

(Hint: Perform an integration by parts and assume that $\chi^*\psi r^2|_{r=0} = 0$.)

20. Show that in cylindrical coordinates ρ, ϕ, z, the kinetic energy of a particle can be expressed as

$$K = \frac{1}{2m}(p_\rho^2 + p_z^2) - \frac{\hbar^2}{8m\rho^2} + \frac{L_z^2}{2m\rho^2}$$

where p_ρ is the radial component of the momentum,

$$p_\rho = \frac{1}{2}\left(\frac{\boldsymbol{\rho}}{\rho}\cdot\mathbf{p} + \mathbf{p}\cdot\frac{\boldsymbol{\rho}}{\rho}\right)$$

Show that this radial component of the momentum and the cylindrical radial coordinate satisfy the standard commutation relation

$$[\rho, p_\rho] = i\hbar$$

8

Central Potentials

As already stated in the preceding chapter, a central potential is a potential $V(r)$ that is a function of only the radial distance r from some suitably chosen origin. Such a potential is also called a spherically symmetric potential. In classical physics, a central potential $V(r)$ is associated with a central force, that is, a force directed toward the origin. Among the important examples of physical systems with central potentials are the isotropic harmonic oscillator $[V(r) \propto r^2]$, the hydrogen atom $[V(r) \propto 1/r]$, and the free particle $[V(r) = 0]$.

The Hamiltonian for a particle moving in a central potential is $H = \mathbf{p}^2/2m + V(r)$. According to the results established in Section 7.2, this Hamiltonian commutes with the angular-momentum operators. In consequence, the angular momentum is conserved, and there exist simultaneous eigenstates of energy and of angular momentum. We will see that the angular-momentum quantum numbers are especially helpful for the classification of the degenerate eigenstates of energy.

8.1 *The Isotropic Harmonic Oscillator*

The isotropic harmonic oscillator is a three-dimensional version of the harmonic oscillator, with equal spring constants for the x, y, and z directions. The Hamiltonian is

$$H = \frac{1}{2m}\left(p_x{}^2 + p_y{}^2 + p_z{}^2\right) + \frac{1}{2}\,m\omega^2(x^2 + y^2 + z^2) \tag{1}$$

Obviously, this Hamiltonian has x, y, and z parts, each of which corresponds to a one-dimensional harmonic oscillator. Therefore, the solution of the eigenvalue problem is trivial: we just have to write down the one-dimensional solution three times, with independent quantum numbers n_1, n_2, and n_3, for the three independent one-dimensional solutions. The energy eigenvalues are the sum of the eigenvalues for the three independent one-dimensional solutions:

$$E_{n_1 n_2 n_3} = \left(n_1 + n_2 + n_3 + \frac{3}{2}\right) \hbar\omega \qquad (2)$$

The eigenfunctions are the product of one-dimensional wavefunctions of the kind discussed in Section 6.1:

$$\psi_{n_1 n_2 n_3}(x, y, z) = \psi_{n_1}(x)\psi_{n_2}(y)\psi_{n_3}(z)$$

Alternatively, we can express the kinetic energy and the potential energy in spherical coordinates. The kinetic energy is $p_r^2/2m + L^2/2mr^2$ [see Eq. (7.121)], and the potential energy is

$$\frac{1}{2} m\omega^2(x^2 + y^2 + z^2) = \frac{1}{2} m\omega^2 r^2$$

Hence the Hamiltonian in spherical coordinates is

$$H = \frac{1}{2m} p_r^2 + \frac{1}{2mr^2} L^2 + \frac{1}{2} m\omega^2 r^2 \qquad (3)$$

Written in this way, the Hamiltonian involves the central potential $V(r) = \frac{1}{2}m\omega^2 r^2$, and the solution of the eigenvalue problem is much harder. However, the advantage of the Hamiltonian (3) is that its eigenstates are also eigenstates of angular momentum. These eigenstates form a very useful complete set.

Since L^2 and L_z commute with H and with each other, we can construct simultaneous eigenstates of H, L^2, and L_z. In our usual notation, we designate such simultaneous eigenstates by $|E,l,m_l\rangle$. The eigenvalues of L_z and L^2 are $m_l\hbar$ and $l(l + 1)\hbar^2$, respectively, and when H acts on $|E,l,m\rangle$, we can effectively replace the operator L^2 contained in H by $l(l + 1)\hbar^2$. The eigenvalue equations for L^2, L_z, and H are then:

$$L^2 |E,l,m_l\rangle = l(l + 1)\hbar^2|E,l,m_l\rangle \qquad (4)$$

$$L_z |E,l,m_l\rangle = m_l\hbar|E,l,m_l\rangle \qquad (5)$$

$$\left(\frac{1}{2m} p_r^2 + \frac{1}{2mr^2} l(l + 1)\hbar^2 + \frac{1}{2} m\omega^2 r^2\right) |E,l,m_l\rangle = E|E,l,m_l\rangle \qquad (6)$$

The operator appearing on the left side of Eq. (6) depends only on r and p_r; thus, the determination of the eigenvalues of this operator reduces to a one-dimensional problem involving only radial variables. We will use the factorization method to find the eigenvalues and eigenvectors of this operator. According to the prescription of Section 6.2, we must find operators η_j that satisfy the recursion relations

$$\eta_1^\dagger \eta_1 + E_1 = \frac{1}{2m} p_r^2 + \frac{1}{2mr^2} l(l+1)\hbar^2 + \frac{1}{2} m\omega^2 r^2 \qquad (7)$$

$$\eta_{j+1}^\dagger \eta_{j+1} + E_{j+1} = \eta_j \eta_j^\dagger + E_j \qquad (8)$$

Instead of η_j we should really write $\eta_j^{(l)}$ in order to indicate that the operator depends on the value of l. But we will for now keep l fixed at one particular value, and no confusion will arise if we designate the operator corresponding to this particular value of l simply by η_j.

Since the right side of Eq. (7) contains a term $\propto r^2$ and a term $\propto 1/r^2$, a likely *Ansatz* for η_j is

$$\eta_j = \frac{1}{\sqrt{2m}} \left[p_r + i \left(b_j r + \frac{c_j}{r} \right) \right] \qquad (9)$$

where b_j and c_j are real constants. With the usual commutation relation for r and p_r [see Eq. (7.115)], this leads to

$$\eta_j^\dagger \eta_j = \frac{1}{2m} \left[p_r^2 + b_j^2 r^2 + b_j(2c_j + \hbar) + \frac{c_j}{r^2}(c_j - \hbar) \right] \qquad (10)$$

and

$$\eta_j \eta_j^\dagger = \frac{1}{2m} \left[p_r^2 + b_j^2 r^2 + b_j(2c_j - \hbar) + \frac{c_j}{r^2}(c_j + \hbar) \right] \qquad (11)$$

Exercise 1. Verify these results.

Substituting $\eta_1^\dagger \eta_1$ into Eq. (7), we obtain

$$\frac{1}{2m} \left[p_r^2 + b_1^2 r^2 + b_1(2c_1 + \hbar) + \frac{c_1}{r^2}(c_1 - \hbar) \right] + E_1$$

$$= \frac{1}{2m} p_r^2 + \frac{1}{2mr^2} l(l+1)\hbar^2 + \frac{1}{2} m\omega^2 r^2 \qquad (12)$$

Comparing the coefficients of equal powers of r on the two sides of this equation, we find

$$\frac{1}{2m} c_1(c_1 - \hbar) = \frac{1}{2m} l(l + 1)\hbar^2$$

$$\frac{1}{2m} b_1{}^2 = \frac{1}{2} m\omega^2$$

$$\frac{1}{2m} b_1(2c_1 + \hbar) + E_1 = 0$$

The solution of these equations that yields the largest value of E_1 is

$$c_1 = (l + 1)\hbar \tag{13}$$

$$b_1 = -m\omega \tag{14}$$

$$E_1 = \hbar\omega \left(l + \frac{3}{2} \right) \tag{15}$$

Next, we must solve the recursion relations for $j \geq 2$. Comparison of equal powers of r on the two sides of Eq. (8) yields

$$\frac{1}{2m} c_{j+1}(c_{j+1} - \hbar) = \frac{1}{2m} c_j(c_j + \hbar) \tag{16}$$

$$\frac{1}{2m} b_{j+1}{}^2 = \frac{1}{2m} b_j{}^2 \tag{17}$$

and

$$\frac{1}{2m} b_{j+1}(2c_{j+1} + \hbar) + E_{j+1} = \frac{1}{2m} b_j(2c_j - \hbar) + E_j \tag{18}$$

The solution of these equations that maximizes the value of E_j is

$$c_j = (l + j)\hbar \tag{19}$$

$$b_j = -m\omega \tag{20}$$

$$E_j = \hbar\omega \left(l + 2j - \frac{1}{2} \right) \tag{21}$$

To establish that the constants E_j are the energy eigenvalues, we must verify that the subsidiary equation $\eta_j | \zeta_j \rangle = 0$ possesses a solution; we will verify this in the next paragraph. Since the energy eigenvalues depend on the angular momentum, we adopt the notation $E_j^{(l)}$ for the jth energy eigenvalue of given angular momentum l:

$$E_j^{(l)} = \hbar\omega \left(l + 2j - \frac{1}{2}\right) \tag{22}$$

The corresponding energy-level diagram is shown in Fig. 8.1. Note that all the states, except the ground state, are degenerate. For a fixed value of the energy, or a fixed value of $l + 2j$, there are several possible values of l and j. Furthermore, for a fixed value of l, there are $2l + 1$ degenerate states with different values of m_l. This degeneracy with respect to the value of m_l is a direct consequence of the spherical symmetry of the potential; obviously, in a spherically symmetric potential there is no preferred direction, and the energy cannot depend on the direction of the angular momentum.

Let us now look for solutions of the subsidiary condition $\eta_j|\zeta_j\rangle = 0$. If we indicate the dependence on l explicitly, Eqs. (19) and (20) give us

$$\eta_j^{(l)} = \frac{1}{\sqrt{2m}} \left[p_r + i\left(-m\omega r + \frac{(l + j)\hbar}{r}\right)\right] \tag{23}$$

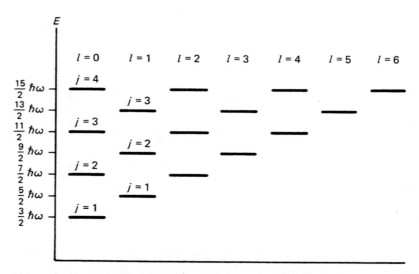

Fig. 8.1 Energy-level diagram of the isotropic harmonic oscillator. The states are labeled with the quantum number l and j, but the quantum number m_l has been omitted.

The subsidiary conditions then becomes

$$\left[p_r + i \left(-m\omega r + \frac{(l+j)\hbar}{r} \right) \right] |\zeta_j^{(l)}\rangle = 0$$

In the position representation, $p_r = (\hbar/i)(d/dr + 1/r)$ and our equation is equivalent to the differential equation

$$\left[\frac{\hbar}{i} \left(\frac{d}{dr} + \frac{1}{r} \right) + i \left(-m\omega r + \frac{(l+j)\hbar}{r} \right) \right] \zeta_j^{(l)}(r) = 0 \tag{24}$$

that is,

$$\left[\frac{d}{dr} + \frac{m\omega}{\hbar} r - \frac{l+j-1}{r} \right] \zeta_j^{(l)}(r) = 0 \tag{25}$$

It is easy to check that this differential equation has the normalizable solution

$$\zeta_j^{(l)}(r) = r^{l+j-1} e^{-(m\omega/2\hbar)r^2} \tag{26}$$

Exercise 2. Check this solution.

We can now construct the energy eigenfunctions:

$$\psi_j^{(l)}(r) \propto \eta_1^{(l)\dagger} \eta_2^{(l)\dagger} \cdots \eta_{j-1}^{(l)\dagger} \zeta_j^{(l)}(r) \tag{27}$$

where

$$\eta_j^{(l)\dagger} = \frac{1}{\sqrt{2m}} \frac{\hbar}{i} \left[\frac{d}{dr} - \frac{m\omega}{\hbar} r + \frac{l+j+1}{r} \right]$$

This leads to the results

$$\psi_1^{(0)}(r) \propto e^{-(m\omega/2\hbar)r^2} \tag{28}$$

$$\psi_1^{(1)}(r) \propto r e^{-(m\omega/2\hbar)r^2} \tag{29}$$

$$\psi_1^{(2)}(r) \propto r^2 e^{-(m\omega/2\hbar)r^2} \tag{30}$$

$$\psi_2^{(0)}(r) \propto \left(\frac{3}{2} - \frac{m\omega r^2}{\hbar} \right) e^{-(m\omega/2\hbar)r^2} \tag{31}$$

Exercise 3. Obtain these results and show that, in general, $\psi_j^{(l)}$ is a polynomial of order $l + 2(j-1)$ multiplied by the exponential function $e^{-(m\omega/2\hbar)r^2}$.

The general expression for $\psi_j^{(l)}(r)$ is customarily written in the form

$$\psi_j^{(l)}(r) \propto u^{l/2} L_{(j-1)}^{l+1/2}(u) e^{-u} \tag{32}$$

where $u = (m\omega/2\hbar)r^2$ and where $L_{(j-1)}^{l+1/2}(u)$ is a polynomial of order $j - 1$ in u, called a *Laguerre polynomial*. Note that $L_{(j-1)}^{l+1/2}(u)$ is a polynomial of order $2(j - 1)$ in r, and therefore the right side of Eq. (32) is a polynomial of order $l + 2(j - 1)$, multiplied by the exponential function, in accord with Exercise 3.

Of course, $\psi_j^{(l)}(r)$ gives only the radial dependence of the wavefunction. The angular dependence must be $Y_l^{m_l}(\theta, \phi)$, since the wavefunction is an eigenfunction of $\mathbf{L}^2$ and L_z [compare Eqs. (4) and (5)]. The complete wavefunction is therefore

$$\psi_{jlm_l}(r, \theta, \phi) = Y_l^{m_l}(\theta, \phi) \psi_j^{(l)}(r) \tag{33}$$

The radial function $\psi_j^{(l)}(r)$ has to be normalized so

$$\int_0^\infty |\psi_j^{(l)}|^2 r^2 \, dr = 1 \tag{34}$$

In view of the normalization of the spherical harmonics [see Eq. (7.83)], this will lead to the correct normalization for the functions ψ_{jlm_l}. The normalization of the radial functions $\psi_j^{(l)}$ can also be obtained from the general results stated in Section 6.3.

The eigenfunctions $\psi_{jlm_l}(r, \theta, \phi)$ in spherical coordinates can be expressed as superpositions of the eigenfunctions $\psi_{n_1 n_2 n_3}(x, y, z)$ in rectangular coordinates, and vice versa. The superposition involves only those eigenfunctions that have the same, given energy. For instance, the spherical functions (29) with $l = 1, j = 1$, and $m_l = -1, 0,$ or $+1$ are superpositions of the rectangular functions $\psi_{100}, \psi_{010},$ and ψ_{001}.

8.2   *The Hydrogen Atom*

Since the nucleus of the hydrogen atom is much heavier than the electron, it is a good approximation to neglect the motion of the nucleus and to pretend that the nucleus remains at a fixed position. We will adopt this approximation for the calculations in the first part of this section, and we will see how to take the motion of the nucleus into account in the second part. If we place our origin of coordinates at the nucleus, the electric potential acting on the electron is, in SI units,

$$V(r) = - \frac{1}{4\pi\varepsilon_0} \frac{e^2}{r} \tag{35}$$

In spherical coordinates, the Hamiltonian for the motion of the electron is then

$$H = \frac{1}{2m} p_r^2 + \frac{1}{2mr^2} \mathbf{L}^2 - \frac{1}{4\pi\varepsilon_0} \frac{e^2}{r} \tag{36}$$

We designate the simultaneous eigenstates of $\mathbf{L}^2$, L_z, and H by $|E,l,m_l\rangle$. As in the case of the isotropic harmonic oscillator, when the operator H acts on such a simultaneous eigenstate, we can effectively replace the operator $\mathbf{L}^2$ by $l(l + 1)\hbar^2$. Our eigenvalue equation for the energy can then be written as

$$\left(\frac{1}{2m} p_r^2 + \frac{1}{2mr^2} l(l + 1)\hbar^2 - \frac{1}{4\pi\varepsilon_0} \frac{e^2}{r} \right) |E,l,m_l\rangle = E|E,l,m_l\rangle \tag{37}$$

Again, the operator appearing on the left side of this eigenvalue equation depends only on r and p_r, and the determination of the eigenvalues of this operator reduces to a one-dimensional problem involving only radial variables. We will use the factorization method to find the eigenvalues and eigenvectors of this operator. For this, we need to solve the recursion relations

$$\eta_1^\dagger \eta_1 + E_1 = \frac{1}{2m} p_r^2 + \frac{1}{2mr^2} l(l + 1)\hbar^2 - \frac{1}{4\pi\varepsilon_0} \frac{e^2}{r} \tag{38}$$

$$\eta_{j+1}^\dagger \eta_{j+1} + E_{j+1} = \eta_j \eta_j^\dagger + E_j \tag{39}$$

We should use the notation $\eta_j^{(l)}$ to indicate that the operator depends on l, but we will, again, omit the superscript $^{(l)}$ and insert it later.

To solve these recursion relations, we make the *Ansatz*

$$\eta_j = \frac{1}{\sqrt{2m}} \left[p_r + i \left(b_j + \frac{c_j}{r} \right) \right] \tag{40}$$

where b_j and c_j are real constants. This leads to

$$\eta_j^\dagger \eta_j = \frac{1}{2m} \left[p_r^2 + b_j^2 + 2b_j \frac{c_j}{r} + \frac{c_j}{r^2} (c_j - \hbar) \right] \tag{41}$$

$$\eta_j \eta_j^\dagger = \frac{1}{2m} \left[p_r^2 + b_j^2 + 2b_j \frac{c_j}{r} + \frac{c_j}{r^2} (c_j + \hbar) \right] \tag{42}$$

Exercise 4. Obtain these expressions for $\eta_j{}^\dagger \eta_j$ and $\eta_j \eta_j{}^\dagger$.

By Eq. (38), we must have

$$\frac{1}{2m}\left[p_r{}^2 + b_1{}^2 + \frac{2b_1 c_1}{r} + \frac{c_1}{r^2}(c_1 - \hbar) \right] + E_1$$

$$= \frac{1}{2m} p_r{}^2 + \frac{1}{2mr^2} l(l + 1)\hbar^2 - \frac{1}{4\pi\varepsilon_0}\frac{e^2}{r} \tag{43}$$

Comparison of the coefficients of equal powers of r on the two sides of this equation yields

$$\frac{1}{2m} c_1(c_1 - \hbar) = \frac{1}{2m} l(l + 1)\hbar^2 \tag{44}$$

$$\frac{b_1 c_1}{m} = -\frac{e^2}{4\pi\varepsilon_0} \tag{45}$$

and

$$\frac{b_1{}^2}{2m} + E_1 = 0 \tag{46}$$

Equation (44) has the solution $c_1 = (l + 1)\hbar$, and then Eq. (45) yields $b_1 = -me^2/4\pi\varepsilon_0(l + 1)\hbar$, and Eq. (46) yields

$$E_1 = -\frac{1}{2m}\left(\frac{1}{4\pi\varepsilon_0}\frac{me^2}{(l + 1)\hbar}\right)^2 \tag{47}$$

This is the first energy eigenvalue.

Exercise 5. Show that the other solution $(c_1 = -l\hbar)$ of Eq. (44) leads to a value of E_1 that is less than the value given in Eq. (47).

The recursion relation (39) yields

$$c_{j+1}(c_{j+1} - \hbar) = c_j(c_j + \hbar) \tag{48}$$

$$b_{j+1} c_{j+1} = b_j c_j \tag{49}$$

$$\frac{b_{j+1}{}^2}{2m} + E_{j+1} = \frac{b_j{}^2}{2m} + E_j \tag{50}$$

One solution of Eq. (48) is $c_{j+1} = c_j + \hbar$, which implies

$$c_j = (j - 1)\hbar + c_1$$

that is,

$$c_j = (l + j)\hbar \tag{51}$$

The recursion relation (49) yields $b_j c_j = b_1 c_1 = -me^2/4\pi\varepsilon_0$, and hence

$$b_j = -\frac{1}{4\pi\varepsilon_0} \frac{me^2}{(l + j)\hbar} \tag{52}$$

Finally, Eq. (50) yields $E_j + b_j{}^2/2m = E_1 + b_1{}^2/2m = 0$. Therefore,

$$E_j = -\frac{b_j{}^2}{2m} = -\frac{1}{2m}\left(\frac{1}{4\pi\varepsilon_0}\frac{me^2}{(l + j)\hbar}\right)^2 \tag{53}$$

Exercise 6. Show that the other solution of the quadratic equation (48) gives a smaller value of the constant E_j.

The constants E_j are the energy eigenvalues. They depend on the angular momentum. For $l = 0$, the energy eigenvalues are

$$-\frac{1}{(4\pi\varepsilon_0)^2}\frac{me^4}{2\hbar^2 j^2}; \quad j = 1, 2, 3, \ldots . \tag{54}$$

For $l = 1$, they are

$$-\frac{1}{(4\pi\varepsilon_0)^2}\frac{me^4}{2\hbar^2(j + 1)^2}; \quad j = 1, 2, 3, \ldots . \tag{55}$$

and so on.

It is convenient to summarize the possible energy eigenvalues by introducing the notation $n \equiv l + j$; the possible energy eigenvalues are then

$$E_n = -\frac{1}{(4\pi\varepsilon_0)^2}\frac{me^4}{2\hbar^2 n^2}; \quad n = 1, 2, 3, \ldots . \tag{56}$$

These energies can also be written as

$$E_n = -\frac{1}{2}\alpha^2\frac{mc^2}{n^2} \tag{57}$$

where

$$\alpha \equiv \frac{1}{4\pi\varepsilon_0} \frac{e^2}{\hbar c} \tag{58}$$

The constant α is called the *fine-structure* constant. This constant is dimensionless and its numerical value is approximately $\alpha \simeq 1/137.036$.

The energy-level diagram is shown in Fig. 8.2. With the exception of the ground state, the energy levels are degenerate. According to the energy-level diagram, the degeneracy of the first excited state is 2, that of the second excited state is 4, and so on. However, the energy-level diagram does not take into account that for each state of given l there are $2l + 1$ states of different values of m_l. Thus, the degeneracy of the first excited state is actually 4, that of the second excited state is 9, and so on.

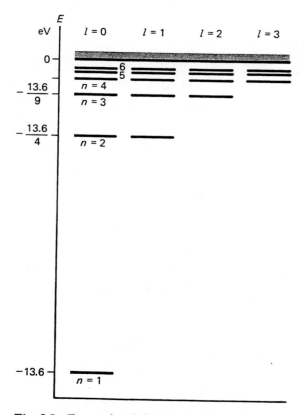

Fig. 8.2 Energy-level diagram for the hydrogen atom. The states are labeled with the quantum numbers n and l. The quantum number m_l has been omitted.

Next, we must construct the solutions of the subsidiary condition $\eta_j|\zeta_j\rangle = 0$. Since the operators η_j depend on l, we will hereafter write them as $\eta_j^{(l)}$; thus

$$\eta_j^{(l)} = \frac{1}{\sqrt{2m}}\left[p_r + i\left(-\frac{1}{4\pi\varepsilon_0}\frac{me^2}{(l+j)\hbar} + \frac{(l+j)\hbar}{r}\right)\right]$$

For convenience, we introduce the Bohr radius

$$a_0 = \frac{4\pi\varepsilon_0\hbar^2}{me^2} \tag{59}$$

and we write

$$\eta_j^{(l)} = \frac{1}{\sqrt{2m}}\left[p_r - \frac{i\hbar}{(l+j)a_0} + \frac{i(l+j)\hbar}{r}\right]$$

The subsidiary condition is then

$$\left[p_r - \frac{i\hbar}{(l+j)a_0} + \frac{i(l+j)\hbar}{r}\right]|\zeta_j^{(l)}\rangle = 0 \tag{60}$$

In the position representation, $p_r = (\hbar/i)(d/dr + 1/r)$, and Eq. (60) becomes the differential equation

$$\left[\frac{d}{dr} + \frac{1}{r} + \frac{1}{(l+j)a_0} - \frac{l+j}{r}\right]\zeta_j^{(l)}(r) = 0 \tag{61}$$

This equation has the normalizable solution

$$\zeta_j^{(l)}(r) = r^{l+j-1}e^{-r/(l+j)a_0}$$

Exercise 7. Verify that this is a solution of Eq. (61)

The eigenfunctions are therefore, except for normalization,

$$\psi_j^{(l)}(r) = \eta_1^{(l)\dagger}\eta_2^{(l)\dagger}\cdots\eta_{j-1}^{(l)\dagger}r^{l+j-1}e^{-r/(l+j)a_0}$$

where

$$\eta_j^{(l)\dagger} = \frac{1}{\sqrt{2m}}\left[\frac{\hbar}{i}\left(\frac{d}{dr} + \frac{1}{r}\right) + \frac{i\hbar}{(l+j)a_0} - \frac{i(l+j)\hbar}{r}\right]$$

The first few eigenfunctions for $l = 0$, $l = 1$ and $l = 2$ are

$$\psi_1^{(0)}(r) = e^{-r/a_0} \tag{62}$$

$$\psi_2^{(0)}(r) = \eta_1^{(0)\dagger}re^{-r/2a_0} \propto \left(1 - \frac{r}{2a_0}\right)e^{-r/2a_0} \tag{63}$$

$$\psi_3{}^{(0)}(r) = \eta_1{}^{(0)\dagger}\eta_2{}^{(0)\dagger}r^2 e^{-r/3a_0} \propto \left(1 - \frac{2}{3}\frac{r}{a_0} + \frac{2}{27}\frac{r^2}{a_0{}^2}\right) e^{-r/3a_0} \qquad (64)$$

$$\psi_1{}^{(1)}(r) = r e^{-r/2a_0} \qquad (65)$$

$$\psi_2{}^{(1)}(r) = \eta_1{}^{(1)\dagger}r^2 e^{-r/3a_0} \propto r\left(1 - \frac{r}{6a_0}\right) e^{-r/3a_0} \qquad (66)$$

$$\psi_1{}^{(2)}(r) = r^2 e^{-r/3a_0} \qquad (67)$$

Exercise 8. Obtain the results stated in Eqs. (66) and (68).

The customary notation for these radial wavefunctions is R_{nl}, where $n = l + j$. With this notation, and with the appropriate normalization factors, Eqs. (62)–(67) become

$$R_{10}(r) = \frac{2}{a_0{}^{3/2}} e^{-r/a_0} \qquad (68)$$

$$R_{20}(r) = \frac{2}{(2a_0)^{3/2}}\left(1 - \frac{r}{2a_0}\right) e^{-r/2a_0} \qquad (69)$$

$$R_{30}(r) = \frac{2}{(3a_0)^{3/2}}\left(1 - \frac{2}{3}\frac{r}{a_0} + \frac{2}{27}\frac{r^2}{a_0{}^2}\right) e^{-r/3a_0} \qquad (70)$$

$$R_{21}(r) = \frac{1}{(2a_0)^{3/2}}\frac{1}{\sqrt{3}}\frac{r}{a_0} e^{-r/2a_0} \qquad (71)$$

$$R_{31}(r) = \frac{1}{(3a_0)^{3/2}}\frac{4\sqrt{2}}{9}\frac{r}{a_0}\left(1 - \frac{r}{6a_0}\right) e^{-r/3a_0} \qquad (72)$$

$$R_{32}(r) = \frac{1}{(3a_0)^{3/2}}\frac{2\sqrt{2}}{27\sqrt{5}}\frac{r^2}{a_0{}^2} e^{-r/3a_0} \qquad (73)$$

The normalization condition satisfied by these radial wavefunctions is similar to Eq. (34):

$$\int_0^\infty (R_{nl})^2 r^2 \, dr = 1$$

The normalization of the wavefunctions R_{nl} can be checked by evaluating this integral. Alternatively, we can use the results of Section 6.3 (see Problem 10).

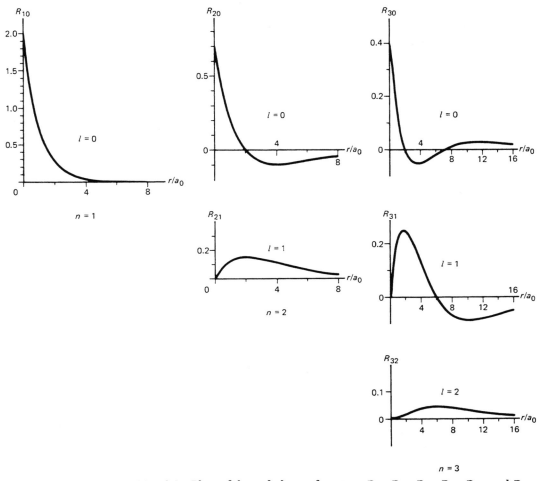

Fig. 8.3 Plots of the radial wavefunctions R_{10}, R_{20}, R_{30}, R_{21}, R_{31}, and R_{32}.

Figure 8.3 shows plots of the radial wavefunctions R_{nl}. All the wavefunctions with $l = 0$ have a maximum at $r = 0$. The other wavefunctions all vanish at $r = 0$. This says that when the electron is in a state of zero angular momentum, the probability for finding the electron in a volume element $dV = dx\ dy\ dz$ is largest at the nucleus.[1] Note, however, that the large probability per unit volume at $r = 0$ does not necessarily mean that this is the most proba-

---

[1] The electron does not interact with the nuclear material except by the Coulomb interaction; thus, the electron can move through the nucleus without hindrance.

ble radius for the electron. If we want the probability for a radial interval dr, we must multiply the probability density by the volume associated with this radial interval; that is, we must multiply by the volume $dV = 4\pi r^2\, dr$ of a thin spherical shell of radius r and thickness dr (see Fig. 8.4). Thus, the probability per unit radial interval is proportional to $|R_{n0}|^2 r^2$, and the most probable radius is determined by the location of the maximum of this function. For instance, the most probable radius for an electron in the ground state is determined by the location of the maximum of $|R_{10}|^2 r^2 \propto r^2 e^{-2r/a_0}$; this maximum is located at the Bohr radius $r = a_0 = 4\pi\varepsilon_0 \hbar^2/me^2$. Likewise, for all the other hydrogen wavefunctions, the most probable radius is different from zero, and it increases with n.

Since our energy eigenfunctions are simultaneous eigenfunctions of $\mathbf{L}^2$ and L_z, their angular dependence is given by the spherical harmonics $Y_l^{m_l}(\theta, \phi)$. The complete wavefunctions, including the radial and the angular dependence, but omitting the time dependence, are then

$$\psi_{nlm_l}(r, \theta, \phi) = R_{nl}(r) Y_l^{m_l}(\theta, \phi) \tag{74}$$

Figure 8.5 displays the probability distributions $|\psi_{nlm_l}(r, \theta, \phi)|^2$ as clouds of dots with a density proportional to the probability.

Our result (56) for the eigenvalues shows that for a given value of n, the energy does not depend on l or m_l. As in the case of the isotropic harmonic oscillator, the degeneracy with respect to m_l is a direct consequence of the spherical symmetry of the potential. But the degeneracy with respect to l is not related to spherical symmetry; rather, it is a peculiar property of the $1/r$ potential, and

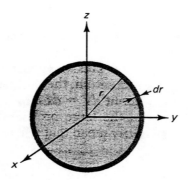

Fig. 8.4   The volume dV associated with a radial interval dr is that of a shell of radius r and thickness dr.

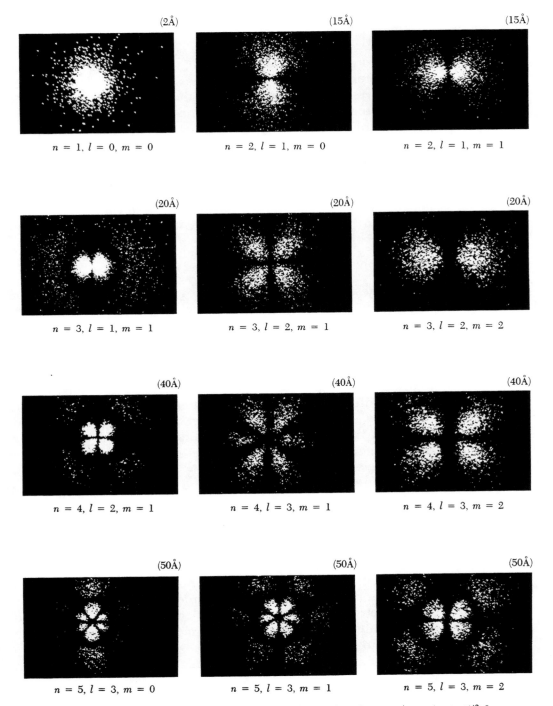

Fig. 8.5 Pictures of the probability distributions $|\psi_{nlm_l}(r, \theta, \phi)|^2$ for some states of the hydrogen atom. The density of dots in these pictures is proportional to the probability. These pictures were generated by a Monte-Carlo computer program that selected points r, θ, ϕ at random, and then decided to plot them or not plot them according to the value of $|\psi_{nlm_l}(r, \theta, \phi)|^2$. (Courtesy A. F. Burr and A. Fisher, New Mexico State University.)

it is called an *accidental degeneracy*. For other spherically symmetric potentials, this degeneracy does not occur. Of course, nothing in physics is an accident, and the "accidental," apparently unmotivated, degeneracy of the energy levels of the $1/r$ potential can be explained in terms of a symmetry of the Hamiltonian with respect to an operation generated by a peculiar combination of the momentum operators.

Let us now count how many states there are with a given energy $-\frac{1}{2}mc^2\alpha^2/n^2$. For a given value of n, l can assume the values $n - 1$, $n - 2$, . . . , 0, and for each of these values of l, m_l can assume any of $2l + 1$ values. Hence the degeneracy of the nth energy level is

$$[2(n - 1) + 1] + [2(n - 2) + 1] + \cdots + 1 = n^2 \qquad (75)$$

Besides the discrete set of negative energy eigenvalues $-\frac{1}{2}mc^2\alpha^2/n^2$ there exists a continuum of positive energy eigenvalues. These positive eigenvalues correspond, of course, to states in which the electron is not bound to the nucleus, but simply moves toward and then away from the nucleus with some positive value of the total energy. The eigenfunctions for this case are very complicated, because even at large distances the electron still feels the effect of the Coulomb potential, that is, the potential has a long range. If we want a complete set of eigenfunctions, we must take both the negative-energy eigenfunctions and the positive-energy eigenfunctions. The negative-energy eigenfunctions ψ_{nlm_l} by themselves do not form a complete set.

If we need the eigenvalues and the eigenfunctions for an electron interacting with a nucleus of charge Ze, then all we have to do is to replace e^2 by Ze^2 everywhere in the above formulas. Instead of the Bohr radius $4\pi\varepsilon_0\hbar^2/me^2$, we must use a new Bohr radius $4\pi\varepsilon_0\hbar^2/Zme^2$. Since the Bohr radius determines the sizes of the "orbits"—or, more precisely, the extent of the wavefunctions—this implies that all the dimensions of the probability distribution are smaller by a factor of Z. For example, the most probable radius for an electron in the ground state is determined by the location of the maximum of $|R_{10}(r)|^2 r^2$, and this most probable radius is proportional to $1/Z$.

Exercise 9. Show that maximum in the radial probability distribution $|R_{10}(r)|^2 r^2$ for a hydrogenic atom with nuclear charge Z occurs at

$$r = \frac{4\pi\varepsilon_0\hbar^2}{Zme^2} \qquad (76)$$

In our treatment of the hydrogen atom we have so far ignored the motion of the nucleus. Let us now take this motion into account. If the subscripts e and n stand for the electron and the nucleus (or the proton), respectively, then the complete Hamiltonian for the system of these two particles is

$$H = \frac{\mathbf{p_e}^2}{2m_e} + \frac{\mathbf{p_n}^2}{2m_n} - \frac{1}{4\pi\varepsilon_0}\frac{e^2}{r} \tag{77}$$

where r is now the distance between the electron and the nucleus, $r = |\mathbf{r_e} - \mathbf{r_n}|$. The electron coordinate and momentum commute with the nucleus coordinate and momentum.

To simplify the Hamiltonian, we proceed as in the classical case, and we separate the motion of the center of mass and the motion of the particles relative to the center of mass. For this purpose, we introduce the quantities

$$\mathbf{P} \equiv \mathbf{p_e} + \mathbf{p_n} \tag{78}$$

$$\mathbf{p} \equiv \frac{1}{M}(m_n\mathbf{p_e} - m_e\mathbf{p_n}) \tag{79}$$

$$m' \equiv \frac{m_e m_n}{m_e + m_n} \tag{80}$$

$$M \equiv m_e + m_n \tag{81}$$

The quantities $\mathbf{P}$, $\mathbf{p}$, m', and M represent, respectively, the total momentum operator, the relative momentum operator, the reduced mass, and the total mass of the system. It is then easy to check that, as in classical mechanics,

$$\frac{\mathbf{p_e}^2}{2m_e} + \frac{\mathbf{p_n}^2}{2m_n} = \frac{\mathbf{P}^2}{2M} + \frac{\mathbf{p}^2}{2m'} \tag{82}$$

Exercise 10. Check this.

By means of Eq. (82) we can separate the Hamiltonian into two terms:

$$H = H_1 + H_2 \tag{83}$$

where

$$H_1 = \frac{\mathbf{P}^2}{2M} \tag{84}$$

and

$$H_2 = \frac{\mathbf{p}^2}{2m'} - \frac{1}{4\pi\varepsilon_0}\frac{e^2}{r} \tag{85}$$

The term H_1 represents the energy of the translational motion of the center of mass of the atom and the term H_2 represents energy of the motion relative to the center of mass. The terms H_1 and H_2 commute with each other and with $\mathbf{P}$.

Exercise 11. Show that H_1 and H_2 commute with each other and with $\mathbf{P}$. (Hint: $\mathbf{p}_e$ and $\mathbf{p}_n$ commute.)

In view of this commutativity, there exist simultaneous eigenstates of H_1, H_2, and $\mathbf{P}$. We can label these states by giving the eigenvalue of $\mathbf{P}$; this fixes the net translational momentum and the net translational energy H_1 of the atom. The other labels describe the internal state of the atom. For these other labels we can use the eigenvalues of H_2, $\mathbf{L}^2$, and L_z, where the angular momentum $\mathbf{L}$ is, of course, the internal angular momentum, $\mathbf{L} = \mathbf{r} \times \mathbf{p}$. A simultaneous eigenstate of $\mathbf{P}$, H_2, $\mathbf{L}^2$, and L_z will be denoted by $|\mathbf{P},E,l,m_l\rangle$. The energy eigenvalues of the Hamiltonian H_1 are simply $\mathbf{P}^2/2M$. The energy eigenvalues of the Hamiltonian H_2 can be readily obtained by comparing it with our previous Hamiltonian for the hydrogen atom without nuclear motion. The relative momentum $\mathbf{p}$ and the relative coordinate $\mathbf{r} = \mathbf{r}_e - \mathbf{r}_n$ satisfy the standard commutation relations

$$[x_j, p_k] = i\hbar\delta_{jk}$$

Exercise 12. Derive these commutation relations from the commutation relations for the electron and proton momenta and coordinates.

Hence, not only is the Hamiltonian H_2 of the same form as our previous Hamiltonian for the hydrogen atom, but the coordinates and momenta appearing in H_2 satisfy the same commutation relations as before. Obviously, the only difference between H_2 and our previous Hamiltonian is that now the reduced mass m' appears in place of the electron mass m. Thus, the energy eigenvalues must be of the same form as previously, but the reduced mass replaces the electron mass. The energy eigenvalues for H_2 are

$$E_n = -\frac{1}{2}\alpha^2\frac{m'c^2}{n^2}$$

Numerically, the energy eigenvalues given by this formula and those given by our previous formula [Eq. (57)] differ by a factor of $m'/m_e = m_n/(m_e + m_n) = 0.99946$. The eigenvalue of the total energy is[2]

$$\frac{\mathbf{P}^2}{2M} - \frac{1}{2} \alpha^2 \frac{m'c^2}{n^2} \tag{86}$$

The state vector that represents the system when in a state in which the position of the center of mass is exactly $\mathbf{R}$ and the relative separation $r_e - r_n$ is exactly r will be designated by $|\mathbf{R}, \mathbf{r}\rangle$. The wavefunction for the system (with time dependence omitted) is a function of these two vector variables $\mathbf{R}$ and $\mathbf{r}$:

$$\psi(\mathbf{R}, \mathbf{r}) = \langle \mathbf{R}, \mathbf{r} | \psi \rangle \tag{87}$$

or, explicitly,

$$\psi(\mathbf{R}, \mathbf{r}) = \frac{1}{(2\pi\hbar)^{3/2}} e^{i\mathbf{P}\cdot\mathbf{R}/\hbar} \psi_{nlm_l}(r, \theta, \phi) \tag{88}$$

Then $|\psi(\mathbf{R}, \mathbf{r})|^2$ gives the probability for finding the center of mass of the atom at the point $\mathbf{R}$ with a relative separation between the electron and the proton of $\mathbf{r}$. The integral

$$\int |\psi(\mathbf{R}, \mathbf{r})|^2 \, d^3r$$

gives the probability for finding the center of mass of the atom at $\mathbf{R}$, irrespective of the relative separation between the electron and the proton. The integral

$$\int |\psi(\mathbf{R}, \mathbf{r})|^2 \, d^3R$$

gives the probability for finding a relative separation $\mathbf{r}$ irrespective of where the center of mass is located. For the wavefunction (88), this statement about the probability interpretation runs into the familiar difficulty that the integral over $\mathbf{R}$ is not convergent (the wavefunction $e^{i\mathbf{P}\cdot\mathbf{R}/\hbar}$ is not normalizable), so we can evaluate only relative probabilities.

---

[2] Note that in this equation $\mathbf{P}$ is not an operator; it is a number (the eigenvalue of the momentum). This should be obvious from the context.

8.3   *The Free Particle*

The energy eigenvectors for a free particle moving in three dimensions have already been discussed in Section 7.1. Taking the Hamiltonian

$$H = \frac{1}{2m}\,(p_1{}^2 + p_2{}^2 + p_3{}^2) \qquad (89)$$

as our starting point, we obtained simultaneous eigenvectors of H, p_1, p_2, and p_3. In spherical coordinates, the Hamiltonian (89) is equal to

$$H = \frac{1}{2m}\,p_r{}^2 + \frac{1}{2mr^2}\,\mathbf{L}^2 \qquad (90)$$

which means that we can regard the free particle as a special case of a central potential with $V(r) = 0$. The expression (90) suggests that we look for a different set of eigenvectors, that is, simultaneous eigenvectors of H, $\mathbf{L}^2$, and L_z. Such a description of the states of a free particle by angular-momentum eigenvectors is very useful in the theory of scattering.

We adopt the notation $|k,l,m_l\rangle$ for the simultaneous eigenvectors of H, $\mathbf{L}^2$, and L_z with eigenvalues $\hbar^2 k^2/2m$, $l(l + 1)\hbar^2$, and $m_l\hbar$, respectively. The constant k is the wave number, $k = \sqrt{2mE}/\hbar$. The eigenvector satisfies the eigenvalue equation

$$\left[\frac{1}{2m}\,p_r{}^2 + \frac{1}{2mr^2}\,l(l + 1)\hbar^2\right] |k,l,m_l\rangle = \frac{\hbar^2 k^2}{2m}\,|k,l,m_l\rangle \qquad (91)$$

The possible eigenvalues are already known: for a free particle, the energy $\hbar^2 k^2/2m$ can assume any positive value between zero and infinity. Thus, the factorization method is not relevant here. However, the wavefunctions $\psi_{klm_l}(r, \theta, \phi)$ remain to be found, and we will see that a construction involving ladder operators is suitable for this purpose.

The angular dependence of the wavefunctions must be $Y_l{}^{m_l}(\theta, \phi)$. We therefore write

$$\psi_{klm_l}(r, \theta, \phi) = j_l(kr)Y_l{}^{m_l}(\theta, \phi)$$

where $j_l(kr)$ gives the radial dependence.[3] In the position repre-

---

[3] We could write $j_l(r)$ in Eq. (91), but $j_l(kr)$ is more convenient, since kr is a dimensionless variable.

sentation, Eq. (91) becomes a differential equation for the function $j_l(kr)$:

$$\left[\frac{1}{2m} p_r^2 + \frac{1}{2mr^2} l(l+1)\hbar^2\right] j_l(kr) = \frac{\hbar^2 k^2}{2m} j_l(kr) \tag{92}$$

where

$$p_r = \frac{\hbar}{i}\left(\frac{d}{dr} + \frac{1}{r}\right)$$

or equivalently,

$$p_r = \frac{\hbar}{ir}\frac{d}{dr} r$$

For the special case $l = 0$, the differential equation (92) reduces to

$$-\frac{\hbar^2}{2m}\left(\frac{1}{r}\frac{d^2}{dr^2} r\right) j_0(kr) = \frac{\hbar^2 k^2}{2m} j_0(kr) \tag{93}$$

This differential equation has two solutions: $(\cos kr)/kr$ and $(\sin kr)/kr$. Of these, only the latter is well behaved at $r = 0$. Adopting this solution, we have

$$j_0(kr) = \frac{\sin kr}{kr} \tag{94}$$

We will show that $j_0(kr)$ can be used to construct $j_l(kr)$ according to the following formula:

$$j_l(kr) = \chi_{l-1}^\dagger \cdots \chi_0^\dagger j_0(kr) \tag{95}$$

where the ladder operators χ_l are

$$\chi_l = \frac{1}{\hbar k}\left[ip_r + \frac{(l+1)\hbar}{r}\right] \tag{96}$$

$$\chi_l^\dagger = \frac{1}{\hbar k}\left[-ip_r + \frac{(l+1)\hbar}{r}\right] \tag{97}$$

The proof of Eq. (95) is very simple. We begin with the expressions

$$\chi_l \chi_l^\dagger = \frac{1}{\hbar^2 k^2}\left[p_r^2 + \frac{l(l+1)\hbar^2}{r^2}\right] \tag{98}$$

$$\chi_l^\dagger \chi_l = \frac{1}{\hbar^2 k^2}\left[p_r^2 + \frac{(l+1)(l+2)\hbar^2}{r^2}\right] \tag{99}$$

Exercise 13. Derive these expressions.

It follows that Eq. (92) can be written as

$$\chi_l \chi_l^\dagger j_l(kr) = j_l(kr) \tag{100}$$

If we multiply by this $\chi_l^\dagger$, we obtain

$$\chi_l^\dagger \chi_l \chi_l^\dagger j_l(kr) = \chi_l^\dagger j_l(kr) \tag{101}$$

which, in view of Eq. (99), is the same as

$$\frac{1}{\hbar^2 k^2}\left[p_r^2 + \frac{(l+1)(l+2)\hbar^2}{r^2}\right] \chi_l^\dagger j_l(kr) = \chi_l^\dagger j_l(kr) \tag{102}$$

This verifies that the function $\chi_l^\dagger j_l(kr)$ satisfies Eq. (92) with l replaced by $l+1$. We conclude that

$$j_{l+1}(kr) = \chi_l^\dagger j_l(kr) \tag{103}$$

This establishes the validity of Eq. (95),

It is convenient to introduce the variable $\rho = kr$ and to write the solution j_0 as

$$j_0(\rho) = \frac{\sin \rho}{\rho} \tag{104}$$

In terms of this variable ρ,

$$\chi_l = \frac{d}{d\rho} + \frac{l+2}{\rho} \tag{105}$$

and

$$\chi_l^\dagger = -\frac{d}{d\rho} + \frac{l}{\rho} \tag{106}$$

Exercise 14. Verify that the expressions (105) and (106) agree with (96) and (97).

Equation (95) then becomes

$$j_l(\rho) = \left(-\frac{d}{d\rho} + \frac{l-1}{\rho}\right)\left(-\frac{d}{d\rho} + \frac{l-2}{\rho}\right)\cdots\left(-\frac{d}{d\rho}\right)\frac{\sin \rho}{\rho} \tag{107}$$

This can also be written in the form

$$j_l(\rho) = \rho^l \left(-\frac{1}{\rho}\frac{d}{d\rho} \right)^l \frac{\sin\rho}{\rho} \qquad (108)$$

Exercise 15. Show, by induction, that

$$\left(-\frac{d}{d\rho} + \frac{l-1}{\rho} \right)\left(-\frac{d}{d\rho} + \frac{l-2}{\rho} \right) \cdots \left(-\frac{d}{d\rho} \right) = \rho^l \left(-\frac{1}{\rho}\frac{d}{d\rho} \right)^l$$

and thereby establish that Eqs. (107) and (108) are equivalent.

The functions $j_l(\rho)$ are called *spherical Bessel functions*. The first few of these functions are

$$j_0(\rho) = \frac{\sin\rho}{\rho}$$

$$j_1(\rho) = \frac{\sin\rho}{\rho^2} - \frac{\cos\rho}{\rho} \qquad (109)$$

$$j_2(\rho) = \left(\frac{3}{\rho^3} - \frac{1}{\rho} \right)\sin\rho - \frac{3}{\rho^2}\cos\rho$$

Figure 8.6 gives plots of these functions.

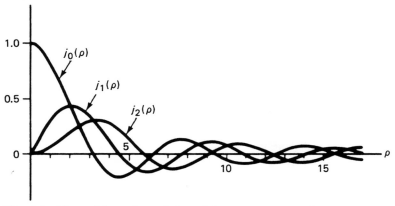

Fig. 8.6 Plots of the spherical Bessel functions j_0, j_1, and j_2.

Exercise 16. Obtain the results for $j_2(\rho)$ and $j_3(\rho)$.

Exercise 17. Show that

$$\chi_{l-1}j_l(\rho) = j_{l-1}(\rho) \tag{110}$$

Exercise 18. By combining Eqs. (103) and (110), show that

$$\frac{1}{\rho}\,j_l(\rho) = \frac{1}{2l+1}\,[j_{l+1}(\rho) + j_{l-1}(\rho)] \tag{111}$$

and that

$$\frac{d}{d\rho}\,j_l(\rho) = -\frac{l+1}{2l+1}\,j_{l+1}(\rho) + \frac{l}{2l+1}\,j_{l-1}(\rho) \tag{112}$$

L_z form a complete set of functions. However, the normalization of these functions runs into the same difficulty as in the case of the momentum eigenfunctions; only a delta-function normalization is possible.

Since these functions form a complete set, it must be possible to express any momentum eigenfunction in terms of them. For instance, suppose that we have a particle whose momentum is in the z direction, so the eigenvalues of the operators p_1, p_2, and p_3 are 0, 0, and $\hbar k$, respectively. The corresponding wavefunction e^{ikz} can be expressed as a superposition of the functions $j_l(kr)Y_l^{m_l}(\theta, \phi)$ as follows:

$$e^{ikz} = \sqrt{4\pi}\sum_{l=0}^{\infty}\sqrt{2l+1}\,\,i^l Y_l^0 j_l(kr) \tag{113}$$

One way to prove this equation is to check that both sides are eigenfunctions of p_3 with the same eigenvalue $\hbar k$, and that the eigenvalues of p_1 and of p_2 are zero [this establishes Eq. (113) to within a constant of proportionality; by evaluating the function at, say, $r = 0$, it is easy to check that this constant has been correctly chosen].

Exercise 19. Show that the right side of Eq. (113) is an eigenfunction of $\partial/\partial z$. [Hint: With $\mu = \cos\theta$, use

$$\frac{\partial}{\partial z} = \mu\frac{\partial}{\partial r} + \frac{1}{r}\,(1 - \mu^2)\frac{\partial}{\partial\mu}$$

to express the derivative with respect to z of the right side of Eq. (113) as

$$\sqrt{4\pi} \sum_l \sqrt{2l+1}\; i^l \left[\mu Y_l^0 \frac{\partial j_l}{\partial r} + (1 - \mu^2) \frac{\partial Y_l^0}{\partial \mu} \frac{j_l}{r} \right]$$

and then apply Eqs. (7.88), (7.89), (111), and (112).]

Exercise 20. Show that the right side of Eq. (113) is an eigenfunction of $\partial/\partial x$ and of $\partial/\partial y$, with eigenvalue zero.

Equation (113) expresses a plane wave as a superposition of spherical waves of definite angular momentum. We will make use of this expression in the theory of scattering.

Exercise 21. Show that for $\rho \to \infty$, $j_l(\rho)$ has the asymptotic behavior

$$j_l(\rho) \to \frac{1}{\rho} (-1)^l \frac{d^l}{d\rho^l} \sin\rho = \frac{1}{\rho} \sin\left(\rho - \frac{l\pi}{2}\right) \qquad (114)$$

[Hint: Use Eq. (108) and omit all terms proportional to $1/\rho^2$, $1/\rho^3$, ...]

PROBLEMS

1. Consider the nth energy level of the isotropic harmonic oscillator, with an energy $(n + \frac{3}{2})\hbar\omega$, where n is an integer. What is the degeneracy of this energy level, that is, how many different eigenvectors have this energy?

2. Obtain the radial function $\psi_3^{(0)}(r)$ for the isotropic harmonic oscillator.

3. Express the wavefunction $\psi_{111}(r, \theta, \phi)$ of the isotropic harmonic oscillator as a superposition of the wavefunctions $\psi_{n_1 n_2 n_3}(x, y, z)$.

4. Obtain the normalization for the radial wavefunctions $\psi_1^{(0)}(r)$, $\psi_1^{(1)}(r)$, $\psi_1^{(2)}(r)$, and $\psi_2^{(0)}(r)$ given in Eqs. (28)–(31).

5. Show that the operator p_r for the isotropic harmonic oscillator can be expressed as

$$p_r = \sqrt{\frac{m}{2}} \left(\eta_1^{(l)} + \eta_1^{(l)\dagger} \right)$$

and use this expression to find the expectation value of p_r in the eigenstate $|E, l, m_l\rangle$.

6. Find the expectation value of p_r^2 in the eigenstate $|E, l, m_l\rangle$ with $j = 1$ for the isotropic harmonic oscillator.

7. Prove the following general virial theorem for a particle in a stationary state in a central potential

$$\langle\psi|\, r\, \frac{\partial V}{\partial r}\,|\psi\rangle = 2\langle\psi|\, K\,|\psi\rangle$$

(Hint: Begin with the expectation value of the commutator $[\mathbf{r}\cdot\mathbf{p}, H]$.)

8. The Hamiltonian for a two-dimensional harmonic oscillator is

$$H = \frac{1}{2m}\,(p_x^2 + p_y^2) + \frac{1}{2}\,m\omega^2(x^2 + y^2)$$

(a) Show that in cylindrical coordinates

$$H = \frac{1}{2m}\,p_\rho^2 - \frac{\hbar^2}{8m\rho^2} + \frac{1}{2m\rho^2}\,L_z^2 + \frac{1}{2}\,m\omega^2\rho^2$$

Show that L_z commutes with H.

(b) Consider an eigenstate of L_z, with eigenvalue $m_z\hbar$. Use the factorization method to show that the energy eigenvalues are

$$E_j = \hbar\omega(|m_z| + 2j - 1) \qquad j = 1, 2, 3, \ldots$$

9. The electron in a hydrogen atom is in the state

$$|\psi(t)\rangle = \frac{1}{\sqrt{3}}\,e^{-iE_1 t/\hbar}\,|E_1,0,0\rangle - \sqrt{\frac{2}{3}}\,e^{-iE_2 t/\hbar}\,|E_2,1,1\rangle$$

Find the expectation values of the operators H, $\mathbf{L}^2$, L_z, and P (where P is the parity operator).

10. By evaluating the integral $\int_0^\infty |\zeta_j^{(l)}(r)|^2 r^2 dr$, show that the functions

$$\zeta_j^{(l)}(r) = \frac{r^{l+j-1}e^{-r/(l+j)a_0}}{\{[2(l+j)]![(l+j)a_0/2]^{2(l+j)+1}\}^{1/2}}$$

are correctly normalized. Then use the results of Section 6.3 to obtain the normalized functions (68)–(73).

11. Show that the operators p_r and $1/r$ for the hydrogen atom can be expressed as

$$p_r = \sqrt{\frac{m}{2}}\,(\eta_1^{(l)} + \eta_1^{(l)\dagger})$$

and

$$\frac{1}{r} = \frac{1}{i(l+1)\hbar}\,\sqrt{\frac{m}{2}}\,(\eta_1^{(l)} - \eta_1^{(l)\dagger}) + \frac{1}{4\pi\varepsilon_0}\,\frac{me^2}{(l+1)^2\hbar^2}$$

Use these expressions to evaluate the expectation values of p_r and of $1/r$ for the eigenstate $|E_n,l,m_l\rangle$. Deduce that the expectation value of the potential energy is $2E_n$, and verify that this is in accord with the general virial theorem stated in Problem 7.

12. Obtain the following results for the expectation values of $1/r^2$ and $1/r^3$ in the eigenstate $|E_n,n-1,m_l\rangle$ of maximum angular momentum of the hydrogen atom:

    (a) $\langle E_n,l,m_l|\dfrac{1}{r^2}|E_n,l,m_l\rangle = \dfrac{2}{(2n-1)n^3a_0^2}$

    (b) $\langle E_n,l,m_l|\dfrac{1}{r^3}|E_n,l,m_l\rangle = \dfrac{2}{(n-1)(2n-1)n^4a_0^3}$

13. Find the expectation value of $\cos^2\theta$ in the eigenstate $|E_n,l,m_l\rangle$ of hydrogen. [Hint: Begin by establishing that

    $$\cos\theta\, Y_l{}^{m_l} = \sqrt{\frac{(l+m_l+1)(l-m_l+1)}{(2l+1)(2l+3)}}\, Y_{l+1}{}^{m_l} + \sqrt{\frac{(l+m_l)(l-m_l)}{(2l-1)(2l+1)}}\, Y_{l-1}{}^{m_l}]$$

14. Show that the most probable radius for the electron in the hydrogen atom in the state $|E_n,n-1,m_l\rangle$ of maximum orbital angular momentum is n^2a_0.

15. Obtain the following results for the expectation values of r and of r^2 in the eigenstate $|E_n,n-1,m_l\rangle$ of maximum orbital angular momentum of the hydrogen atom:

    (a) $\langle E_n,n-1,m_l|r|E_n,n-1,m_l\rangle = n(n+\tfrac{1}{2})a_0$
    (b) $\langle E_n,n-1,m_l|r^2|E_n,n-1,m_l\rangle = (n+1)(n+\tfrac{1}{2})n^2a_0^2$
    (c) $\Delta r = n\sqrt{\tfrac{1}{2}(n+\tfrac{1}{2})}\,a_0$

16. Show that all the hydrogen wavefunctions with $l=0$ have a maximum at $r=0$ and that the wavefunctions with $l\neq 0$ vanish at $r=0$.

17. Find the locations of the maxima in the probability distribution $|\psi_{311}(r,\theta,\phi)|^2$ of the hydrogen atom. Find the nodal surfaces in this probability distribution. Compare your results with Fig. 8.5.

18. When the motion of the nucleus is taken into account, the state of the hydrogen atom can be represented by a wavefunction $\psi(\mathbf{R},\mathbf{r})$. Suppose that a hydrogen atom is in a state such that the total momentum has equal probabilities for the values $\mathbf{P}$ and $-\mathbf{P}$ and such that the internal states $|E_1,0,0\rangle$ and $|E_2,1,1\rangle$ have probabilities of 1/4 and 3/4, respectively. These latter probabilities are uncorrelated with the total momentum. Except for phase factors, what is the wavefunction for the atom? What is the expectation value of the total energy?

19. Consider a particle of orbital angular momentum $l = 0$ in the central potential

$$V(r) = -\frac{V_0}{e^{\kappa r} - 1}$$

called Hulthen's potential. Find the first energy eigenvalue for this particle. [Hint: Try

$$\eta_j = \frac{1}{\sqrt{2m}}\left(p_r + ib_j + \frac{ic_j}{e^{\kappa r} - 1}\right)]$$

9

Spin and the Exclusion Principle

Throughout the previous chapters we have assumed that the state vector of a particle is uniquely specified by the eigenvalues of the three components of the momentum or, in the case of a particle in a central potential, by the eigenvalues of the orbital energy and the orbital angular momentum. Thus, we used the state vector $|p_1, p_2, p_3\rangle$ to describe the state of a free particle and the state vector $|E_n, l, m_l\rangle$ to describe the state of a particle moving in a central potential. However, if the particle in question is an electron, then three eigenvalues, or three quantum numbers, are not sufficient to specify the state vector uniquely. For instance, for an electron in a hydrogen atom, the quantum numbers $n = 1$, $l = 0$, $m_l = 0$ do not specify a single, unique state; rather, they specify a pair, or doublet, of states, differing slightly in energy (the energy difference between the two states in this doublet is only about 10^{-5} eV). Likewise, for a free electron, the eigenvalues p_1, p_2, and p_3 of the components of the momentum do not specify a single, unique state, but a pair of states. Experimentally, the two states of a free electron associated with a given momentum can be distinguished by sending the electron through an inhomogeneous magnetic field (Stern–Gerlach experiment[1]); the deflection suffered by the electron is different for the two states.

---

[1] A Stern–Gerlach experiment cannot be performed directly with free electrons, because the uncertainty in the position of the electron is always at least as large as the deflection attainable by the interaction of the magnetic moment of the electron with the magnetic field (see Problem 7). However, in principle, an indirect Stern–Gerlach experiment is feasible: first capture the electron in an ionized atom, then perform a Stern–Gerlach experiment on this atom, and then ionize the electron off the atom (without changing its direction of spin).

This characteristic two-valuedness of the electron states is due to the *spin,* or intrinsic angular momentum. The spin quantum number of the electron is $s = \frac{1}{2}$, and the quantum numbers for the z component of spin are $m_s = +\frac{1}{2}$ and $m_s = -\frac{1}{2}$. Thus, for every orbital state, there are two possible spin states.

It is tempting to picture the spin as due to some kind of rotational motion in the interior of the electron. But such a naive picture is untenable—scattering experiments have established that the electron is a truly pointlike particle, with no discernible internal structure. For an understanding of the mechanism that gives rise to the spin of the electron we must appeal to relativistic wave mechanics, and we must examine the properties of the relativistic wave equation for the electron, called the *Dirac equation.* According to the picture that emerges from relativistic wave mechanics, the spin is not due to an internal rotation of the electron, but rather to a circulating flow of energy in the wave field. We can understand this picture of spin by analogy with the angular momentum found in a circularly polarized electromagnetic wave. Examination of the Poynting vector of such a wave reveals that there is not only an energy flow along the direction of propagation of the wave, but there is also an energy flow in circles around the direction of propagation. This circulating flow of energy amounts to a rotational motion, and therefore gives rise to an angular momentum. This angular momentum of the circularly polarized electromagnetic wave accounts for the spin of the photon. A similar mechanism accounts for the spin of the electron. Examination of the "Poynting" vector for the relativistic electron wave reveals a similar circulating flow of energy, which gives rise to an angular momentum. Thus, the origin of the spin is not found in the internal structure of the electron, but in the structure of its wave field.

In this chapter, we will deal only with the mathematical properties of spin, without regard to the mechanism that causes it. We will concentrate on electrons, but the mathematical results apply also to other particles of spin quantum number $s = \frac{1}{2}$, such as protons or neutrons. However, the mechanism underlying the spin of protons, neutrons, and other baryons differs somewhat from that of electrons. Protons, neutrons, and other baryons are composite bodies, with a finite size and an internal structure; they are made of quarks orbiting around each other. Thus, the spin of a proton, a neutron, or some other baryon is in part due to the orbital angular momentum of the quarks, and in part to the intrinsic spin ($s = \frac{1}{2}$) of the quarks. This means that the spin of a baryon is analogous to the net angular momentum of an atom or a molecule, which is in part due to the orbital angular momentum of the electrons and in part to the spins of the electrons.

9.1 *The Spin of the Electron*

In general, we can define the spin of a particle (or of a system of particles, such as an atom) as that part of the angular momentum that is not associated with the momentum of the center of mass. Thus, the spin is the remainder we obtain when we subtract from the total angular momentum the orbital angular momentum $\mathbf{L} = \mathbf{R} \times \mathbf{P}$ of the motion of the center of mass. More simply, the spin is the angular momentum of the particle in its own rest frame (in which the momentum $\mathbf{P}$ is zero).

As for every observable, we want to associate a hermitian operator with the spin. We will designate this operator by $\mathbf{S}$. Since spin is a form of angular momentum, it must have three components S_x, S_y, and S_z, and we will postulate that these three components obey exactly the same commutation relations as L_x, L_y, and L_z:

$$[S_x, S_y] = i\hbar S_z$$

$$[S_y, S_z] = i\hbar S_x \tag{1}$$

$$[S_z, S_x] = i\hbar S_y$$

It follows that $\mathbf{S}^2$ and S_z commute and that they have simultaneous eigenvectors with eigenvalues $s(s + 1)\hbar^2$ and $m_s\hbar$:

$$\mathbf{S}^2|s,m_s\rangle = s(s + 1)\hbar^2|s,m_s\rangle \tag{2}$$

$$S_z|s,m_s\rangle = m_s\hbar|s,m_s\rangle \tag{3}$$

We have seen that in the case of orbital angular momentum we can accept only integer values of l, because the half-integer values of l lead to inconsistencies in the equations that the wavefunctions $Y_l^{m_l}(\theta, \phi)$ ought to obey. The spin has nothing to do with position coordinates and with translational momentum. Thus, we never represent the spin states by functions of θ and ϕ, and any trouble with the wavefunctions $Y_l^{m_l}(\theta, \phi)$ is entirely irrelevant to the spin. There is then no reason to reject half-integer values of the angular-momentum quantum number. The possible values of s are then

$$s = 0, \tfrac{1}{2}, 1, \tfrac{3}{2}, 2 \ldots \tag{4}$$

and the possible values of m_s are

$$m_s = -s, -s + 1, \ldots, s - 1, s \tag{5}$$

Table 9.1 lists the values of the spin quantum number s for a few particles. The last three particles in this table (photon, neutrino, graviton) have rest mass zero, and they always move at the speed of light; a full description of the spin of these particles requires relativistic quantum mechanics. Particles with integer spin are called bosons; those with half-integer spin are called fermions.

We will now treat the case of the electron spin in detail. The spin quantum number s for the electron is $s = \frac{1}{2}$ and the two possible values of m_s are $+\frac{1}{2}$ and $-\frac{1}{2}$. We designate the possible spin states of the electron by $|\frac{1}{2},+\frac{1}{2}\rangle$ and $|\frac{1}{2},-\frac{1}{2}\rangle$:

$$\mathbf{S}^2|\tfrac{1}{2},\pm\tfrac{1}{2}\rangle = \tfrac{3}{4}\hbar^2|\tfrac{1}{2},\pm\tfrac{1}{2}\rangle \tag{6}$$

$$S_z|\tfrac{1}{2},\pm\tfrac{1}{2}\rangle = \pm\tfrac{1}{2}\hbar|\tfrac{1}{2},\pm\tfrac{1}{2}\rangle \tag{7}$$

Here, we have ignored the quantum numbers needed to specify the orbital state of the electron. For convenience, we will adopt the concise notation $|+\rangle$ for the state $|\frac{1}{2},+\frac{1}{2}\rangle$ and $|-\rangle$ for the state $|\frac{1}{2},-\frac{1}{2}\rangle$, and we will call these the spin "up" and spin "down" states, respectively. Since $\mathbf{S}$ is assumed to be a hermitian operator, $|+\rangle$ and $|-\rangle$ are orthogonal, and we assume that they are normalized:

$$\langle+|+\rangle = \langle-|-\rangle = 1$$

$$\langle+|-\rangle = 0 \tag{8}$$

TABLE 9.1 Spin Quantum Numbers

Particle	Spin Quantum Number, s
Electron (e)	1/2
Proton (p)	1/2
Neutron (n)	1/2
Pi meson (π^+, π^-, π^0)	0
Omega (Ω^-, Ω^+)	3/2
Photon(γ)	1
Neutrino (ν)	1/2
Graviton (Γ)*	2

*No experimental measurement has been performed on the graviton; the value of the spin is obtained from theory.

If we indicate the orbital quantum numbers explicitly, then the spin up and down states are $|q,+\rangle$ and $|q,-\rangle$, where q stands for whatever orbital quantum numbers we are dealing with. For example, q could stand for n, l, m_l (if we are dealing with an electron in an atom) or q could stand for $\mathbf{p}$ (if we are dealing with a free electron). An arbitrary state will then be some superposition of orbital and spin states:

$$|\psi\rangle = \sum_q c_q^{(+)}|q,+\rangle + \sum_q c_q^{(-)}|q,-\rangle \tag{9}$$

We will denote the state in which the electron is localized at $\mathbf{r}$ and has spin up or down by $|\mathbf{r},+\rangle$ and by $|\mathbf{r},-\rangle$, respectively. Out of the state vector (9), we can then form *two* wavefunctions in the position representation:

$$\psi_+(\mathbf{r}) = \langle \mathbf{r},+|\psi\rangle = \sum_q c_q^{(+)} \langle \mathbf{r},+|q,+\rangle \tag{10}$$

and

$$\psi_-(\mathbf{r}) = \langle \mathbf{r},-|\psi\rangle = \sum_q c_q^{(-)} \langle \mathbf{r},-|q,-\rangle \tag{11}$$

The first gives the probability amplitude for finding the electron at $\mathbf{r}$ with spin up; the second with spin down. Note that the probability for finding the electron at $\mathbf{r}$ irrespective of spin is $|\psi_+(\mathbf{r})|^2 + |\psi_-(\mathbf{r})|^2$, and the probability for finding it with, say, spin up irrespective of position is

$$\int |\psi_+(\mathbf{r})|^2 \, d^3r$$

It is convenient to combine ψ_+ and ψ_- into a single *two-component* object. We write this object as a column vector, or a 2×1 matrix:

$$\begin{pmatrix} \psi_+(\mathbf{r}) \\ \psi_-(\mathbf{r}) \end{pmatrix} \tag{12}$$

Such a two-component object is called a *spinor*. For example, suppose that the electron is in the ground state of a hydrogen atom and that it has spin up. Then the spinor (12) becomes

$$\begin{pmatrix} \dfrac{2}{a_0^{3/2}} \, e^{-r/a_0} Y_0^0(\theta, \phi) \\ 0 \end{pmatrix}$$

Alternatively, we can write Eq. (12) as

$$\psi_+(\mathbf{r}) \begin{pmatrix} 1 \\ 0 \end{pmatrix} + \psi_-(\mathbf{r}) \begin{pmatrix} 0 \\ 1 \end{pmatrix} \tag{13}$$

Each of these terms consists of the product of a purely orbital part and a spin part. We can regard the unit vectors

$$\chi_+ = \begin{pmatrix} 1 \\ 0 \end{pmatrix} \quad \text{and} \quad \chi_- = \begin{pmatrix} 0 \\ 1 \end{pmatrix} \tag{14}$$

respectively, as representations of the spin-up state and the spin-down state. If we adopt this representation, we will have to represent operators, such as $\mathbf{S}$, that act on the spin states by 2×2 matrices. Thus,

$$S_z = \begin{pmatrix} (S_z)_{11} & (S_z)_{12} \\ (S_z)_{21} & (S_z)_{22} \end{pmatrix} \tag{15}$$

To find the matrix elements, we can use Eq. (5.43):

$$(S_z)_{11} = \langle +|S_z|+\rangle = \langle +|\tfrac{1}{2}\hbar|+\rangle = \tfrac{1}{2}\hbar$$

$$(S_z)_{12} = \langle +|S_z|-\rangle = \langle +|(-\tfrac{1}{2}\hbar)|-\rangle = 0$$

$$(S_z)_{21} = \langle -|S_z|+\rangle = \langle -|\tfrac{1}{2}\hbar|+\rangle = 0$$

$$(S_z)_{22} = \langle -|S_z|-\rangle = \langle -|(-\tfrac{1}{2}\hbar)|-\rangle = -\tfrac{1}{2}\hbar$$

Hence the 2×2 matrix that represents S_z is

$$S_z = \begin{pmatrix} \langle +|S_z|+\rangle & \langle +|S_z|-\rangle \\ \langle -|S_z|+\rangle & \langle -|S_z|-\rangle \end{pmatrix} = \begin{pmatrix} \tfrac{1}{2}\hbar & 0 \\ 0 & -\tfrac{1}{2}\hbar \end{pmatrix}$$

$$= \tfrac{1}{2}\hbar \begin{pmatrix} 1 & 0 \\ 0 & -1 \end{pmatrix} \tag{17}$$

We can also find the 2×2 matrices that represent S_x and S_y. For this purpose, we recall what the raising and lowering operators do when applied to the state vectors $|+\rangle$ and $|-\rangle$ with $m_s = \tfrac{1}{2}$ and $m_s = -\tfrac{1}{2}$:

$$(S_x + iS_y)|+\rangle = 0$$

$$(S_x - iS_y)|-\rangle = 0$$

$$(S_x + iS_y)|-\rangle = \hbar|+\rangle$$

$$(S_x - iS_y)|+\rangle = \hbar|-\rangle \tag{18}$$

Combining these equations, we readily find

$$S_x|+\rangle = \tfrac{1}{2}\hbar|-\rangle$$
$$S_x|-\rangle = \tfrac{1}{2}\hbar|+\rangle \tag{19}$$

Exercise 1. Obtain these results.

From Eq. (19), we find

$$S_x = \begin{pmatrix} \langle+|S_x|+\rangle & \langle+|S_x|-\rangle \\ \langle-|S_x|+\rangle & \langle-|S_x|-\rangle \end{pmatrix} = \begin{pmatrix} 0 & \tfrac{1}{2}\hbar \\ \tfrac{1}{2}\hbar & 0 \end{pmatrix}$$

$$= \tfrac{1}{2}\hbar \begin{pmatrix} 0 & 1 \\ 1 & 0 \end{pmatrix} \tag{20}$$

By a similar argument, we find that the 2×2 matrix representing S_y is

$$S_y = \tfrac{1}{2}\hbar \begin{pmatrix} 0 & -i \\ i & 0 \end{pmatrix} \tag{21}$$

Exercise 2. Show this.

The matrices appearing in these equations are called the *Pauli spin matrices*, designated by σ_x, σ_y, and σ_z:

$$\sigma_x = \begin{pmatrix} 0 & 1 \\ 1 & 0 \end{pmatrix} \quad \sigma_y = \begin{pmatrix} 0 & -i \\ i & 0 \end{pmatrix} \quad \sigma_z = \begin{pmatrix} 1 & 0 \\ 0 & -1 \end{pmatrix} \tag{22}$$

In terms of these matrices, we can write the spin operator as

$$\mathbf{S} = \tfrac{1}{2}\hbar\boldsymbol{\sigma} \tag{23}$$

where the symbol $\boldsymbol{\sigma}$ stands for the vector operator with components σ_x, σ_y, and σ_z. Note that the Pauli matrices are hermitian, which implies that $\mathbf{S}$ is hermitian, as it should be. The matrices σ_x, σ_y, σ_z in conjunction with the unit matrix

$$\mathbf{1} = \begin{pmatrix} 1 & 0 \\ 0 & 1 \end{pmatrix} \tag{24}$$

form a complete set of 2×2 matrices, that is, any 2×2 matrix can be written as a superposition of these four matrices.

Exercise 3. Show this.

The Pauli matrices satisfy several algebraic relations. By comparison with the angular-momentum commutation relations or by direct multiplication, we can check that the commutation relations of the Pauli matrices are

$$[\sigma_x, \sigma_y] = 2i\sigma_z$$

$$[\sigma_y, \sigma_z] = 2i\sigma_x \qquad (25)$$

$$[\sigma_z, \sigma_x] = 2i\sigma_y$$

Let us define the *anticommutator* of two operators A and B as

$$[A, B]_+ \equiv AB + BA \qquad (26)$$

By direct multiplication of the Pauli matrices we can then verify that

$$[\sigma_x, \sigma_x]_+ = [\sigma_y, \sigma_y]_+ = [\sigma_z, \sigma_z]_+ = 2 \cdot \mathbf{1}$$

$$[\sigma_x, \sigma_y]_+ = [\sigma_x, \sigma_z]_+ = [\sigma_y, \sigma_z]_+ = 0$$

Instead of σ_x, σ_y, and σ_z we sometimes will write σ_1, σ_2, and σ_3. With this notation, the anticommutation relations of the Pauli matrices take the concise form

$$[\sigma_r, \sigma_s]_+ = 2\delta_{rs}\mathbf{1} \qquad r, s = 1, 2, 3 \qquad (27)$$

Accordingly, the square of each of the Pauli matrices equals the unit matrix:

$$\sigma_r^2 = \mathbf{1} \qquad (28)$$

and consequently,

$$\sigma_1^2 + \sigma_2^2 + \sigma_3^2 = 3 \cdot \mathbf{1} \qquad (29)$$

If $\mathbf{A}$ is any vector object with three components A_x, A_y, and A_z, we can formally define the dot product of $\mathbf{A}$ and $\boldsymbol{\sigma}$ in the usual way

$$\mathbf{A} \cdot \boldsymbol{\sigma} = A_x\sigma_x + A_y\sigma_y + A_z\sigma_z \qquad (30)$$

In particular,

$$\boldsymbol{\sigma} \cdot \boldsymbol{\sigma} = \sigma_x\sigma_x + \sigma_y\sigma_y + \sigma_z\sigma_z \qquad (31)$$

Thus, we can write Eq. (29) as

$$\boldsymbol{\sigma} \cdot \boldsymbol{\sigma} = 3 \cdot \mathbf{1} \qquad (32)$$

In the our discussion of spin states we have focused on the eigenstates $|+\rangle$ and $|-\rangle$ of S_z. According to the general principles governing measurements of observables in quantum mechanics, the electron will be thrown into one of these spin states if we perform a measurement of the z component of the spin, for instance, with a Stern–Gerlach experiment. However, if we perform a measurement of, say, the x component of the spin, then the electron will be thrown into an eigenstate of S_x. Obviously, the eigenvalues of S_x are $\pm\frac{1}{2}\hbar$, the same as the eigenvalues of S_z. To find the eigenvector of S_x of eigenvalue $+\frac{1}{2}\hbar$, we must solve the equation

$$S_x \begin{pmatrix} a \\ b \end{pmatrix} = \tfrac{1}{2}\hbar \begin{pmatrix} a \\ b \end{pmatrix}$$

With the matrix (20) for S_x, this becomes

$$\tfrac{1}{2}\hbar \begin{pmatrix} b \\ a \end{pmatrix} = \tfrac{1}{2}\hbar \begin{pmatrix} a \\ b \end{pmatrix}$$

The solution of this equation requires $a = b$. The normalized eigenvector is then

$$\frac{1}{\sqrt{2}} \begin{pmatrix} 1 \\ 1 \end{pmatrix} \tag{33}$$

Likewise, we can show that the eigenvector of S_x of eigenvalue $-\frac{1}{2}\hbar$ is

$$\frac{1}{\sqrt{2}} \begin{pmatrix} 1 \\ -1 \end{pmatrix} \tag{34}$$

We will designate the spinors appearing in Eqs. (33) and (34) by χ_+' and χ_-', respectively:

$$\chi_+' = \frac{1}{\sqrt{2}} \begin{pmatrix} 1 \\ 1 \end{pmatrix} \quad \text{and} \quad \chi_-' = \frac{1}{\sqrt{2}} \begin{pmatrix} 1 \\ -1 \end{pmatrix} \tag{35}$$

Of course, these spinors can still be multiplied by an overall orbital function $\psi(\mathbf{r})$.

To gain some insight on the physical relationships between the spinors χ_+, χ_-, χ_+', and χ_-', let us suppose that we initially have an electron with spin up, that is, an electron with a spinor

$$\chi_+ = \begin{pmatrix} 1 \\ 0 \end{pmatrix} \tag{36}$$

What is the probability that a measurement of the x component of the spin will yield the value $\frac{1}{2}\hbar$? According to our general rules for computing probabilities, the answer is the square of the scalar product of the spinors χ_+' and χ_+:

$$|\chi_+'^* \cdot \chi_+|^2 = \left| \frac{1}{\sqrt{2}} \times 1 + \frac{1}{\sqrt{2}} \times 0 \right|^2 = \frac{1}{2} \tag{37}$$

Likewise, the probability that the measurement of the x component of the spin will yield the value $-\frac{1}{2}\hbar$ is the square of the scalar product of the spinors χ_-' and χ_+:

$$|\chi_-'^* \cdot \chi_+|^2 = \left| \frac{1}{\sqrt{2}} \times 1 - \frac{1}{\sqrt{2}} \times 0 \right|^2 = \frac{1}{2} \tag{38}$$

Thus, the probabilities for the two eigenvalues $\pm\frac{1}{2}\hbar$ of the x component of the spin are equal. This can be also made obvious by writing the spinor (36) as follows:

$$\begin{pmatrix} 1 \\ 0 \end{pmatrix} = \frac{1}{\sqrt{2}} \left[\frac{1}{\sqrt{2}} \begin{pmatrix} 1 \\ 1 \end{pmatrix} + \frac{1}{\sqrt{2}} \begin{pmatrix} 1 \\ -1 \end{pmatrix} \right]$$

which expresses χ_+ as a superposition of χ_+' and χ_-' with equal amplitudes. Such a superposition of spin states provides an exceptionally simple example of how a system that is in an eigenstate of one observable can, at the same time, be in a superposition of several eigenstates of another observable.

Example 4. Show that χ_+ and χ_- form a complete set of two-component vectors and show that χ_+' and χ_-' also form a complete set.

9.2 *The Magnetic Moment*

Besides spin, or intrinsic angular momentum, the electron also has an intrinsic magnetic moment. The operator associated with the magnetic moment is

$$\boldsymbol{\mu} = -\frac{e}{m}\,\mathbf{S} \tag{39}$$

This relation between the spin and the magnetic moment can be deduced for relativistic quantum mechanics.[2] Note that classical mechanics tells us that for a rigidly rotating body with a charge distribution proportional to the mass distribution, the magnetic moment is related to the angular momentum of rotation by $\boldsymbol{\mu} = -(e/2m)\mathbf{S}$; this differs by a factor of two from the relation (39) for the electron.

The eigenstates of $\boldsymbol{\mu}^2$ and of μ_z are simply those of $\mathbf{S}^2$ and S_z:

$$\boldsymbol{\mu}^2|\pm\rangle = \frac{3}{4}\frac{e^2\hbar^2}{m^2}|\pm\rangle \tag{40}$$

$$\mu_z|\pm\rangle = \mp\frac{e\hbar}{2m}|\pm\rangle \tag{41}$$

The magnitude of the z component of the magnetic moment of the electron is called a *Bohr magneton*, $\mu_B = e\hbar/2m = 9.27 \times 10^{-24}$ J/T.

If the electron is placed in a uniform external magnetic field, the interaction of the magnetic moment with this magnetic field adds an extra term $-\boldsymbol{\mu}\cdot\mathbf{B}$ to the Hamiltonian:

$$H = H_{orb} - \boldsymbol{\mu}\cdot\mathbf{B}$$

where the orbital part H_{orb} of the Hamiltonian stands for the usual kinetic plus potential energy which depends only on the orbital coordinates and momentum (however, in a magnetic field, the orbital Hamiltonian must also include the effect of the magnetic field on the orbital motion).

Suppose that the electron is in some given orbital state. We assume it to be an energy eigenstate, so $H_{orb}\psi(\mathbf{r}, t) = E_{orb}\psi(\mathbf{r}, t)$. Then the complete spinor wavefunction is

$$\psi(\mathbf{r}, t)\begin{pmatrix}a(t)\\b(t)\end{pmatrix} \tag{42}$$

and it satisfies the Schrödinger equation

$$(H_{orb} - \boldsymbol{\mu}\cdot\mathbf{B})\left[\psi(\mathbf{r}, t)\begin{pmatrix}a\\b\end{pmatrix}\right] = i\hbar\frac{d}{dt}\left[\psi(\mathbf{r}, t)\begin{pmatrix}a\\b\end{pmatrix}\right] \tag{43}$$

---

[2] Actually, the factor appearing in Eq. (39) is not exactly e/m, but $e/m \times 1.0011$ . . . ; the extra factor is a "radiative correction," arising from the interaction of the electron with itself and with the quanta of the electromagnetic field.

or

$$H_{\text{orb}}\psi(\mathbf{r},\ t)\begin{pmatrix}a\\b\end{pmatrix} - \psi(\mathbf{r},\ t)\boldsymbol{\mu}\cdot\mathbf{B}\begin{pmatrix}a\\b\end{pmatrix} = i\hbar\left(\frac{d}{dt}\ \psi(\mathbf{r},\ t)\right)\begin{pmatrix}a\\b\end{pmatrix}$$

$$+ i\hbar\psi(\mathbf{r},\ t)\frac{d}{dt}\begin{pmatrix}a\\b\end{pmatrix} \quad (44)$$

If we use the orbital Schrödinger equation

$$H_{\text{orb}}\psi(\mathbf{r},\ t) = i\hbar\ \frac{d}{dt}\ \psi(\mathbf{r},\ t)$$

we obtain

$$-\boldsymbol{\mu}\cdot\mathbf{B}\begin{pmatrix}a\\b\end{pmatrix} = i\hbar\ \frac{d}{dt}\begin{pmatrix}a\\b\end{pmatrix}$$

that is,

$$\frac{e}{m}\ \mathbf{B}\cdot\mathbf{S}\begin{pmatrix}a\\b\end{pmatrix} = i\hbar\ \frac{d}{dt}\begin{pmatrix}a\\b\end{pmatrix} \quad (45)$$

This shows that the orbital and the spin motions are independent. This independence hinges on the assumption that the only interaction of the spin is with the *external* magnetic field.[3]

We will now solve Eq. (45). Suppose that the magnetic field is along the z axis. Then

$$\mathbf{B}\cdot\mathbf{S} = BS_z = \tfrac{1}{2}B\hbar\sigma_z = \tfrac{1}{2}B\hbar\begin{pmatrix}1 & 0\\0 & -1\end{pmatrix} \quad (46)$$

and Eq. (45) becomes

$$\frac{eB\hbar}{2m}\begin{pmatrix}1 & 0\\0 & -1\end{pmatrix}\begin{pmatrix}a\\b\end{pmatrix} = i\hbar\begin{pmatrix}da/dt\\db/dt\end{pmatrix} \quad (47)$$

that is,

$$\frac{eB}{2m}\begin{pmatrix}a\\-b\end{pmatrix} = i\begin{pmatrix}da/dt\\db/dt\end{pmatrix} \quad (48)$$

---

[3] Our calculation ignores spin-orbit coupling (see Section 9.3).

Obviously, this spinor equation is equivalent to a pair of separate equations:

$$\frac{eB}{2m} a = i \frac{da}{dt}$$

and

$$-\frac{eB}{2m} b = i \frac{db}{dt}$$

One solution of this pair of equations is

$$a(t) = e^{-i(eB/2m)t} \qquad b(t) = 0 \tag{49}$$

Another solution is

$$a(t) = 0 \qquad b(t) = e^{+i(eB/2m)t} \tag{50}$$

If the orbital wavefunction has a time dependence $\psi(\mathbf{r}, t) = \psi(\mathbf{r})e^{-iE_{\text{orb}}t/\hbar}$, then the complete spinor wavefunctions, which include both the orbital part and the spinor part, are

$$\psi(\mathbf{r})e^{-i(E_{\text{orb}}+eB\hbar/2m)t/\hbar} \begin{pmatrix} 1 \\ 0 \end{pmatrix} \tag{51}$$

and

$$\psi(\mathbf{r})e^{-i(E_{\text{orb}}-eB\hbar/2m)t/\hbar} \begin{pmatrix} 0 \\ 1 \end{pmatrix} \tag{52}$$

These two solutions describe stationary states. The solutions show that if the electron is initially in the state of spin up (or down), it will remain in this state. The energy for the spin up (or down) state is $E_{\text{orb}} + eB\hbar/2m$ (or $E_{\text{orb}} - eB\hbar/2m$).

If the electron is initially in a superposition of the two eigenstates (51) and (52) with, say, equal amplitudes, then the spinor wavefunction has a time dependence

$$\psi(\mathbf{r}, t) \begin{pmatrix} a(t) \\ b(t) \end{pmatrix} = \frac{1}{\sqrt{2}} \psi(\mathbf{r})e^{-iEt/\hbar} \begin{pmatrix} e^{-i(eB/2m)t} \\ e^{i(eB/2m)t} \end{pmatrix} \tag{53}$$

Such a spinor does not correspond to a stationary state, and the expectation values of S_x and of S_y are time dependent:

$$\langle S_x \rangle = \frac{1}{2} \hbar \cos \frac{eB}{m} t \tag{54}$$

$$\langle S_y \rangle = \frac{1}{2} \hbar \, \sin \frac{eB}{m} \, t \qquad (55)$$

These equations indicate that, on the average, the spin precesses around the z axis with an angular frequency eB/m. This frequency coincides with the classical precession frequency expected for a spinning rigid body of angular momentum $\frac{1}{2}\hbar$ and magnetic moment $e\hbar/2m$.

Exercise 5. Obtain Eqs. (54) and (55).

Exercise 6. Show that for the state (53), the uncertainty of the energy is

$$\Delta E = \frac{eB\hbar}{2m}$$

and that the time required for $\langle S_x \rangle$ or $\langle S_y \rangle$ to change appreciably is

$$\Delta t \simeq \frac{m}{eB} \frac{\pi}{2}$$

which is consistent with the uncertainty relation $\Delta E \, \Delta t \simeq \hbar$.

9.3   *Addition of Angular Momenta*

The total angular momentum of a physical system is often the sum of several angular momenta of individual parts of the system. For example, the total angular momentum of an atom is the sum of the orbital and the spin angular momenta of all the electrons in the atom. The total angular momentum is, of course, quantized, and the eigenstates of the total angular momentum are combinations of the eigenstates of the individual angular momenta. However, the rules for constructing the former eigenstates from the latter eigenstates are fairly complicated. We can understand the reasons for these complications by recalling that according to the vector model (see Section 7.2), each angular-momentum eigenstate is described graphically by a vector whose direction is spread out over a cone. The combination of the eigenstates of the individual angular momenta therefore involves the combination of several such vectors of uncertain directions, and the combination must be finely tuned so the resultant is, again, a vector spread out over a cone.

In this section we will consider only the addition of two angular momenta: the orbital angular momentum of an electron and its spin angular momentum. The procedure we will develop for the addition of two angular momenta can be readily generalized to the addition of more than two angular momenta.

The total angular-momentum operator for the electron is the sum of the orbital and the spin angular-momentum operators:

$$\mathbf{J} = \mathbf{L} + \mathbf{S} \tag{56}$$

The orbital and spin operators commute with each other, and the commutation relations for the total angular momentum are therefore completely determined by the commutation relations for the orbital and spin angular momenta. For instance,

$$
\begin{aligned}
[J_x, J_y] &= [(L_x + S_x), (L_y + S_y)] \\
&= [L_x, L_y] + [L_x, S_y] + [S_x, L_y] + [S_x, S_y] \\
&= i\hbar L_z + 0 + 0 + i\hbar S_z \\
&= i\hbar J_z
\end{aligned}
$$

Likewise,

$$[J_y, J_z] = i\hbar J_x$$

$$[J_z, J_x] = i\hbar J_y$$

Thus, the commutation relations for $\mathbf{J}$ are of the same form as those for $\mathbf{L}$ or $\mathbf{S}$.

We recall from Chapter 7 that the quantization rules for the angular momentum are a direct consequence of the commutation relations. The quantization rules for the total angular momentum are therefore the same as those for the orbital angular momentum or the spin. The eigenvalues of $\mathbf{J}^2$ are $j(j + 1)\hbar^2$ and the eigenvalues of J_z are $m_j\hbar$:

$$\mathbf{J}^2|j,m_j\rangle = j(j + 1)\hbar^2|j,m_j\rangle \tag{57}$$

$$J_z|j,m_j\rangle = m_j|j,m_j\rangle \tag{58}$$

where the possible values of j and m_j are

$$j = 0, \tfrac{1}{2}, 1, \tfrac{3}{2}, 2, \ldots \tag{59}$$

$$m_j = -j, -j + 1, \ldots, j - 1, j \tag{60}$$

The crucial question in the addition of angular momenta is then: If we start with the eigenstates $|l,m_l\rangle$ and $|s,m_s\rangle$ of some given values of l and s, how do we combine these to form the eigenstates $|j,m_j\rangle$, and what values of j will emerge? To answer this question, let us consider a concrete example involving an electron with orbital quantum number $l = 1$. The individual eigenstates of orbital angular momentum and of spin are then

$$|1,+1\rangle, \quad |1,0\rangle, \quad |1,-1\rangle \tag{61}$$

and

$$|+\rangle, \quad |-\rangle \tag{62}$$

where, as in Section 9.1, we use the abbreviated notation $|\pm\rangle$ for the spin eigenstates $|\tfrac{1}{2},\pm\tfrac{1}{2}\rangle$. The possible spin-and-orbital states are then constructed by taking all possible products of the states listed in (61) and (62). Altogether, there are six such possible products:[4]

$$|1,-1\rangle|-\rangle \qquad\qquad\qquad (m_j = -\tfrac{3}{2}) \tag{63}$$

$$|1,-1\rangle|+\rangle, \quad |1,0\rangle|-\rangle \qquad (m_j = -\tfrac{1}{2}) \tag{64}$$

$$|1,+1\rangle|-\rangle, \quad |1,0\rangle|+\rangle \qquad (m_j = +\tfrac{1}{2}) \tag{65}$$

$$|1,+1\rangle|+\rangle \qquad\qquad\qquad (m_j = +\tfrac{3}{2}) \tag{66}$$

Here, the products have been grouped on separate lines according to the value of $m_l + m_s$; the values of $m_l + m_s$ for the first, second, third, and fourth lines are $-\tfrac{3}{2}$, $-\tfrac{1}{2}$, $+\tfrac{1}{2}$, and $+\tfrac{3}{2}$, respectively. Since $J_z = L_z + S_z$, an eigenstate of L_z and of S_z is also eigenstate of J_z, and the eigenvalue m_j is the sum of the eigenvalues m_l and m_s,

$$m_j = m_l + m_s \tag{67}$$

Thus, the values of m_j for the first, second, third, and fourth lines are also $-\tfrac{3}{2}$, $-\tfrac{1}{2}$, $+\tfrac{1}{2}$, and $+\tfrac{3}{2}$, respectively. These values have been listed on the far right on each of the lines (63)–(66).

The largest and the smallest values of m_j occur for the states $|1,+1\rangle|+\rangle$ and $|1,-1\rangle|-\rangle$, which have $m_j = \pm\tfrac{3}{2}$, respectively. These states must have $j = \tfrac{3}{2}$, since a lower value of j is incompatible with $m_j = \tfrac{3}{2}$, and a higher j value is incompatible with the absence of higher values of m_j in our list of products (63)–(66).

---

[4] We could equally well use the position representation and write (63)–(66) as

$$Y_1^{-1}(\theta, \phi)\chi_-, \; Y_1^{-1}(\theta, \phi)\chi_+, \; Y_1^{0}(\theta, \phi)\chi_-, \ldots.$$

But since we will not make any use of the dependence of the wavefunctions on θ and ϕ, we prefer the ket notation.

Thus, the states (63) and (66) are eigenstates of the total angular momentum:

$$|j=\tfrac{3}{2}, m_j=+\tfrac{3}{2}\rangle = |1,+1\rangle|+\rangle \tag{68}$$

and

$$|j=\tfrac{3}{2}, m_j=-\tfrac{3}{2}\rangle = |1,-1\rangle|-\rangle \tag{69}$$

In conjunction with these eigenstates of maximum or minimum m_j, there must be eigenstates of the same value of j, but intermediate values of m_j, that is, $m_j = \tfrac{1}{2}$ and $m_j = -\tfrac{1}{2}$. Obviously, the eigenstate $|j=\tfrac{3}{2}, m_j=+\tfrac{1}{2}\rangle$ must be some superposition of the two states listed in (65), and the eigenstate $|j = \tfrac{3}{2}, m_j = -\tfrac{1}{2}\rangle$ must be some superposition of the two states listed in (64). But since (65) includes two independent states, we can form two independent superpositions, and only one of these can belong to $j = \tfrac{3}{2}$. The other must belong to a different value of j. Since $m_j = \tfrac{1}{2}$ for the states (65), the only other possible value of j is $\tfrac{1}{2}$. Likewise, $m_j = -\tfrac{1}{2}$ for the states (64), and one of the two independent superpositions we can form with these states belongs to $j = \tfrac{3}{2}$, and the other belongs to $j = \tfrac{1}{2}$.

This exhausts all the possible states we have available in our list (63)–(66), and it therefore exhausts the possible eigenstates $|j, m_j\rangle$ we can construct. We see that the six possible products listed in (63)–(66) give rise to six eigenstates of the total angular momentum; four of these have $j = \tfrac{3}{2}$ and two of these have $j = \tfrac{1}{2}$. In the vector model, these two sets of states are said to correspond to "parallel" spin and orbital angular momentum and "antiparallel" spin and orbital angular momentum (see Fig. 9.1).

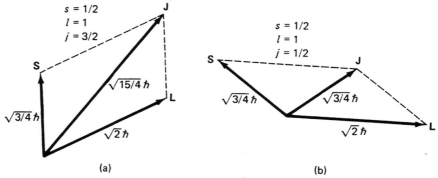

(a)   (b)

Fig. 9.1 Graphical representation of the addition of spin and orbital angular momenta ($s = \tfrac{1}{2}$ and $l = 1$) according to the vector model. (a) If the spin and angular momentum vectors are as nearly parallel as possible, the result is a state with $j = \tfrac{3}{2}$. (b) If the spin and angular momentum vectors are as nearly antiparallel as possible, the result is a state with $j = \tfrac{1}{2}$.

Note that in this example of the addition of the spin and orbital angular momentum of an electron, the largest possible value of j is $j = l + s$, and the smallest possible value is $j = l - s$. This illustrates a general rule for the addition of two arbitrary angular momenta: When we add two angular momenta of quantum numbers j_1 and j_2, the maximum possible quantum number of the resulting total angular momentum is $j_1 + j_2$, the minimum possible quantum number is $|j_1 - j_2|$, and all the other possible intermediate quantum numbers differ from these by integers. For instance, if $j_1 = 1$ and $j_2 = 3$, then the possible values of the total angular-momentum quantum number are 4, 3, and 2. This rule is applicable to any two (commuting) angular momenta, for example, the orbital angular momentum and the spin of one particle, or the two spin angular momenta of two particles, or the two orbital angular momenta of two particles, etc. The proof of this rule is a simple generalization of the arguments given above.

To finish our example of the addition of the spin and orbital angular momenta of the electron, let us explicitly construct the superpositions involved in the eigenstates with $m_j = \pm\frac{1}{2}$. This construction can be conveniently performed by means of the raising operators for angular momentum. The raising operator for the total angular momentum is

$$J_+ = J_x + iJ_y \tag{70}$$

which can also be written as

$$J_+ = L_+ + S_+ \tag{71}$$

To construct $|j=\frac{3}{2}, m_j=-\frac{1}{2}\rangle$, we begin with the eigenstate (69) and apply the raising operator J_+:

$$J_+|j=\frac{3}{2}, m_j=-\frac{3}{2}\rangle = (L_+ + S_+)|1,-1\rangle|-\rangle$$
$$= (L_+|1,-1\rangle)\,|-\rangle + |1,-1\rangle\,(S_+|-\rangle) \tag{72}$$

But according to Eq. (7.51), J_+ acting on an eigenstate $|j,m_j\rangle$ yields

$$J_+|j,m_j\rangle = \hbar\sqrt{j(j+1) - m_j(m_j+1)}\,|j,m_j+1\rangle \tag{73}$$

Inserting this and similar expressions for $L_+|1,-1\rangle$ and $S_+|-\rangle$ into Eq. (72), we obtain, with appropriate choices for the numerical coefficients,

$$\hbar\sqrt{\tfrac{15}{4} - \tfrac{3}{4}}\,|j = \tfrac{3}{2}, m_j=-\tfrac{1}{2}\rangle$$
$$= (\hbar\sqrt{2 - 0}\,|1,0\rangle)\,|-\rangle + |1,-1\rangle\,(\hbar\sqrt{\tfrac{3}{4} + \tfrac{1}{4}}\,|+\rangle) \tag{74}$$

that is,

$$|j = \tfrac{3}{2}, m_j = -\tfrac{1}{2}\rangle = \frac{1}{\sqrt{3}} \left(\sqrt{2}\,|1,0\rangle|-\rangle + |1,-1\rangle|+\rangle \right) \tag{75}$$

By applying the raising operator J_+ once more, we obtain

$$|j = \tfrac{3}{2}, m_j = +\tfrac{1}{2}\rangle = \frac{1}{\sqrt{3}} \left(|1,+1\rangle|-\rangle + \sqrt{2}\,|1,0\rangle|+\rangle \right) \tag{76}$$

Exercise 7. Obtain the result (76) by applying the raising operator J_+ to (75); alternatively, obtain this result by applying the lowering operator J_- to $|j=\tfrac{3}{2}, m_j=+\tfrac{3}{2}\rangle$.

As expected from our arguments above, the eigenstate (75) is a superposition of the two states (64). To construct the "other" independent superposition, which belongs to $j = \tfrac{1}{2}$, we must find a state vector orthogonal to (75). Obviously, the vector orthogonal to (75) is, with a choice of phase,

$$\frac{1}{\sqrt{3}} \left(|1,0\rangle|-\rangle - \sqrt{2}\,|1,-1\rangle|+\rangle \right) \tag{77}$$

and thus

$$|j = \tfrac{1}{2}, m_j = -\tfrac{1}{2}\rangle = \frac{1}{\sqrt{3}} \left(|1,0\rangle|-\rangle - \sqrt{2}\,|1,-1\rangle|+\rangle \right) \tag{78}$$

The other eigenvector belonging to $j = \tfrac{1}{2}$ can be constructed by applying the raising operator J_+ to (78):

$$|j = \tfrac{1}{2}, m_j = +\tfrac{1}{2}\rangle = \frac{1}{\sqrt{3}} \left(\sqrt{2}\,|1,+1\rangle|-\rangle - |1,0\rangle|+\rangle \right) \tag{79}$$

Although we have here performed the construction of the eigenvectors of the total angular momentum only for the special case $l = 1$, $s = \tfrac{1}{2}$, a similar construction can be performed for any other two values of the individual angular-momentum quantum numbers l and s or, more generally, j_1 and j_2. The coefficients appearing in superpositions such as (75), (76), (78), and (79) are called *Clebsch–Gordan coefficients*. Tables of values of these coefficients, for different pairs of values of j_1 and j_2, are available in handbooks (see Appendix 2).

The construction of eigenstates of total angular momentum we have performed in this section does not hinge on any physical coupling, or interaction, of the spin and the orbital angular momentum. The addition of these angular momenta can be viewed as a purely geometrical exercise, which involves only the mathematical properties of the angular-momentum eigenstates. For an electron without interactions, the eigenstates of the individual spin and orbital angular momenta [such as the eigenstates (63)–(66)] and the eigenstates of the total angular momentum are both physically attainable. Which of these two kinds of states is realized for the electron depends on what measurement we perform. The observables L^2, S^2, L_z, and S_z are one set of commuting observables, and L^2, S^2, J^2, and J_z are another set (note that not all the observables in the second set commute with those in the first). If we measure the orbital angular momentum and the spin separately, the electron will be thrown into a state such as listed in Eqs. (63)–(66); if we measure the total angular momentum, then the electron will be thrown into a state such as given in Eq. (75). For an electron without interactions, both kinds of angular-momentum eigenstates are simultaneous eigenstates of the energy, and both are equally suitable for the specification of the state of the electron.

However, for an electron with interactions, such as the electron in a hydrogen atom, one kind of angular-momentum eigenstate is usually more suitable than the other. The spin and the orbital angular momentum of an electron in an atom interact by an effective spin-orbit coupling, which contributes a term proportional to $L \cdot S$ to the Hamiltonian. This coupling term arises indirectly, in part from the interaction of the moving spin with the electric field, and in part from a relativistic precession (Thomas precession) of the spin brought about by the accelerated motion of the electron. Because of the spin-orbit coupling term, both the orbital angular momentum and the spin fail to commute with the Hamiltonian, and therefore these individual angular momenta are not conserved. Hence we cannot use the quantum numbers m_l and m_s to specify the stationary states of the electron. However, the total angular momentum still commutes with the Hamiltonian, and hence we can use the quantum numbers j and m_j. This means that the simultaneous eigenstates of energy and angular momentum are states of the kind given in Eq. (75).

Exercise 8. Suppose that the Hamiltonian for the hydrogen atom contains an extra term of the form $f(r)L \cdot S$. Show that this term commutes with the total angular momentum J, but not with the individual angular momenta L and S.

Although for most problems in atomic physics, the eigenstates of total angular momentum prove to be the most suitable set of states, under exceptional circumstances the eigenstates of the individual spin and orbital angular momenta prove to be better. Such exceptional circumstances occur if the atom is immersed in a very strong external magnetic field (strong-field Zeeman effect). Then the spin-orbit coupling becomes negligible compared with the individual interactions of the spin and orbital magnetic moments with the external magnetic field, and the quantum numbers m_l and m_s provide a better classification scheme for the stationary states than the quantum numbers j and m_j.

9.4   *Fermions and Bosons; The Exclusion Principle*

All electrons are identical—they all have exactly the same mass, electric charge, spin, magnetic moment, and every other physical attribute. In classical physics, such an identity among particles does not imply that the particles are indistinguishable. In classical physics, we can distinguish between two particles by continually observing them as they move along their separate trajectories. Even if their paths were to cross, and the two particles were to collide, we could still tell which is which because classical physics allows us to predict the trajectories after the collision. However, in quantum physics the situation is fundamentally different: the identity of electrons implies their indistinguishability because the quantum uncertainties prevent us from "tracking" electrons. Whenever two electrons collide or approach closely, so their wavefunctions overlap, we cannot be sure which is which after the collision.

The indistinguishability of electrons has drastic consequences for the construction of the wavefunction for a system of several electrons. Here we will deal with the case of the wavefunction of two electrons; our results can be generalized to more than two electrons. The spinor wavefunction for two electrons depends on two orbital variables $(\mathbf{r}_1, \mathbf{r}_2)$ and on two spin variables (m_{s1}, m_{s2}). We will write the general wavefunction for two electrons as[5]

$$\psi(\mathbf{r}_1, \mathbf{r}_2; m_{s1}, m_{s2}) \qquad (80)$$

Here the spin quantum numbers m_{s1} and m_{s2} take their usual values $\pm\frac{1}{2}$. However, we prefer not to specify the numerical values of

---

[5] For the sake of simplicity, we omit the time dependence of the wavefunction.

m_{s1} and m_{s2} in (80), and we prefer to regard ψ as a function of m_{s1} and m_{s2}. The advantage of this is that the general notation (80) can then serve equally well for a state of definite values of the spins and for a state containing a superposition of different values of the spins, such as the superposition used in Section 9.3 in the construction of eigenstates of the total angular momentum. In contrast, our previous notation with subscripts $\pm$ for the spin [see, for example, Eqs. (10) and (11)] would require us to write such a superposition as an explicit sum of several terms.

The probability for finding the first electron in the volume dV_1 with spin m_{s1} and the second in the volume dV_2 with spin m_{s2} is

$$|\psi(\mathbf{r}_1, \mathbf{r}_2; m_{s1}, m_{s2})|^2 \, dV_1 dV_2 \tag{81}$$

and the probability for finding the first electron in the volume dV_2 with spin m_{s2} and the second in the volume dV_1 with spin m_{s1} is

$$|\psi(\mathbf{r}_2, \mathbf{r}_1; m_{s2}, m_{s1})|^2 \, dV_1 dV_2 \tag{82}$$

Since the electrons are indistinguishable, it makes no difference which electron we regard as "first" and which as "second." Thus, the two expressions (81) and (82) must be equal—each merely gives the probability of finding *some* electron in the volume dV_1 with spin m_{s1} and *some* electron in dV_2 with spin m_{s2}. From the equality

$$|\psi(\mathbf{r}_1, \mathbf{r}_2; m_{s1}, m_{s2})|^2 = |\psi(\mathbf{r}_2, \mathbf{r}_1; m_{s2}, m_{s1})|^2 \tag{83}$$

we deduce that the two wavefunctions differ at most by a phase factor:

$$\psi(\mathbf{r}_1, \mathbf{r}_2; m_{s1}, m_{s2}) = e^{i\delta}\psi(\mathbf{r}_2, \mathbf{r}_1; m_{s2}, m_{s1}) \tag{84}$$

The phase δ must be a universal constant, that is, it must be independent of the details of the wavefunction. We recognize the necessity for this universality if we contemplate the superposition of two wavefunctions, say, $\psi(\mathbf{r}_1, \mathbf{r}_2; m_{s1}, m_{s2}) + \xi(\mathbf{r}_1, \mathbf{r}_2; m_{s1}, m_{s2})$. Unless both ψ and ξ acquire the same phase factor upon exchange of $(\mathbf{r}_1, m_{s1})$ and $(\mathbf{r}_2, m_{s2})$, this superposition would change by *more* than a phase factor, in violation of Eq. (84). We can now evaluate the phase factor by a trick: exchange $(\mathbf{r}_1, m_{s1})$ and $(\mathbf{r}_2, m_{s2})$ once more on the right side of Eq. (84); this yields once more the same phase factor,

$$\psi(\mathbf{r}_1, \mathbf{r}_2; m_{s1}, m_{s2}) = e^{i\delta}\psi(\mathbf{r}_2, \mathbf{r}_1; m_{s2}, m_{s1})$$

$$= e^{2i\delta}\psi(\mathbf{r}_1, \mathbf{r}_2; m_{s1}, m_{s2}) \tag{85}$$

From this we see that $e^{2i\delta} = 1$ and that

$$e^{i\delta} = \pm 1 \tag{86}$$

Thus, the wavefunctions of indistinguishable particles are either symmetric functions under exchange of their variables,

$$\psi(\mathbf{r}_1, \mathbf{r}_2; m_{s1}, m_{s2}) = +\psi(\mathbf{r}_2, \mathbf{r}_1; m_{s2}, m_{s1}) \tag{87}$$

or else antisymmetric functions,

$$\psi(\mathbf{r}_1, \mathbf{r}_2; m_{s1}, m_{s2}) = -\psi(\mathbf{r}_2, \mathbf{r}_1; m_{s2}, m_{s1}) \tag{88}$$

For the general mathematical treatment of indistinguishable particles, it is useful to define an exchange operator P_{ex} that exchanges the particles; that is, the operator P_{ex} exchanges the variables $(\mathbf{r}_1, m_{s1})$ and $(\mathbf{r}_2, m_{s2})$ in the wavefunction:

$$P_{ex}\psi(\mathbf{r}_1, \mathbf{r}_2; m_{s1}, m_{s2}) = \psi(\mathbf{r}_2, \mathbf{r}_1; m_{s2}, m_{s1}) \tag{89}$$

Since two exchanges in succession produce no change, we have the identity

$$P_{ex}{}^2 = 1 \tag{90}$$

Obviously, this identity implies that the eigenvalues of P_{ex} are $+1$ and -1. According to Eqs. (87) and (88), the symmetric and the antisymmetric wavefunctions are eigenfunctions of P_{ex} with eigenvalues $+1$ and -1, respectively.

If two particles are indistinguishable, their energy must remain unchanged when we exchange the particles. This implies that the exchange operator P_{ex} has no effect on the Hamiltonian of the two-particle system; that is, P_{ex} commutes with the Hamiltonian:

$$[H, P_{ex}] = 0 \tag{91}$$

As we know from Chapter 5, the vanishing of the commutator of H and any operator implies that the expectation value of the operator is constant in time:

$$\langle\psi|P_{ex}|\psi\rangle = [\text{constant}] \tag{92}$$

This means that if a two-particle state is initially symmetric or antisymmetric, it remains that way forever. Thus, symmetry or antisymmetry under exchange is a permanent property of the two-particle state, and it can be regarded as an attribute of the particles themselves. Particles whose wavefunctions are symmetric are called *bosons;* particles whose wavefunctions are antisymmetric are called *fermions.*

Empirical observations establish that whether a particle is a boson or a fermion is directly correlated with its spin. Electrons and all other particles of spin $\frac{1}{2}$, or $\frac{3}{2}$, or $\frac{5}{2}$, . . . are fermions, whereas photons and all particles of spin 0, or 1, or 2, . . . are bosons. Although we will here accept this correlation between the spin and the boson or fermion character of a particle as an empirical rule, in relativistic quantum mechanics this rule can be established as a theorem (the spin-statistics theorem), which can be deduced from fundamental theoretical principles.

As an example of the implementation of the antisymmetry requirement, let us consider the wavefunction for the two electrons in the ground state of the helium atom. For the sake of simplicity, let us ignore the Coulomb interaction between the electrons. This means that the only interaction is that of each electron with the helium nucleus. If there were only one electron in the helium atom (as in singly ionized helium), its wavefunction would be

$$\psi_0(\mathbf{r}_1)|+\rangle \quad \text{or} \quad \psi_0(\mathbf{r}_1)|-\rangle \tag{93}$$

where $\psi_0(\mathbf{r}_1)$ is the usual wavefunction for the hydrogen-like ground state. Since we are ignoring the interaction between the electrons, the presence of the second electron does not alter the dependence of the wavefunction $\psi_0(\mathbf{r}_1)$ on its radial coordinate. Likewise, the presence of the first electron does not alter the dependence of the wavefunction $\psi_0(\mathbf{r}_2)$ on its radial coordinate. Thus, the wavefunction for the two-electron system is simply a product of the form

$$\psi_0(\mathbf{r}_1)|\pm\rangle \, \psi_0(\mathbf{r}_2)|\pm\rangle \tag{94}$$

or a superposition of several such products. Since the wavefunction (94) is symmetric under the exchange of $\mathbf{r}_1$ and $\mathbf{r}_2$, the antisymmetry requirement for the two-electron system must be met by making some suitable choice of superpositions of products of the spin states. Obviously, the only antisymmetric superposition we can construct with $|\pm\rangle|\pm\rangle$ is, with the appropriate normalization,

$$\frac{1}{\sqrt{2}} \left(|+\rangle|-\rangle - |-\rangle|+\rangle \right) \tag{95}$$

Thus, the wavefunction for our two-electron system must be

$$\psi(\mathbf{r}_1, \mathbf{r}_2; m_{s1}, m_{s2}) = \frac{1}{\sqrt{2}} \psi_0(\mathbf{r}_1)\psi_0(\mathbf{r}_2)(|+\rangle|-\rangle - |-\rangle|+\rangle) \qquad (96)$$

It is easy to check that the spin state (95) corresponds to net spin zero.

Exercise 9. Verify that (95) is an eigenstate of $(\mathbf{S}_1 + \mathbf{S}_2)^2$ and of $S_{z1} + S_{z2}$ with eigenvalues zero.

We therefore see that two electrons with a symmetric orbital state must be in a state of zero net spin, that is, a state of opposite spins. This conclusion remains valid even if the electrons interact with each other. When there is interaction, the orbital part of the wavefunction cannot be expressed as a product $\psi_0(\mathbf{r}_1)\psi_0(\mathbf{r}_2)$ of individual wavefunctions; instead, it becomes some new function $\psi(\mathbf{r}_1, \mathbf{r}_2)$ in which the variables $\mathbf{r}_1$ and $\mathbf{r}_2$ are entangled in some complicated way. However, if the two electrons are in the *same* orbital state, then $\psi(\mathbf{r}_1, \mathbf{r}_2)$ is necessarily symmetric under the exchange of $\mathbf{r}_1$ and $\mathbf{r}_2$, and therefore the spin part of the wavefunction necessarily must be antisymmetric, as in Eq. (95).

From this argument we reach the general conclusion that whenever two electrons are in the same orbital state, their spins must be opposite. If we include the spin as part of the specification of a state, then we can rephrase this as follows: *No more than one electron can be in any given orbital and spin state.* This statement is the *exclusion principle*, which was discovered by Pauli. The exclusion principle is valid not only for a system with two electrons, but also for a system with more than two electrons, and it plays a crucial role in determining the arrangement of electrons in the states of atoms.

For electrons in different orbital states, the antisymmetry requirement for the wavefunction does not impose restrictions on the orientations of the spins; such electrons can have opposite spins or parallel spins. If they have parallel spins, the spin part of the wavefunction is symmetric, and the orbital part of the wavefunction must then be antisymmetric. Note that the antisymmetry of the orbital part of the wavefunction implies that the wavefunction vanishes for $\mathbf{r}_1 = \mathbf{r}_2$; thus, the probability for finding the two electrons at the same place is zero. This is a characteristic property of the wavefunction for electrons with parallel spins.

The zero probability for finding two electrons at the same place can be interpreted to mean that electrons of parallel spins tend to avoid one another. In contrast, electrons of opposite spins have no such tendency [their wavefunction usually does not vanish for $r_1 = r_2$; see Eq. (96)]. This correlation between the spin orientation and the probability distribution has an important effect on the Coulomb interaction between the electrons. If the electrons tend to avoid one another, then their mutual Coulomb energy will tend to be low. Thus, we see that electrons of parallel spins tend to have a lower Coulomb energy than electrons of opposite spins. This difference in Coulomb energy is effectively equivalent to a spin-spin interaction which tends to align the spins of the electrons. This effective spin-spin interaction (produced indirectly by the Coulomb interaction) plays an important role in atoms, where it is much stronger than the ordinary magnetic interaction of the spins (produced directly by the interaction of the spin magnetic moments).

PROBLEMS   1. An electron is in the spin state $1/3 \,|+\rangle + 2\sqrt{2}/3 \,|-\rangle$.

(a) What is the probability that a measurement of the z component of spin will result in $\frac{1}{2}\hbar$? In $-\frac{1}{2}\hbar$?

(b) What is the expectation value of S_z?

(c) What is the rms uncertainty in S_z?

2. At one instant, the electron in a hydrogen atom is in the state

$$|\psi\rangle = \frac{1}{\sqrt{5}} \,|E_2,1,-1,+\rangle + \frac{2}{\sqrt{5}} \,|E_1,0,0,-\rangle$$

(a) What are the expectation values of the operators

$$H, \; p_x, \; p_y, \; p_z, \; \mathbf{L}^2, \; L_x, \; L_y, \; L_z, \; \mathbf{S}^2, \; S_x, \; S_y, \; S_z, \; \mathbf{L}\cdot\mathbf{S}, \text{ and } P \text{ (parity)?}$$

(b) List those operators for which $|\psi\rangle$ is an eigenstate.

(c) Express the state $|\psi\rangle$ in the position representation, as a spinor wavefunction.

3. (a) Find the eigenvectors of S_y. Express them as spinors.

(b) Suppose that an electron is in the spin state

$$\frac{1}{\sqrt{5}} \begin{pmatrix} 2 \\ -1 \end{pmatrix}$$

If we measure the y component of the spin, what is the probability for finding a value of $+\frac{1}{2}\hbar$?

4. Prove the following identity

$$(\boldsymbol{\sigma}\cdot\mathbf{A})(\boldsymbol{\sigma}\cdot\mathbf{B}) = \mathbf{A}\cdot\mathbf{B} + i\boldsymbol{\sigma}\cdot(\mathbf{A} \times \mathbf{B})$$

where **A** and **B** are any two vectors which commute with $\boldsymbol{\sigma}$.

5. (a) Show that for the eigenstates $|+\rangle$ and $|-\rangle$ of S_z the expectation values of S_x and of S_y are zero:

$$\langle+|S_x|+\rangle = \langle+|S_y|+\rangle = \langle-|S_x|-\rangle = \langle-|S_y|-\rangle = 0$$

   (b) Show that there exists no spin $\frac{1}{2}$ state for which the expectation values of S_x, S_y, and S_z are all zero simultaneously.

6. Suppose that instead of measuring the z component of the spin, we measure the component of the spin along an axis inclined at an angle θ with respect to the z axis, in the x-z plane. Find the eigenvectors for this component of the spin, and express them as superpositions of the spinors χ_+ and χ_-. [Hint: The component of the spin vector **S** along this direction is $\frac{1}{2}\hbar(\sigma_z \cos \theta + \sigma_x \sin \theta)$; why?]

7. Show that the magnetic moment of a free electron cannot be measured in a Stern–Gerlach experiment, by the following argument:

   (a) The magnetic field on the centerline of the magnet is vertical, in the z direction, and it has a vertical gradient $\partial B_z/\partial z$ (see Fig. 9.2). The vertical force on the electron is therefore $-\mu_z \partial B_z/\partial z$.

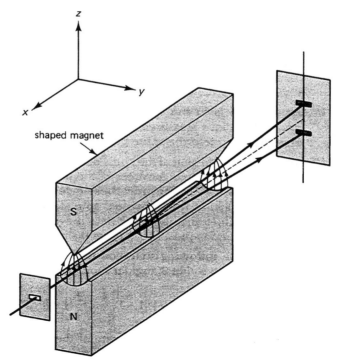

Fig. 9.2 Magnet used for the Stern–Gerlach experiment.

(b) On the exact centerline, the horizontal magnetic field is zero. However, if the electron has a horizontal uncertainty Δy, the horizontal magnetic field acting on the electron has an uncertainty $\Delta y \, \partial B_y / \partial y$, and the vertical force has an uncertainty $|ev_x \, \Delta y \, \partial B_y / \partial y|$, where v_x is the velocity of the electron (before deflection). Show that this equals $|ev_x \, \Delta y \, \partial B_z / \partial z|$, because $\nabla \cdot \mathbf{B} = 0$.

(c) By comparing the results obtained in (a) and (b), show that the uncertainty in the vertical force always exceeds the magnitude of the vertical force, and that therefore the vertical deflection of the electron is not detectable.

(d) Show that if the electron is attached to an atom of mass $M \gg m_e$, then its magnetic moment can be measured in such a Stern–Gerlach experiment.

8. For two angular momenta of quantum numbers j_1 and j_2, there are $(2j_1 + 1) \times (2j_2 + 1)$ possible products $|j_1, m_{j_1}\rangle|j_2, m_{j_2}\rangle$ of eigenstates of the individual angular momenta. Count all the possible eigenstates $|j, m_j\rangle$ of the total angular momentum, and show that there are exactly $(2j_1 + 1) \times (2j_2 + 1)$ such eigenstates.

9. Consider the addition of three (commuting) angular momenta with quantum numbers j_1, j_2, and j_3, where $j_1 \geq j_2 \geq j_3$. What is the maximum possible value of the quantum number j characterizing the total angular momentum? What is the minimum value? (Hint: Add the first two angular momenta and then add the third to this sum.)

10. A particle has spin $s = 1$ and orbital angular momentum $l = 2$. What are the possible values of the quantum number j for the total angular momentum? For each of these values of j draw the angular-momentum vectors $\mathbf{S}$, $\mathbf{L}$, and $\mathbf{J}$ according to the vector model.

11. An electron is in the spin-and-orbital state $|1, -1\rangle|+\rangle$. What is the probability that a measurement of the total angular momentum of this electron will yield the value $j = \frac{3}{2}$?

12. Consider the addition of the spins of two electrons, each of which has $s = \frac{1}{2}$. The possible eigenstates of the total angular momentum then have $j = 1$ or $j = 0$. Construct these eigenstates as superpositions of products $|\pm\rangle|\pm\rangle$ of the individual spin states of the first and second electrons. Show that the $j = 1$ state is symmetric under exchange of the two electron spins, and the $j = 0$ state is antisymmetric.

13. Consider the following state constructed out of products of eigenstates of two individual angular momenta with $j_1 = \frac{3}{2}$ and $j_2 = 1$:

$$\sqrt{\frac{3}{5}} \, |\tfrac{3}{2}, -\tfrac{1}{2}\rangle|1, -1\rangle + \sqrt{\frac{2}{5}} \, |\tfrac{3}{2}, -\tfrac{3}{2}\rangle|1, 0\rangle$$

(a) Show that this is an eigenstate of the total angular momentum. What are the values of j and m_j for this state?

(b) Construct a (normalized) state of the same j, but a value of m_j larger by 1.

14. Consider two free electrons, with single-particle wavefunctions $e^{i\mathbf{p}_1 \cdot \mathbf{r}_1/\hbar}|\pm\rangle$ and $e^{i\mathbf{p}_2 \cdot \mathbf{r}_2/\hbar}|\pm\rangle$.

    (a) Construct the antisymmetric two-electron wavefunction of net spin zero.

    (b) Construct the antisymmetric two-electron wavefunction of net spin one. Assume that both spins are up.

15. Suppose that ψ is a function of $\mathbf{r}_1$, $\mathbf{r}_2$, m_{s1}, and m_{s2} that is neither symmetric nor antisymmetric. Show that then $(1 + P_{ex})\psi$ is symmetric and $(1 - P_{ex})\psi$ is antisymmetric.

16. Suppose that five electrons are placed in a one-dimensional infinite potential well of length L. What is the energy of the ground state of this system of five electrons? What is the net spin of the ground state? Take the exclusion principle into account, and ignore the Coulomb interaction of the electrons with each other.

17. A beam of protons is scattered off a target of protons. In most of the scattering events, the incident proton is in a state of zero orbital angular momentum relative to the target proton, since this permits the closest approach. What is the net spin of the incident proton and the target proton for this state?

18. Three electrons of parallel spins (say, spin up) are placed in three different orbital states, ψ_1, ψ_2, and ψ_3. Since the spin part of the wavefunction is symmetric, the orbital part must be antisymmetric.

    (a) Construct a superposition of products $\psi_1(\mathbf{r}_1)\psi_2(\mathbf{r}_2)\psi_3(\mathbf{r}_3)$, $\psi_1(\mathbf{r}_2)\psi_2(\mathbf{r}_1)\psi_3(\mathbf{r}_3)$, ... such that it is antisymmetric under the exchange of any two of the variables $\mathbf{r}_1$, $\mathbf{r}_2$, and $\mathbf{r}_3$.

    (b) Show that except for a normalization factor, your antisymmetric superposition equals the determinant

    $$\begin{vmatrix} \psi_1(\mathbf{r}_1) & \psi_1(\mathbf{r}_2) & \psi_1(\mathbf{r}_3) \\ \psi_2(\mathbf{r}_1) & \psi_2(\mathbf{r}_2) & \psi_2(\mathbf{r}_3) \\ \psi_3(\mathbf{r}_1) & \psi_3(\mathbf{r}_2) & \psi_3(\mathbf{r}_3) \end{vmatrix}$$

    (c) If the wavefunctions ψ_1, ψ_2, and ψ_3 are normalized, what normalization factor is required with this determinant?

10

Perturbation Theory

Exact solutions of the Schrödinger equation are known only for a select group of problems that have sufficiently simple Hamiltonians. More often than not, we have to be satisfied with approximate solutions. Perturbation theory is a powerful method for obtaining approximate solutions. This method hinges on the assumption that the Hamiltonian H of the system under consideration does not differ very much from some other Hamiltonian H_0 with known eigenvectors. The difference between the actual Hamiltonian and the other Hamiltonian, with known eigenvectors, is called the perturbation, $H' = H - H_0$. The perturbation may be time independent or time dependent. We will deal with these two cases separately.

10.1 *Time-Independent Perturbations*

As an example of a time-independent perturbation, consider a hydrogen atom immersed in a uniform external electric field, say, an electric field $\mathscr{E}$ in the z direction. The potential energy of an electron immersed in such an electric field is $e\mathscr{E}z$. The Hamiltonian for the electron must then include both the potential energy contributed by this uniform electric field and the potential energy contributed by the Coulomb field of the nucleus:

$$H = \frac{\mathbf{p}^2}{2m} - \frac{1}{4\pi\varepsilon_0}\frac{e^2}{r} + e\mathscr{E}z \tag{1}$$

and the eigenvalue equation is

$$\left(\frac{\mathbf{p}^2}{2m} - \frac{1}{4\pi\varepsilon_0}\frac{e^2}{r} + e\mathscr{E}z\right)|\psi\rangle = E|\psi\rangle \tag{2}$$

(here, as elsewhere, we are neglecting the spin-orbit coupling of the electron). The presence of the term $e\mathscr{E}z$ in the Hamiltonian destroys the spherical symmetry and makes the solution of Eq. (2) extremely difficult.[1] It is therefore advisable to seek an approximate solution that exploits what we already know about the eigenvectors of the unperturbed hydrogen atom. Thus, we regard $H_0 = \mathbf{p}^2/2m - 1/4\pi\varepsilon_0 \, e^2/r$ as the unperturbed Hamiltonian and we regard $H' = e\mathscr{E}z$ as the perturbation.

In this section, we will develop a general method of successive approximations for the determination of the eigenvectors and eigenvalues of some given Hamiltonian. Suppose that, as in the example,

$$H = H_0 + H' \tag{3}$$

where H_0 is a Hamiltonian whose eigenvalues and eigenvectors are known exactly and H' is the perturbation. For the general discussion of the approximation method, we do not need to specify the detailed form of H_0 or H', but we need to assume that the perturbation H' is "small" compared with H_0, in a sense to be spelled out below.

The eigenvalue equation we want to solve is

$$(H_0 + H')|\psi_n\rangle = E_n|\psi_n\rangle \tag{4}$$

Since H' is "small" compared with H_0, we can write H' in the form

$$H' = \lambda W \tag{5}$$

where λ is a numerical, dimensionless parameter chosen so $\lambda \ll 1$, and W is an operator of a magnitude roughly comparable to H_0. Just what it means to compare the magnitudes of two operators is not quite clear at this stage, but our approximation method does not require a precise definition of these magnitudes. All that is required is that the numerical parameter λ give a rough characterization of the smallness of the perturbation. The definition of λ will vary from case to case. For instance, for the hydrogen atom immersed in the uniform external electric field, we can take $\lambda = \mathscr{E}/(e/4\pi\varepsilon_0 a_0^2)$, which is the ratio of the external electric field $\mathscr{E}$ and the typical electric field produced by the nucleus of the atom. Since $e/4\pi\varepsilon_0 a_0^2 = 5 \times 10^{11}$ V/m, the parameter λ will be small whenever the external electric field is small compared with $5 \times$

---

[1] The eigenvalue equation (2) can actually be solved exactly by introducing parabolic coordinates. But even so, a simple approximate solution is welcome.

10^{11} V/m. With this choice of λ, the operator W has the form $W = (e^2/4\pi\varepsilon_0)(z/a_0^2)$.

By hypothesis, we know the exact solutions of

$$H_0|\psi_k^{(0)}\rangle = E_k^{(0)}|\psi_k^{(0)}\rangle \tag{6}$$

The superscript $^{(0)}$ indicates that this is the equation that holds when $\lambda = 0$. The eigenvectors $|\psi_k^{(0)}\rangle$ form a complete set. We can therefore write any eigenvector $|\psi_n\rangle$ of the perturbed Hamiltonian as a superposition of the eigenvectors $|\psi_k^{(0)}\rangle$:

$$|\psi_n\rangle = \sum_k c_{kn}|\psi_k^{(0)}\rangle \tag{7}$$

Substituting this into Eq. (4), we obtain

$$(H_0 + \lambda W)\sum_k c_{kn}|\psi_k^{(0)}\rangle = E_n\sum_k c_{kn}|\psi_k^{(0)}\rangle \tag{8}$$

that is,

$$\sum_k (E_k^{(0)} + \lambda W)c_{kn}|\psi_k^{(0)}\rangle = E_n\sum_k c_{kn}|\psi_k^{(0)}\rangle$$

or

$$\sum_k c_{kn}\lambda W|\psi_k^{(0)}\rangle = \sum_k c_{kn}(E_n - E_k^{(0)})|\psi_k^{(0)}\rangle \tag{9}$$

Taking the scalar product of Eq. (9) with $|\psi_m^{(0)}\rangle$, we find that

$$\sum_k c_{kn}\langle\psi_m^{(0)}|\lambda W|\psi_k^{(0)}\rangle = (E_n - E_m^{(0)})c_{mn} \tag{10}$$

We can rewrite this equation in the compact form

$$\lambda\sum_k W_{mk}c_{kn} = (E_n - E_m^{(0)})c_{mn} \tag{11}$$

where the "matrix element" W_{mk} is defined as

$$W_{mk} = \langle\psi_m^{(0)}|W|\psi_k^{(0)}\rangle \tag{12}$$

Equation (11) is exact. But we will now introduce our approximation. For this purpose we assume that both c_{kn} and E_n can be written as power series in λ,

$$c_{kn} = c_{kn}^{(0)} + \lambda c_{kn}^{(1)} + \lambda^2 c_{kn}^{(2)} + \cdots$$

$$= \delta_{kn} + \lambda c_{kn}^{(1)} + \lambda^2 c_{kn}^{(2)} + \cdots \tag{13}$$

$$E_n = E_n^{(0)} + \lambda E_n^{(1)} + \lambda^2 E_n^{(2)} + \cdots \tag{14}$$

In Eq. (13) we have used the condition that for $\lambda = 0$ (no perturbation), the coefficient c_{kn} must reduce to δ_{kn} [see Eq. (7)].

We now substitute these power series into Eq. (11), with the result

$$\lambda \sum_k W_{mk}(\delta_{kn} + \lambda c_{kn}{}^{(1)} + \lambda^2 c_{kn}{}^{(2)} + \cdots)$$

$$= (E_n{}^{(0)} - E_m{}^{(0)} + \lambda E_n{}^{(1)} + \lambda^2 E_n{}^{(2)} + \cdots)$$

$$(\delta_{mn} + \lambda c_{mn}{}^{(1)} + \lambda^2 c_{mn}{}^{(2)} + \cdots) \qquad (15)$$

If we collect all the terms of order 1, λ, $\lambda^2, \ldots$, this equation becomes

$$0 = (E_n{}^{(0)} - E_m{}^{(0)})\delta_{mn} + \lambda[\delta_{mn}E_n{}^{(1)} + c_{mn}{}^{(1)}(E_n{}^{(0)} - E_m{}^{(0)}) - W_{mn}]$$

$$+ \lambda^2[\delta_{mn}E_n{}^{(2)} + c_{mn}{}^{(1)}E_n{}^{(1)} + c_{mn}{}^{(2)}(E_n{}^{(0)} - E_m{}^{(0)})$$

$$- \sum_k c_{kn}{}^{(1)}W_{mk}] + \cdots \qquad (16)$$

We can regard the parameter λ as a variable, which may assume any value between 0 and some upper limit. If Eq. (16) is to remain valid for all values of λ, we must demand that the coefficient of each power of λ be zero. The constant term, or zero-power term, in Eq. (16) is identically zero, since $E_n{}^{(0)} - E_m{}^{(0)} = 0$ for $m = n$ and $\delta_{mn} = 0$ for $m \neq n$. If we demand that the coefficient of λ be zero, we obtain

$$\delta_{mn}E_n{}^{(1)} + c_{mn}{}^{(1)}(E_n{}^{(0)} - E_m{}^{(0)}) - W_{mn} = 0 \qquad (17)$$

For $m \neq n$, this equation yields

$$c_{mn}{}^{(1)} = \frac{W_{mn}}{E_n{}^{(0)} - E_m{}^{(0)}} \qquad (18)$$

For $m = n$, it yields

$$E_n{}^{(1)} = W_{nn} \qquad (19)$$

To first order in λ, our approximate solution is then, with $H'_{kn} = \lambda W_{kn} = \langle \psi_k{}^{(0)} | H' | \psi_n{}^{(0)} \rangle$,

$$|\psi_n\rangle = |\psi_n{}^{(0)}\rangle + \sum_k \lambda c_{kn}{}^{(1)} |\psi_k{}^{(0)}\rangle$$

$$= |\psi_n{}^{(0)}\rangle + \lambda c_{nn}{}^{(1)} |\psi_n{}^{(0)}\rangle + \sum_{k \neq n} \frac{H'_{kn}}{E_n{}^{(0)} - E_k{}^{(0)}} |\psi_k{}^{(0)}\rangle \qquad (20)$$

and

$$E_n = E_n^{(0)} + \lambda E_n^{(1)} = E_n^{(0)} + H'_{nn} \tag{21}$$

Equation (21) shows that, to first order, the change in the energy produced by the perturbation is simply the expectation value of the perturbation.

Note that Eqs. (18) and (20) rely on the implicit supposition that $E_n^{(0)} - E_m^{(0)} \neq 0$ if $n \neq m$, that is, the unperturbed states are supposed to be nondegenerate. The modifications in the approximation method needed to handle degenerate states will be discussed in the next section.

To finish our first-order solution, we still have to find $c_{nn}^{(1)}$. The eigenvectors (20) must be orthogonal to each other. This implies

$$\langle \psi_n | \psi_m \rangle = \langle \psi_n^{(0)} | \psi_m^{(0)} \rangle + \lambda c_{mm}^{(1)} \langle \psi_n^{(0)} | \psi_m^{(0)} \rangle$$

$$+ \lambda c_{nn}^{(1)*} \langle \psi_n^{(0)} | \psi_m^{(0)} \rangle + \sum_{k \neq m} \frac{\lambda W_{km}}{E_m^{(0)} - E_k^{(0)}} \langle \psi_n^{(0)} | \psi_k^{(0)} \rangle$$

$$+ \sum_{k \neq n} \frac{\lambda W_{kn}^*}{E_n^{(0)} - E_k^{(0)}} \langle \psi_k^{(0)} | \psi_m^{(0)} \rangle + O(\lambda^2) \tag{22}$$

Since we are dealing with the first-order approximation, we can neglect terms of order λ^2 in Eq. (22). Taking into account the orthonormality of the vectors $|\psi_n^{(0)}\rangle$, we see that Eq. (22) reduces to

$$\langle \psi_n | \psi_m \rangle = \delta_{mn} + \lambda(c_{mm}^{(1)} + c_{nn}^{(1)*})\delta_{mn} + \frac{\lambda(W_{nm} - W_{mn}^*)}{E_m^{(0)} - E_n^{(0)}} \tag{23}$$

The perturbation H' is, of course, a hermitian operator. This implies that $W_{nm} = W_{mn}^*$, and thus the last term in Eq. (23) is zero. If the eigenvectors are to be orthonormal, Eq. (23) demands, for $m = n$,

$$c_{nn}^{(1)} + c_{nn}^{(1)*} = 0 \tag{24}$$

Thus, c_{nn} is *purely imaginary*. We will show that this entitles us to assume that $c_{nn}^{(1)} = 0$. The argument is as follows: If we ignore terms of order λ^2,

$$|\psi_n\rangle = |\psi_n^{(0)}\rangle + \sum_{k \neq n} \lambda c_{kn}^{(1)} |\psi_k^{(0)}\rangle + \lambda c_{nn}^{(1)} |\psi_n^{(0)}\rangle + O(\lambda^2)$$

$$= (1 + \lambda c_{nn}^{(1)}) \left(|\psi_n^{(0)}\rangle + \sum_{k \neq n} \lambda c_{kn}^{(1)} |\psi_n^{(0)}\rangle \right) + O(\lambda^2) \tag{25}$$

But, if we ignore terms of order λ^2, then $(1 + \lambda c_{nn}^{(1)}) = e^{\lambda c_{nn}^{(1)}}$, and this is a phase factor since $c_{nn}^{(1)}$ is purely imaginary. Thus, the choice of $c_{nn}^{(1)}$ amounts to no more than an irrelevant choice of phase for the vectors $|\psi_n\rangle$, and we may as well assume that $c_{nn}^{(1)}$ is zero. We can then write Eq. (20) as

$$|\psi_n\rangle = |\psi_n^{(0)}\rangle + \sum_{k \neq n} \frac{H'_{kn}}{E_n^{(0)} - E_k^{(0)}} |\psi_k^{(0)}\rangle \qquad (26)$$

As next step in our approximation scheme, we demand that the coefficient of λ^2 in Eq. (16) be zero, and we obtain

$$\delta_{mn} E_n^{(2)} + c_{mn}^{(1)} E_n^{(1)} + c_{mn}^{(2)}(E_n^{(0)} - E_m^{(0)}) - \sum_k c_{kn}^{(1)} W_{mk} = 0 \qquad (27)$$

For $m = n$, this yields

$$E_n^{(2)} = -c_{nn}^{(1)} E_n^{(1)} + \sum_k c_{kn}^{(1)} W_{nk}$$

$$= \sum_{k \neq n} \frac{W_{kn} W_{nk}}{E_n^{(0)} - E_k^{(0)}} \qquad (28)$$

For $m \neq n$, it yields an equation for $c_{mn}^{(2)}$; we will not bother with this equation (see, however, Problem 13). Combining Eqs. (21) and (28), we obtain an expression for the energy, to second order in λ:

$$E_n = E_n^{(0)} + H'_{nn} + \sum_{k \neq n} \frac{H'_{kn} H'_{nk}}{E_n^{(0)} - E_k^{(0)}} \qquad (29)$$

If these results are to provide a satisfactory approximation, H' must be "small" compared with H_0. We can now state this requirement more precisely. The second term in Eq. (26) must be small compared with the first, that is,

$$|H'_{kn}| \ll |E_n^{(0)} - E_k^{(0)}| \qquad (30)$$

This says that the matrix element of the perturbation must be small compared with the energy difference between any two unperturbed states. In particular, if a state is degenerate in energy with some other state, then our results are inapplicable to this state, no matter how small the perturbation (however, our results remain applicable to any other, nondegenerate states of the system).

Let us employ these results to calculate the change of energy of the ground state of the hydrogen atom immersed in an external electric field. The unperturbed Hamiltonian is

$$H_0 = \frac{\mathbf{p}^2}{2m} - \frac{1}{4\pi\varepsilon_0} \frac{e^2}{r}$$

and the perturbation is

$$H' = e\mathscr{E}z \tag{31}$$

For consistency with the notation used in this chapter, we adopt the notation $|\psi_0^{(0)}\rangle$ for the ground state. According to Eq. (21), the energy change of the ground state is, in first order,

$$\langle\psi_0^{(0)}|e\mathscr{E}z|\psi_0^{(0)}\rangle$$

But this is zero, since the expectation value, or the average value, of the z coordinate of the electron is zero. Thus, in first order, there is no energy change of the ground state. To obtain a nontrivial result, we must examine the energy change in second order. According to Eq. (29), we must then evaluate the infinite sum

$$\sum_{k\neq 0} \frac{H'_{k0}H'_{0k}}{E_0^{(0)} - E_k^{(0)}}$$

In this sum, the index k ranges over *all* the excited states of the hydrogen atom, that is, states of negative energy *and* states of positive energy. In general, such infinite sums are quite difficult to evaluate; but in our case the evaluation can be performed by a devious trick based on the identity

$$z|\psi_0^{(0)}\rangle = (GH_0 - H_0G)|\psi_0^{(0)}\rangle \tag{32}$$

where

$$G = -\frac{ma_0}{2\hbar^2}(r + 2a_0)z$$

Exercise 1. Verify this identity. (Hint: Use the explicit form of the ground-state wavefunction.)

This identity permits us to cancel the denominator in the infinite sum, which makes the evaluation easy. We begin by using the identity (32) in H'_{k0}:

$$H'_{k0} = \langle\psi_k^{(0)}|H'|\psi_0^{(0)}\rangle$$

$$= e\mathscr{E}\,\langle\psi_k^{(0)}|z|\psi_0^{(0)}\rangle$$

$$= e\mathscr{E}\,\langle\psi_k^{(0)}|(GH_0 - H_0G)|\psi_0^{(0)}\rangle$$

$$= e\mathscr{E}\,\langle\psi_k^{(0)}|(GE_0^{(0)} - E_k^{(0)}G)|\psi_0^{(0)}\rangle$$

$$= e\mathscr{E}(E_0^{(0)} - E_k^{(0)})\,\langle\psi_k^{(0)}|G|\psi_0^{(0)}\rangle$$

With this, our infinite sum becomes

$$\sum_{k \neq 0} \frac{1}{E_0^{(0)} - E_k^{(0)}} \times e\mathscr{E}(E_0^{(0)} - E_k^{(0)}) \langle \psi_k^{(0)} | G | \psi_0^{(0)} \rangle \langle \psi_0^{(0)} | e\mathscr{E}z | \psi_k^{(0)} \rangle$$

$$= (e\mathscr{E})^2 \langle \psi_0^{(0)} | z \left(\sum_{k \neq 0} | \psi_k^{(0)} \rangle \langle \psi_k^{(0)} | \right) G | \psi_0^{(0)} \rangle$$

$$= (e\mathscr{E})^2 \langle \psi_0^{(0)} | z \left(\mathbf{1} - | \psi_0^{(0)} \rangle \langle \psi_0^{(0)} | \right) G | \psi_0^{(0)} \rangle$$

$$= (e\mathscr{E})^2 \langle \psi_0^{(0)} | zG | \psi_0^{(0)} \rangle - (e\mathscr{E})^2 \langle \psi_0^{(0)} | z | \psi_0^{(0)} \rangle \langle \psi_0^{(0)} | G | \psi_0^{(0)} \rangle$$

Here we have used the completeness relation for the eigenstates of hydrogen to eliminate the infinite sum. Since the average value of z for the ground state is zero, our expression for the energy change reduces to the single term:

$$(e\mathscr{E})^2 \langle \psi_0^{(0)} | zG | \psi_0^{(0)} \rangle = -(e\mathscr{E})^2 \frac{ma_0}{2\hbar^2} \langle \psi_0^{(0)} | z^2(r + 2a_0) | \psi_0^{(0)} \rangle$$

The calculation of the expectation value of $z^2(r + 2a_0)$ is straightforward, and yields the final result $-9\pi\varepsilon_0 a_0^3 \mathscr{E}^2$ for the energy change of the ground state of hydrogen.

Exercise 2. Calculate the expectation value of $z^2(r + 2a_0)$ for the ground state of hydrogen. (Hint: In this state, the expectation values of z^2r, y^2r, and x^2r are equal. Why?)

Experimentally, this energy change can be detected by the consequent shifts of the spectral lines of the atom ("quadratic Stark effect"). The electric field of course also produces energy changes in the excited states of hydrogen, but these states are degenerate, and therefore the energy changes cannot be calculated with the formulas of this section. We will see how to calculate the energy changes of the excited states in the next section.

We now turn to another example, again involving the ground state of hydrogen. In our treatment of hydrogen in Chapter 8 we assumed that the nucleus is pointlike, with a Coulomb potential $-(1/4\pi\varepsilon_0)(e^2/r)$, and we found the exact eigenstates for an electron subjected to this potential. However, the nucleus is actually a charge distribution of finite size, and we will now use perturbation theory to calculate the effect on the ground-state energy of this finite size of the nucleus. We will pretend that the nucleus is a uniformly charged sphere of radius R. The potential energy for an electron inside or outside such a sphere is

$$
V = \begin{cases}
-\dfrac{1}{4\pi\varepsilon_0}\dfrac{e^2}{r} & \text{for } r \geq R \\[2ex]
-\dfrac{1}{4\pi\varepsilon_0}\dfrac{e^2}{R}\left(\dfrac{3}{2}-\dfrac{1}{2}\dfrac{r^2}{R^2}\right) & \text{for } r \leq R
\end{cases}
$$

To treat this problem by perturbation theory, we separate the Hamiltonian $H = \mathbf{p}^2/2m + V(r)$ into an unperturbed part H_0 and a perturbation H':

$$
H_0 = \frac{\mathbf{p}^2}{2m} - \frac{1}{4\pi\varepsilon_0}\frac{e^2}{r}
$$

and

$$
H' = \begin{cases}
-\dfrac{1}{4\pi\varepsilon_0}\dfrac{e^2}{R}\left(\dfrac{3}{2}-\dfrac{1}{2}\dfrac{r^2}{R^2}\right)+\dfrac{1}{4\pi\varepsilon_0}\dfrac{e^2}{r} & \text{for } r \leq R \\[2ex]
0 & \text{for } r \geq R
\end{cases}
\tag{33}
$$

Note that the perturbation H' is simply the difference between the actual potential energy and the Coulomb potential energy.

According to Eq. (21), the energy change of the ground state is $\langle\psi_0^{(0)}|H'|\psi_0^{(0)}\rangle$. For the evaluation of this matrix element, it is convenient to adopt the position representation:

$$
\int_0^R \int d\Omega \; r^2 \, dr \; \psi_{100}(r)^* \left[-\frac{1}{4\pi\varepsilon_0}\frac{e^2}{R}\left(\frac{3}{2}-\frac{1}{2}\frac{r^2}{R^2}\right) + \frac{1}{4\pi\varepsilon_0}\frac{e^2}{r} \right] \psi_{100}(r)
$$

Over the short range of integration, we can neglect the exponential factor in the wavefunction $\psi_{100}(r)$, and we can approximate this wavefunction by $\psi_{100}(r) \simeq 2/\sqrt{4\pi}a_0^{3/2}$. This makes the integration trivial and leads to the result

$$
\langle\psi_0^{(0)}|H'|\psi_0^{(0)}\rangle = \frac{2}{5}\frac{e^2}{4\pi\varepsilon_0}\frac{R^2}{a_0^3}
\tag{34}
$$

Exercise 3. Obtain this result.

Equation (34) indicates that the energy of the ground state is slightly larger than the unperturbed value $-\tfrac{1}{2}mc^2\alpha^2$ we obtained in Chapter 7. We can understand this increase qualitatively, since the electron experiences a reduced potential when it penetrates the nucleus. With $R \simeq 10^{-15}\,m$, the numerical value of the energy correction for the ground state of hydrogen is only about

$$\frac{2}{5} \frac{e^2}{4\pi\varepsilon_0} \frac{R^2}{a_0{}^3} \simeq 10^{-8} \text{ eV}$$

However, for other atoms, with a nuclear charge Ze, the energy corrections can be much larger. For such atoms, Eq. (34) is multiplied by Z^4 (since e^2 must be replaced by Ze^2, and a_0 must be replaced by a_0/Z); furthermore, the radii of such nuclei are larger.

10.2 *Degenerate Perturbations*

Suppose that the unperturbed system has a group of energy eigenstates that are degenerate. To calculate the effect of a perturbation in this case, we have to return to our exact equation (11). Suppose that the group of degenerate states consists of j states; for convenience we label the eigenvectors $|\psi_1{}^{(0)}\rangle, |\psi_2{}^{(0)}\rangle, \ldots, |\psi_j{}^{(0)}\rangle$. Equation (26) suggests that the coefficients c_{kn}, with k and $n = 1, 2, \ldots, j$, are *not* of the form $\delta_{kn} +$ (small quantity) we assumed earlier; that is, the power series (13) is not valid for these coefficients. So let us assume that the coefficients c_{kn}, with k and $n = 1, 2, \ldots, j$ are of the order of magnitude of 1. If we then retain in Eq. (11) the *largest* terms, we obtain

$$\lambda \sum_{k=1}^{j} W_{mk} c_{kn} = (E_n - E_m{}^{(0)}) c_{mn} \qquad m, n = 1, 2, \ldots, j \qquad (35)$$

We expect that the solutions of this equation will serve as a reasonable first approximation to the solutions of the exact equation (11).

For each fixed value of n, Eq. (35) is a linear homogeneous system of equations for the j unknowns c_{kn}, $k = 1, 2, \ldots, j$. Let us rewrite this system of equations in the standard form

$$\sum_{k=1}^{j} [H'_{mk} - (E_n - E_k{}^{(0)})\delta_{mk}] c_{kn} = 0 \qquad (36)$$

It is well known that such a homogeneous system of equations has a solution if and only if the determinant of the coefficients multiplying the unknowns vanishes:[2]

$$|H'_{mk} - (E_n - E_k{}^{(0)})\delta_{mk}| = 0 \qquad (37)$$

---

[2] The notation $|A_{mk}|$ stands for the determinant of the matrix A_{mk}. Note that in Eq. (37), n is held fixed. The determinant is formed with respect to the indices m and k.

This equation is a polynomial of order j in E_n. The roots of this polynomial give us the energy eigenvalues. Since this polynomial has j roots, we obtain not only one solution of Eqs. (36) and (37), but a set of j separate solutions. Furthermore, we obtain the same set of j solutions no matter which value of n we choose, because Eq. (37) is an equation for the unknown E_n, and the roots of this equation do not depend on whether we call this unknown E_1, E_2, E_3, ...

The fact that each value of n leads to the same set of solutions means that there is no unique way of establishing a correspondence between perturbed and unperturbed states. We start with j degenerate unperturbed eigenvectors. Solving our equations, we obtain j perturbed eigenvectors and eigenvalues. But we cannot tell which of the perturbed eigenvectors corresponds to which of the unperturbed eigenvectors. If we look at one of the perturbed eigenvectors and take the limit as the perturbation goes to zero, we will usually not obtain one of the eigenvectors $|\psi_n^{(0)}\rangle$, but rather some superposition of *several* degenerate eigenvectors $|\psi_n^{(0)}\rangle$. This is quite different from the case of nondegenerate perturbation theory, where, in the limit of zero perturbation, we obtain one and only one of the unperturbed eigenvectors (see the results of the preceding section).

Since the possible solutions of Eqs. (36) and (37) are independent of the choice of n, we could actually omit this label from the equations. We will not do this, but instead give n a new meaning by using it to distinguish between the different possible solutions of Eqs. (36) and (37). Thus, E_1 will designate the, say, lowest root of Eq. (37), E_2 the next root, etc.; and c_{k1}, c_{k2}, . . . will designate the corresponding solutions of Eq. (36). It can, of course, happen that some of the roots of Eq. (37) are equal, so some states remain degenerate even in the presence of the perturbation; in this case, we must adopt some additional conventions for assigning the labels n.

The calculation of the eigenvectors in the first and higher approximations and the calculation of the energy in the *second* approximation[3] can now proceed as in the nondegenerate case. We have only to pretend that the vectors

$$\sum_{k=1}^{j} c_{kn}|\psi_k^{(0)}\rangle \tag{38}$$

---

[3] The solution of Eq. (36) gives us the energy correct to first order, and it gives us the coefficients c_{kn} with $k,n = 1, 2, . . . , j$. But the coefficients with $k > j$ still remain to be calculated in first order.

[with c_{kn} as given by Eq. (36)] are the "unperturbed" eigenvectors. The vanishing denominators in Eqs. (26) and (28) will not give us any trouble, because the matrix elements of H' taken between any two different "unperturbed" eigenvectors are zero, and therefore the terms with vanishing denominators drop out of our equations [this can best be seen from Eq. (16) before division by $E_n^{(0)} - E_m^{(0)}$].

Note that the method used for degenerate states can and should also be used for states that are not degenerate, but that are such that the condition (30) fails. "Degenerate" perturbation theory is to be used whenever the perturbation energy is not small compared with the energy differences between unperturbed states.

As an example, we return to the hydrogen atom immersed in a uniform electric field, and we calculate the energy changes of the first excited state. The first excited state of hydrogen is degenerate: the eigenvectors $|\psi_{200}\rangle$, $|\psi_{210}\rangle$, $|\psi_{211}\rangle$, and $|\psi_{21-1}\rangle$ all have the same energy.[4] For convenience, we designate these eigenvectors by $|\psi_1^{(0)}\rangle$, $|\psi_2^{(0)}\rangle$, $|\psi_3^{(0)}\rangle$, and $|\psi_4^{(0)}\rangle$, respectively. Equation (36) for $m = 1$ reads (omitting the label n)

$$H'_{11}c_1 + H'_{12}c_2 + H'_{13}c_3 + H'_{14}c_4 - (E - E_1^{(0)})c_1 = 0 \qquad (39)$$

and for $m = 2$,

$$H'_{21}c_1 + H'_{22}c_2 + H'_{23}c_3 + H'_{24}c_4 - (E - E_2^{(0)})c_2 = 0 \qquad (40)$$

Here, of course, $E_1^{(0)} = E_2^{(0)} = -\tfrac{1}{8}\alpha^2 mc^2$. A straightforward calculation shows that all the matrix elements of H', except H'_{12} and H'_{21}, are zero.

Exercise 4. Show that

$$H'_{11} = H'_{22} = H'_{13} = H'_{14} = H'_{23} = H'_{24} = 0 \qquad (41)$$

Exercise 5. Show that

$$H'_{12} = H'_{21} = \langle \psi_1^{(0)} | e\mathcal{E}z | \psi_2^{(0)} \rangle = 3a_0 e\mathcal{E} \qquad (42)$$

---

[4] If we take into account the spin-orbit coupling, then the energy of the first of these states does not coincide with that of the others. But the energy differences are small, and, as mentioned above, "degenerate" perturbation theory is to be used whenever the energy differences are small.

Equations (39) and (40) now simplify to

$$-(E + \tfrac{1}{8}\alpha^2 mc^2)c_1 + 3a_0 e \mathscr{E} c_2 = 0 \qquad (43)$$

$$3a_0 e \mathscr{E} c_1 - (E + \tfrac{1}{8}\alpha^2 mc^2)c_2 = 0 \qquad (44)$$

and Eq. (37) becomes

$$\begin{vmatrix} -(E + \tfrac{1}{8}\alpha^2 mc^2) & 3a_0 e \mathscr{E} \\ 3a_0 e \mathscr{E} & -(E + \tfrac{1}{8}\alpha^2 mc^2) \end{vmatrix} = 0 \qquad (45)$$

This is a second-order polynomial, with roots

$$E_1 = -\tfrac{1}{8}\alpha^2 mc^2 - 3a_0 e \mathscr{E} \qquad (46)$$

$$E_2 = -\tfrac{1}{8}\alpha^2 mc^2 + 3a_0 e \mathscr{E} \qquad (47)$$

These are the perturbed energy eigenvalues.

Exercise 6. Check that the expressions for E_1 and E_2 are the roots of the polynomial (45).

Inserting these eigenvalues into Eqs. (43) and (44), we obtain the solutions for the coefficients c_1 and c_2. With our convention, a second subscript is to be attached to these coefficients to indicate to which of the two eigenvalues the coefficient corresponds. The normalized solutions are then

$$c_{11} = \frac{1}{\sqrt{2}} \qquad c_{21} = -\frac{1}{\sqrt{2}} \qquad \text{for } E_1 \qquad (48)$$

and

$$c_{12} = \frac{1}{\sqrt{2}} \qquad c_{22} = \frac{1}{\sqrt{2}} \qquad \text{for } E_2 \qquad (49)$$

Note that as $\mathscr{E} \to 0$, the coefficients c_{12} and c_{21} do not vanish (as they would in the nondegenerate case). This means that if the atom is in an eigenstate of energy in the presence of the electric field $\mathscr{E}$ (say, the state of energy E_1), and then the field $\mathscr{E}$ is gradually decreased to zero, the atom will be left in the state

$$|\psi_1\rangle = \frac{1}{\sqrt{2}}(|\psi_1^{(0)}\rangle - |\psi_2^{(0)}\rangle) = \frac{1}{\sqrt{2}}(|\psi_{200}\rangle - |\psi_{210}\rangle) \qquad (50)$$

that is, the limit of zero perturbation gives a superposition of unperturbed states.

Exercise 7. Show that to first approximation, the energy of the states $|\psi_3^{(0)}\rangle$ and $|\psi_4^{(0)}\rangle$ are unchanged.

Exercise 8. Show that for an atom in the state given by Eq. (50), the expectation value of the dipole moment is

$$\langle\psi_1|(-ez)|\psi_1\rangle = 3a_0e \tag{51}$$

The energy $3a_0e\mathscr{E}$ of this atom in the electric field [see Eq. (46)] is therefore simply the dipole energy, (dipole moment) $\times$ $\mathscr{E}$.

10.3   *Time-Dependent Perturbations*

We now want to find a method for the approximate solution of the time-dependent Schrödinger equation

$$H|\psi(t)\rangle = i\hbar\,\frac{\partial}{\partial t}\,|\psi(t)\rangle \tag{52}$$

We suppose that $H = H_0 + H'(t)$, where H_0 is time independent, and the perturbation $H'(t)$ may be, but need not be, time dependent. An example of a system with a time-dependent Hamiltonian is a hydrogen atom immersed in an oscillating electric field, say, the electric field of a radio wave. When subjected to such a time-dependent perturbation, the atom will not remain in a stationary state, but instead will oscillate from one state to another in response to the oscillating perturbation caused by the radio wave. However, even if the perturbation is time independent, we have to deal with the time-dependent Schrödinger equation whenever we want to examine how the system evolves in response to a perturbation that is "switched on" at some initial time. For instance, if we want to calculate the rate of a reaction (or the cross section for a reaction), we need to know what becomes of the initial state after it has been subjected to a steady perturbation for a while.

To deal with Eq. (52) we, again, express the state vector $|\psi\rangle$ as a superposition of eigenvectors $|\psi_n^{(0)}\rangle$ of the unperturbed Hamiltonian H_0. In contrast to the superposition in Section 10.1, we now must include a time dependence in the coefficients c_n:

$$|\psi(t)\rangle = \sum_n e^{-iE_n^{(0)}t/\hbar}c_n(t)\,|\psi_n^{(0)}\rangle \tag{53}$$

The exponential time dependence has been factored out, so in the absence of perturbation the coefficients $c_n(t)$ reduce to constants.

Exercise 9. Check that if the perturbation is absent, then Eq. (53) with constant coefficients c_n is a solution of the Schrödinger equation.

If we substitute Eq. (53) into Eq. (52), we obtain

$$\sum_n (E_n^{(0)}|\psi_n^{(0)}\rangle + H'|\psi_n^{(0)}\rangle)e^{-iE_n^{(0)}t/\hbar}c_n(t)$$

$$= i\hbar \sum_n |\psi_n^{(0)}\rangle e^{-iE_n^{(0)}t/\hbar}\left(-\frac{iE_n^{(0)}}{\hbar}c_n + \frac{dc_n}{dt}\right)$$

that is,

$$\sum_n H'|\psi_n^{(0)}\rangle e^{-iE_n^{(0)}t/\hbar}c_n(t) = i\hbar \sum_n |\psi_n^{(0)}\rangle e^{-iE_n^{(0)}t/\hbar}\frac{dc_n}{dt} \qquad (54)$$

We can take the scalar product of this equation with $|\psi_k^{(0)}\rangle$:

$$\sum_n \langle\psi_k^{(0)}|H'|\psi_n^{(0)}\rangle e^{-iE_n^{(0)}t/\hbar}c_n(t) = i\hbar \sum_n \langle\psi_k^{(0)}|\psi_n^{(0)}\rangle e^{-iE_n^{(0)}t/\hbar}\frac{dc_n}{dt} \qquad (55)$$

Since $\langle\psi_k^{(0)}|\psi_n^{(0)}\rangle = \delta_{kn}$ and $\langle\psi_k^{(0)}|H'|\psi_n^{(0)}\rangle = H'_{kn}$, this equation becomes

$$\frac{dc_k}{dt} = \frac{1}{i\hbar}\sum_n H'_{kn}e^{-i(E_n^{(0)}-E_k^{(0)})t/\hbar}c_n(t) \qquad (56)$$

This is a linear system of coupled differential equations of first order for the coefficients $c_n(t)$. These equations are exact. We will now seek an approximate solution of these equations. Suppose that the initial conditions are $c_n(0) = \delta_{ns}$; that is, all the coefficients c_n are initially zero, except c_s. This means that the system is initially known to be in the state $|\psi_s^{(0)}\rangle$. If there were no perturbation, the system would of course stay in the state $|\psi_s^{(0)}\rangle$, which would be a stationary state. But because of the perturbation, the coefficients c_n vary with time, and the system is likely to be found in a different state at a later time. Our equation determines the time dependence of the coefficients c_n.

If the perturbation is small, we can assume that c_s remains near 1, and that all the other coefficients c_n with $n \neq s$ remain very small compared with c_s. It is therefore a satisfactory approximation to neglect all the terms in the summation in Eq. (56) except the term with $n = s$, for which term $c_n = c_s = 1$:

$$\frac{dc_k}{dt} = \frac{1}{i\hbar}H'_{ks}e^{-i(E_s^{(0)}-E_k^{(0)})t/\hbar} \qquad (57)$$

This approximate differential equation for the time dependence of the coefficients has the obvious solutions

$$c_k(t) = \frac{1}{i\hbar} \int_0^t H'_{ks}(t) e^{-i(E_s^{(0)} - E_k^{(0)})t/\hbar} \, dt \qquad \text{for } k \neq s \qquad (58)$$

$$c_s(t) = 1 + \frac{1}{i\hbar} \int_0^t H'_{ss}(t) \, dt \qquad\qquad \text{for } k = s \qquad (59)$$

Exercise 10. Verify that these expressions are solutions of Eq. (57).

To perform the integrations in Eqs. (58) and (59), we need to specify the dependence of H'_{ks} on time. We first consider the effects of a constant perturbation. We suppose that H' acts during the time interval from 0 to t, and that it has some constant magnitude during this time interval (see Fig. 10.1). What H' does before and after this time interval need not concern us. The integration in Eq. (58) is then trivial:

$$c_k(t) = \frac{1}{i\hbar} H'_{ks} \int_0^t e^{-i(E_s^{(0)} - E_k^{(0)})t/\hbar} \, dt = H'_{ks} \frac{e^{-i(E_s^{(0)} - E_k^{(0)})t/\hbar} - 1}{E_s^{(0)} - E_k^{(0)}} \qquad (60)$$

Initially, the system is in the state $|\psi_s^{(0)}\rangle$, and the probabilities for all other states are zero, that is, the coefficients c_n with $k \neq s$ are zero. At a later time t, the system has finite probabilities for other states, that is, the coefficients c_k are not zero. The probability for the state $|\psi_k^{(0)}\rangle$ is $|c_k(t)|^2$:

$$|c_k(t)|^2 = 4|H'_{ks}|^2 \frac{\sin^2(E_s^{(0)} - E_k^{(0)})t/2\hbar}{(E_s^{(0)} - E_k^{(0)})^2} \qquad (61)$$

This is called the *transition probability.*

The result (61) is valid as long as $|c_k(t)|^2 \ll 1$. If we want this

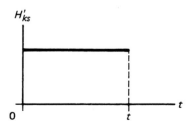

Fig. 10.1 A constant perturbation H'_{ks}.

to hold at all times, we must require

$$|H'_{ks}|^2 \ll |E_s^{(0)} - E_k^{(0)}|^2 \tag{62}$$

This is the familiar condition for the validity of nondegenerate perturbation theory. Let us suppose that this condition is satisfied. Equation (61) then shows that the probability oscillates permanently with a frequency $(E_s^{(0)} - E_k^{(0)})/2\pi\hbar$ (see Fig. 10.2).

Exercise 11. Use the results of *time-independent* nondegenerate perturbation theory to find the coefficient $c_k(t)$. (Hint: Find the perturbed eigenstates, express the initial state as a superposition of these, and consider the evolution in time.)

Let us next suppose that the condition (62) is *not* satisfied, that is, suppose that H'_{ks} is of the order of $|E_s^{(0)} - E_k^{(0)}|$. Then, to keep $|c_k(t)|$ small, we must demand that t be small,

$$\frac{|E_s^{(0)} - E_k^{(0)}|t}{\hbar} \ll 1 \tag{63}$$

and

$$\frac{|H'_{ks}|t}{\hbar} \ll 1 \tag{64}$$

For states $|\psi_k^{(0)}\rangle$ and times t such that these conditions are satisfied, we obtain a transition probability

$$|c_k(t)|^2 = 4|H'_{ks}|^2 \frac{t^2}{4\hbar^2} \tag{65}$$

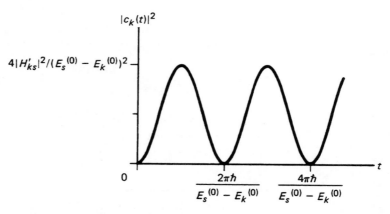

Fig. 10.2 Transition probability $|c_k(t)|^2$ vs. time.

This transition probability increases in proportion to the square of the time; however, this increase is not valid indefinitely, since the time t is subject to the conditions (63) and (64).

Figure 10.3 shows plots of the transition probability (61) vs. the energy difference $(E_s^{(0)} - E_k^{(0)})$, for two different values of the time t. The central peak displays the growth $\propto t^2$ specified by Eq. (65). Outside the central peak, the transition probability oscillates in time.

The presence of the peak centered about the energy $E_k^{(0)} = E_s^{(0)}$ shows that there is a preference for transitions to occur to states of energy $E_k^{(0)}$ in the interval $E_s^{(0)} - 2\pi\hbar/t < E_k^{(0)} < E_s^{(0)} + 2\pi\hbar/t$. This preference becomes more and more marked as time goes on, that is, the peak grows and becomes more narrow. In the limit of large t, transitions to states of energy $E_k^{(0)} \simeq E_s^{(0)}$ are by far the most likely. The reciprocal relationship between the elapsed time and the width ΔE of the preferred energy interval is an instance of the energy–time uncertainty relation.

Note that a transition from one state to another involves a change in the energy of the system. The energy change is accounted for by the work done by the perturbation acting on the system.

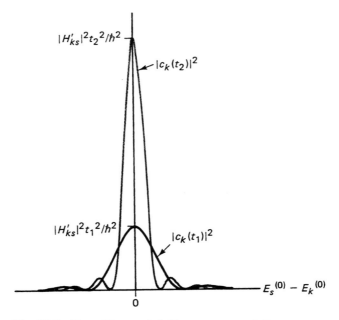

Fig. 10.3 Transition probability vs. energy difference. The gray curve is plotted for a time twice as large as the black curve.

Next, we consider the effects of a perturbation that varies periodically in time. For instance, such a perturbation is provided by an electromagnetic wave incident on an atom. If the frequency of oscillation of the perturbation is ω, then

$$H'_{ks}(t) = H_{ks}(0) \cos \omega t$$

$$= H'_{ks}(0) \frac{e^{i\omega t} + e^{-i\omega t}}{2} \tag{66}$$

Integration of Eq. (58) with this perturbation yields

$$c_k(t) = \frac{1}{2i\hbar} H'_{ks}(0) \int_0^t \left(e^{-i(E_s^{(0)} - E_k^{(0)} - \hbar\omega)t/\hbar} + e^{-i(E_s^{(0)} - E_k^{(0)} + \hbar\omega)t/\hbar} \right) dt$$

$$= \frac{1}{2} H'_{ks}(0) \left(\frac{e^{-i(E_s^{(0)} - E_k^{(0)} - \hbar\omega)t/\hbar} - 1}{E_s^{(0)} - E_k^{(0)} - \hbar\omega} + \frac{e^{-i(E_s^{(0)} - E_k^{(0)} + \hbar\omega)t/\hbar} - 1}{E_s^{(0)} - E_k^{(0)} + \hbar\omega} \right) \tag{67}$$

Obviously, the preferred transitions are those for which $E_s^{(0)} - E_k^{(0)} = \pm\hbar\omega$. For these transitions, $|c_k(t)|^2$ increases as t^2. Transitions with $E_k^{(0)} = E_s^{(0)} + \hbar\omega$ correspond to the *absorption* of energy quanta $\hbar\omega$ from the perturbing field. Transitions with $E_k^{(0)} = E_s^{(0)} - \hbar\omega$ correspond to *stimulated emission*, or induced emission, of quanta. In both cases, the system is in resonance with the applied perturbation.

10.4 *Transitions to the Continuum*

An important problem is the calculation of the transition probability from some initial state to a set of final states that are continuously distributed over a range of energies, that is, a set of final states lying in the continuum. For instance, the initial state might be a bound state of an atom and the final state an unbound state, of positive energy. This situation arises in many scattering problems, where an incident electromagnetic wave or an incident beam of particles of some kind perturbs an atom and causes an electron to make a transition from a bound state to an unbound state, with a loss of energy for the incident wave. The transition probabilities for the final states are then characterized by a set of coefficients c_k with a continuously variable index k, instead of a discrete index, and the existence of a continuous distribution of final states requires some tricky modifications of our formula for the transition probability.

Let us consider transitions to a set of final states of energy near $E_k^{(0)}$, say, energy in an interval from $E_k^{(0)} - \frac{1}{2}\varepsilon$ to $E_k^{(0)} + \frac{1}{2}\varepsilon$, where

ε is small. We can define a *density of states* as

$$\frac{dN}{dE_k{}^{(0)}} = \frac{1}{\varepsilon} \text{ [number of states in energy interval } \varepsilon\text{]}$$

$$= \text{[number of states per unit energy interval]}$$

This density of states is usually a function of the energy $E_k{}^{(0)}$. We will work out an explicit formula for the density of states for a free particle in the next section.

First, we consider transitions generated by a constant perturbation H'. The transition probability from the initial state s to the set of states in the energy interval ε is

$$P(t) \atop {s \to \text{set}} = \sum_{E_k{}^{(0)}=E_s{}^{(0)}-\frac{1}{2}\varepsilon}^{E_k{}^{(0)}=E_s{}^{(0)}+\frac{1}{2}\varepsilon} |c_k(t)|^2$$

$$= \sum_{E_k{}^{(0)}=E_s{}^{(0)}-\frac{1}{2}\varepsilon}^{E_k{}^{(0)}=E_s{}^{(0)}+\frac{1}{2}\varepsilon} 4|H'_{ks}|^2 \frac{\sin^2(E_s{}^{(0)} - E_k{}^{(0)})t/2\hbar}{(E_s{}^{(0)} - E_k{}^{(0)})^2}$$

Here the final energy interval is assumed to be centered around the initial energy E_s; according to the discussion following Eq. (65), this is the preferred energy interval in the limit of large t. Since we are dealing with a continuous distribution of states, the sum in these equations is a continuous sum, that is, an integral:

$$P(t) \atop {s \to \text{set}} = \int_{E_k{}^{(0)}=E_s{}^{(0)}-\frac{1}{2}\varepsilon}^{E_k{}^{(0)}=E_s{}^{(0)}+\frac{1}{2}\varepsilon} 4|H'_{ks}|^2 \frac{\sin^2(E_s{}^{(0)} - E_k{}^{(0)})t/2\hbar}{(E_s{}^{(0)} - E_k{}^{(0)})^2} \, dN$$

$$= \int_{E_k{}^{(0)}=E_s{}^{(0)}-\frac{1}{2}\varepsilon}^{E_k{}^{(0)}=E_s{}^{(0)}+\frac{1}{2}\varepsilon} 4|H'_{ks}|^2 \frac{\sin^2(E_s{}^{(0)} - E_k{}^{(0)})t/2\hbar}{(E_s{}^{(0)} - E_k{}^{(0)})^2} \frac{dN}{dE_k{}^{(0)}} \, dE_k{}^{(0)} \tag{68}$$

Provided that ε is small enough, so H'_{ks} and $dN/dE_k{}^{(0)}$ do not vary appreciably over the range of energies ε, we can treat both of these quantities as constants, and we find

$$P(t) \atop {s \to \text{set}} = 4|H'_{ks}|^2 \frac{dN}{dE_k{}^{(0)}} \int_{E_k{}^{(0)}=E_s{}^{(0)}-\frac{1}{2}\varepsilon}^{E_k{}^{(0)}=E_s{}^{(0)}+\frac{1}{2}\varepsilon} \frac{\sin^2(E_s{}^{(0)} - E_k{}^{(0)})t/2\hbar}{(E_s{}^{(0)} - E_k{}^{(0)})^2} \, dE_k{}^{(0)}$$

$$= 4|H'_{ks}|^2 \frac{dN}{dE_k{}^{(0)}} \frac{t}{2\hbar} \int_{-t\varepsilon/2\hbar}^{t\varepsilon/2\hbar} \frac{\sin^2\xi}{\xi^2} \, d\xi \tag{69}$$

where $\xi \equiv t(E_k{}^{(0)} - E_s{}^{(0)})/2\hbar$. Next, we assume that $t\varepsilon/\hbar \gg 1$, which means that we look at the transition probability after a fairly long time has elapsed. Since most of the contribution to the integral comes from the interval $-\pi < \xi < \pi$, the value of the integral is not sensitive to the exact value of the limits of integration, as long as they are large compared with π. We can then just as well replace the large limits of integration by $\pm\infty$, without significant error:

$$
\begin{aligned}
P(t) &= 4|H'_{ks}|^2 \frac{dN}{dE_k^{(0)}} \frac{t}{2\hbar} \int_{-\infty}^{\infty} \frac{\sin^2\xi}{\xi^2} \, d\xi \\
&\xrightarrow{s \to \text{set}}
\end{aligned}
$$

$$
= 4|H'_{ks}|^2 \frac{dN}{dE_k^{(0)}} \frac{t\pi}{2\hbar} \tag{70}
$$

This shows that the transition probability increases in direct proportion to the time. We will designate the transition probability per unit time, or the *transition rate*, by $W_{s \to k}$:

$$
W_{s \to k} = \frac{2\pi}{\hbar} |H'_{ks}|^2 \frac{dN}{dE_k^{(0)}} \tag{71}
$$

This equation for the transition rate is called *Fermi's Golden Rule*.

Note that the proportionality of the transition probability to t arises from a combination of two factors: the probability of transitions to states of energy very near $E_s^{(0)}$ increases as t^2, but the number of such states decreases as $1/t$ [in Fig. 10.3, the height of the peak increases as t^2, but the width of the peak along the energy axis decreases as $1/t$; the width of the peak is proportional to the number of states, $dN = (dN/dE_k) \, \Delta E$].

The preceding calculation assumed a constant perturbation. Next, we consider an oscillating perturbation, of the form given in Eq. (66). To obtain a relatively large transition rate, one or the other of the denominators in Eq. (67) must be at or near zero, that is, the energy $E_k^{(0)}$ must be at or near $E_s^{(0)} + \hbar\omega$ or $E_s^{(0)} - \hbar\omega$. Let us suppose that the energy is at or near $E_s^{(0)} + \hbar\omega$, which corresponds to an upward transition, with absorption of a quantum $\hbar\omega$. Then the second term in parentheses in Eq. (67) is dominant, and the first term can be neglected. By a calculation similar to that for a constant perturbation, we find that the transition rate is

$$
W_{s \to k} = \frac{2\pi}{\hbar} \left| \frac{1}{2} H'_{ks}(0) \right|^2 \frac{dN}{dE_k^{(0)}} \tag{72}
$$

Exercise 12. Supply the steps of the calculation leading to Eq. (72).

We can interpret the result (72) as follows: In the perturbation

$$H'_{ks}(t) = \frac{1}{2} H'_{ks}(0) \, e^{i\omega t} + \frac{1}{2} H'_{ks}(0) \, e^{-i\omega t}$$

the *second* term produces upward transitions to continuum states of energy $E_k^{(0)} = E_s^{(0)} + \hbar\omega$ with a matrix element given by the coefficient of $e^{-i\omega t}$. The *first* term produces downward transitions to continuum states (if any) of energy $E_k^{(0)} = E_s^{(0)} - \hbar\omega$, again with a matrix element equal to the coefficient of $e^{i\omega t}$. The two terms behave (approximately) as independent perturbations.

10.5   *The Density of States*

Before we can apply the Golden Rule to a problem, we need to find the density of states, $dN/dE_k^{(0)}$. For the sake of simplicity, we will assume that the final states are those of a single particle, say, an electron, and that the interaction of the particle with the remainder of the system can be neglected for these final states. The final states are then free-particle states, that is, they are momentum eigenstates with wavefunctions $e^{i\mathbf{p}\cdot\mathbf{r}/\hbar}$. These wavefunctions cannot be normalized to unit probability. In earlier chapters we introduced a normalizing factor $1/(2\pi\hbar)^{3/2}$ for these wavefunctions. For our present purposes, this is somewhat inconvenient, because it leaves the wavefunctions with the wrong dimensions. To avoid this difficulty, we will use a trick: instead of normalizing the wavefunction by integration over all space, we integrate only over a box of volume V, which we choose to be very large. "Normalized" in this way, our momentum wavefunctions are

$$\psi_{\mathbf{p}}(\mathbf{r}) = \frac{1}{\sqrt{V}} \, e^{i\mathbf{p}\cdot\mathbf{r}/\hbar} \tag{73}$$

At the end of our calculations we will take the limit $V \to \infty$; this leads to no trouble, because, as we will see, the volume V always cancels out.

From the wavefunction (73) we can deduce the density of free-particle states in the volume V. We note that

$$\frac{V}{h^3} \int \psi_{\mathbf{p}}^*(\mathbf{r})\psi_{\mathbf{p}}(\mathbf{r}') \, d^3p = \frac{V}{\hbar^3} \frac{1}{(2\pi)^3} \int \frac{1}{V} e^{(-i\mathbf{p}\cdot\mathbf{r} + i\mathbf{p}\cdot\mathbf{r}')/\hbar} \, d^3p$$

$$= \delta(\mathbf{r} - \mathbf{r}') \tag{74}$$

Let us compare this with the completeness relation, written as a sum:

$$\sum_p \psi_p^*(\mathbf{r})\psi_p(\mathbf{r}') = \delta(\mathbf{r} - \mathbf{r}') \tag{75}$$

The sum can be converted into an integral by introducing the number $dN(\mathbf{p})$ of states in the momentum interval d^3p:

$$\sum_p \psi_p^*(\mathbf{r})\psi_p(\mathbf{r}') = \int \psi_p^*(\mathbf{r})\psi_p(\mathbf{r}') \, dN(\mathbf{p}) \tag{76}$$

Thus, comparison of Eqs. (74) and (76) tells us that the number of states is

$$dN(\mathbf{p}) = \frac{V}{h^3} \, d^3p \tag{77}$$

Expressed in words, this says that

[number of momentum states]

$$= \frac{[\text{volume in space}] \times [\text{volume in momentum space}]}{h^3} \tag{78}$$

In polar coordinates, we can describe the momentum vector $\mathbf{p}$ by its magnitude p and the polar angles θ and ϕ. The volume element d^3p then becomes $p^2 \, dp \sin\theta \, d\theta \, d\phi$, or $p^2 \, dp \, d\Omega$, where $d\Omega$ is the solid angle, and Eq. (77) becomes

$$dN = \frac{V}{h^3} \, p^2 \, dp \, d\Omega$$

But the momentum is related to the energy by $p^2 = 2mE$; hence $p^2 \, dp = 2mE \, m \, dE/p = m\sqrt{2mE} \, dE$ and

$$\frac{dN}{dE} = \frac{V}{h^3} \, m\sqrt{2mE} \, d\Omega \tag{79}$$

We have here obtained the density of states by examining the normalization of the momentum wavefunctions. Alternatively, the density of states can be obtained by directly counting the number of possible standing waves within a box of volume V (see Problem 18).

10.6 *The Photoelectric Effect*

An electromagnetic wave incident on an atom will perturb the electrons and sometimes eject an electron from the atom. We can use perturbation theory to calculate the transition rate and the cross section for this process. Suppose that we are dealing with a hydrogen atom initially in its ground state and that the electron is ejected from the atom with an energy large compared with 13.6 eV (but not so large that the electron becomes relativistic). The ejected electron can then be treated as approximately free, that is, we can neglect the residual interaction of the unbound electron with the nucleus.

We will treat the incident electromagnetic wave classically. Its electric and magnetic fields are derivable from a vector potential $\mathbf{A}$ according to the familiar formulas:

$$\mathbf{E} = -\frac{\partial \mathbf{A}}{\partial t}$$

$$\mathbf{B} = \nabla \times \mathbf{A} \tag{80}$$

The interaction of the wave with the electron is described by a Hamiltonian that includes the vector potential:

$$H = \frac{1}{2m}\left(\mathbf{p}_{op} + e\mathbf{A}\right)\cdot\left(\mathbf{p}_{op} + e\mathbf{A}\right) - \frac{1}{4\pi\varepsilon_0}\frac{e^2}{r} \tag{81}$$

This Hamiltonian is obtained from the usual one by replacing $\mathbf{p}_{op}$ by $\mathbf{p}_{op} + e\mathbf{A}$. This rule can be justified by examining the equation of motion of a classical charged particle in an electromagnetic field.

Exercise 13. Show that with

$$H = \frac{1}{2m}\left(\mathbf{p} - q\mathbf{A}\right)\cdot\left(\mathbf{p} - q\mathbf{A}\right) \tag{82}$$

the Hamiltonian equation $dx_n/dt = \partial H/\partial p_n$ yields a canonical momentum

$$\mathbf{p} = m\mathbf{v} + q\mathbf{A} \tag{83}$$

and the Hamiltonian equation $dp_n/dt = -\partial H/\partial x_n$ yields the classical equation of motion

$$m\frac{d^2 x_n}{dt^2} = qE_n + q(\mathbf{v} \times \mathbf{B})_n \tag{84}$$

[It is important to keep in mind that, according to Eq. (83), the canonical momentum $\mathbf{p}$ differs from the kinetic momentum $m\mathbf{v}$.]

The components of the canonical momentum $\mathbf{p}_{op}$ obey the standard commutation relation with x, y, and z. Hence, in the position representation, the canonical momentum operator takes the form

$$\mathbf{p}_{op} = \frac{\hbar}{i}\,\boldsymbol{\nabla} \tag{85}$$

Note that $\mathbf{p}_{op}$ does not commute with $\mathbf{A}$, because the vector potential depends on $\mathbf{r}$. In Eq. (81), the product of $\mathbf{p}_{op}$ and $\mathbf{A}$ enters symmetrically; the symmetrization of the product of $\mathbf{p}_{op}$ and $\mathbf{A}$ is required to make the Hamiltonian hermitian. If we multiply out the square in Eq. (81), we obtain

$$H = \frac{1}{2m}\,\mathbf{p}_{op}{}^2 + \frac{e}{2m}\,(\mathbf{p}_{op}\cdot\mathbf{A} + \mathbf{A}\cdot\mathbf{p}_{op}) + \frac{e^2}{2m}\,\mathbf{A}^2 - \frac{1}{4\pi\varepsilon_0}\frac{e^2}{r} \tag{86}$$

For reasonable values of the vector potential, the third term, proportional to $\mathbf{A}^2$, is small compared with the second term; we will hereafter neglect the third term. Thus, the perturbation associated with the vector potential $\mathbf{A}$ is approximately

$$H' = \frac{e}{2m}\,(\mathbf{p}_{op}\cdot\mathbf{A} + \mathbf{A}\cdot\mathbf{p}_{op}) \tag{87}$$

or, in the position representation,

$$H' = \frac{e\hbar}{2im}\,(\boldsymbol{\nabla}\cdot\mathbf{A} + \mathbf{A}\cdot\boldsymbol{\nabla}) \tag{88}$$

For a plane electromagnetic wave of frequency ω incident from the negative y direction, with an electric field in the z direction, the vector potential is

$$\mathbf{A} = A_0\hat{\mathbf{z}}\,\cos(\omega t - ky) \tag{89}$$

where $k = \omega/c$.

Exercise 14. Show that the electric and magnetic fields in this wave are

$$\mathbf{E} = A_0\omega\hat{\mathbf{z}}\,\sin(\omega t - ky) \tag{90}$$

$$\mathbf{B} = A_0k\hat{\mathbf{x}}\,\sin(\omega t - ky) = A_0\,\frac{\omega}{c}\,\hat{\mathbf{x}}\,\sin(\omega t - ky) \tag{91}$$

The vector potential given by Eq. (89) has zero divergence, and hence the operator $\boldsymbol{\nabla}\cdot$ commutes with $\mathbf{A}$:

$$\boldsymbol{\nabla}\cdot\mathbf{A} = \mathbf{A}\cdot\boldsymbol{\nabla} \tag{92}$$

Exercise 15. Show this.

Thus, the perturbation H' becomes

$$H' = \frac{e\hbar}{im} \mathbf{A} \cdot \nabla \tag{93}$$

or

$$H' = \frac{e\hbar}{2im} A_0(e^{i(\omega t - ky)} + e^{-i(\omega t - ky)}) \frac{\partial}{\partial z} \tag{94}$$

We will assume that the wavelength of the electromagnetic wave is much larger than the Bohr radius a_0. Then e^{iky} does not vary appreciably over the volume of the atom and can be replaced by its value 1 at the origin (which is assumed to coincide with the center of the atom). Furthermore, since we want to consider upward transitions, with absorption of energy, only the second exponential in Eq. (94) contributes. The transition rate from the ground state to a state of momentum $\mathbf{p}$ is then

$$W_{100 \rightarrow \mathbf{p}} = \frac{2\pi}{\hbar} \left| \langle \psi_{\mathbf{p}} | \frac{e\hbar}{2im} A_0 \frac{\partial}{\partial z} | \psi_{100} \rangle \right|^2 \frac{dN(E)}{dE} \tag{95}$$

where $E = \hbar\omega - 13.6$ eV $\simeq \hbar\omega$.

To complete the calculation of this transition rate, we need the matrix element of $\partial/\partial z$:

$$\langle \psi_{\mathbf{p}} | \frac{\partial}{\partial z} | \psi_{100} \rangle = \int \frac{e^{-i\mathbf{p} \cdot \mathbf{r}/\hbar}}{\sqrt{V}} \frac{\partial}{\partial z} \frac{2}{a_0^{3/2} \sqrt{4\pi}} e^{-r/a_0} d^3r$$

$$= -\frac{2}{a_0^{3/2} \sqrt{4\pi V}} \int \left(\frac{\partial}{\partial z} e^{-i\mathbf{p} \cdot \mathbf{r}/\hbar} \right) e^{-r/a_0} d^3r$$

$$= \frac{ip_z}{\hbar} \frac{1}{a_0^{3/2} \sqrt{\pi V}} \int \exp(-i\mathbf{p} \cdot \mathbf{r}/\hbar - r/a_0) d^3r \tag{96}$$

For the evaluation of this integral, we use a new coordinate system x', y', z' with the z' axis along the direction of the momentum $\mathbf{p}$. Then the integral becomes

$$\int \exp(-i\mathbf{p} \cdot \mathbf{r}/\hbar - r/a_0) d^3r$$

$$= \int \int \int \exp\left[-\frac{i}{\hbar} pr \cos \theta' - r/a_0 \right] r^2 dr \sin \theta' d\theta' d\phi'$$

$$= \frac{8\pi/a_0}{\left(\frac{1}{a_0^2} + \frac{p^2}{\hbar^2} \right)^2} \tag{97}$$

Exercise 16. Perform the integrations over ϕ', θ', and r, and obtain the result (97).

We also need the density of states

$$\frac{dN(E)}{dE} = \frac{V}{h^3} \, m \sqrt{2mE} \, d\Omega = \frac{V}{h^3} \, m \sqrt{2m\hbar\omega} \, d\Omega \qquad (98)$$

Combining Eqs. (95), (96), and (97), we obtain

$$W_{100 \to \mathbf{p}} = \frac{2\pi}{\hbar} \left(\frac{4eA_0}{m} \right)^2 \frac{\pi}{a_0^5} \frac{p_z^2}{\left(\dfrac{1}{a_0^2} + \dfrac{p^2}{\hbar^2} \right)^4} \frac{m \sqrt{2m\hbar\omega}}{h^3} \, d\Omega$$

or with $p_z = p \cos\theta = \sqrt{2m\hbar\omega} \cos\theta$,

$$W_{100 \to \mathbf{p}} = \frac{4}{\pi} \frac{e^2 A_0^2}{m\hbar^4 a_0^5} \frac{\cdot (2m\hbar\omega)^{3/2}}{\left(\dfrac{1}{a_0^2} + \dfrac{2m\omega}{\hbar} \right)^4} \cos^2\theta \, d\Omega \qquad (99)$$

Note that the volumes V have canceled out. The transition rate is zero for electrons with $p_z = 0$. This can be understood by looking at the electric field (90); it points in the z direction, and therefore can eject electrons only in that direction.

We now define a *differential cross section* $d\sigma$ such that

$d\sigma \times$ [incoming photon flux]
    = [electron emission rate into solid angle $d\Omega$ at θ, ϕ] (100)

that is,

$$d\sigma = \frac{W_{100 \to \mathbf{p}}}{[\text{photon flux}]} \qquad (101)$$

We can calculate the photon flux from the energy flux, or the Poynting vector:

$$[\text{photon flux}] = \frac{[\text{energy flux}]}{\hbar\omega} = \frac{\overline{\mathbf{E}^2}}{c\mu_0\hbar\omega} \qquad (102)$$

where $\overline{\mathbf{E}^2}$ is the time-average value of $\mathbf{E}^2$. According to Eq. (90), this time-average value is $\omega^2 A_0^2/2$. The photon flux is then $\omega A_0^2/2c\mu_0\hbar$, and

$$\frac{d\sigma}{d\Omega} = \frac{8\mu_0 c}{\pi} \frac{e^2}{m\omega\hbar^3 a_0^5} \frac{(2m\hbar\omega)^{3/2}}{\left(\dfrac{1}{a_0^2} + \dfrac{2m\omega}{\hbar} \right)^4} \cos^2\theta \qquad (103)$$

Note that in obtaining this result we have used $\hbar\omega \gg 13.6$ eV and $\lambda = 2\pi c/\omega \gg a_0$. This implies that $\hbar\omega$ must be in the range 10^4 eV $\gg \hbar\omega \gg 10$ eV.

PROBLEMS   1. Consider a particle of mass m in an infinite square potential well that extends in one dimension from $x = 0$ to $x = L$. What is the first-order correction to the ground-state energy and wavefunction if this potential is subjected to a perturbation $H' = \frac{1}{2}kx^2$? What condition must be imposed on k if perturbation theory is to yield a good approximation?

2. Consider a particle of mass m in an infinite square potential well that extends in one dimension from $x = 0$ to $x = L$. Suppose that the potential is perturbed by an extra potential (see Fig. 10.4)

$$H' = \begin{cases} -V_0 & \text{for } x < \dfrac{L}{2} \\[2ex] 0 & \text{for } x > \dfrac{L}{2} \end{cases}$$

What is the first-order correction to the energy eigenvalue of the nth eigenstate? What is the first-order correction to the wavefunction of the nth eigenstate? What condition must be imposed on V_0 if perturbation theory is to yield a good approximation?

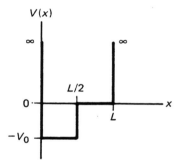

Fig. 10.4   An infinite potential well with a perturbation $-V_0$.

3. A harmonic oscillator consists of a particle of mass m moving along the x axis with a potential $V(x) = \frac{1}{2}m\omega^2 x^2$. The particle has a charge e and is subjected to a perturbing electric field $\mathscr{E}$ in the x direction.

   (a) To first order in $\mathscr{E}$, what is the energy of the nth excited state?

   (b) To first order in $\mathscr{E}$, what are the coefficients c_{mn}?

   (c) Find the *exact* perturbed energy eigenvalues and the *exact* wavefunctions. (Hint: Consider what happens to the potential $\frac{1}{2}m\omega^2 x^2$ if you replace x by $x - b$, which is a shift of the origin of coordinates.)

4. A hydrogen atom is placed in a uniform electric field directed along the z axis. This electric field produces a perturbation $e\mathscr{E}z$ in the Hamiltonian.

(a) Is $\mathbf{L}^2$ a constant of the motion?

(b) Is L_z a constant of the motion?

(c) Are the eigenstates of the perturbed Hamiltonian eigenstates of parity?

5. Consider an anharmonic oscillator consisting of a particle of mass m moving along the x axis in a potential $V(x) = \frac{1}{2}m\omega^2 x^2 + bx^4$. Treating the term bx^4 as a perturbation, find the energy of the ground state, in first order.

6. A particle of mass m is confined to a one-dimensional infinite square potential well that extends from $x = 0$ to $x = L$. The energy eigenvalues for the (nonrelativistic) Hamiltonian are (see Section 3.2)

$$E_n = \frac{n^2\pi^2\hbar^2}{2mL^2}$$

If the mass of the particle is small or if the length L is small, the energy eigenvalues will be large, and the particle may become relativistic (this happens if the energy is comparable with or larger than mc^2). The relativistic Hamiltonian is

$$H = \sqrt{p^2c^2 + m^2c^4} - mc^2$$

(a) Use first-order perturbation theory to find the new energy eigenvalues that correspond to this relativistic Hamiltonian.

(b) The energy eigenvalues obtained by first-order perturbation theory are actually the *exact* eigenvalues for the relativistic Hamiltonian. Explain carefully why this is so.

7. Because of the relativistic increase of mass, the usual (nonrelativistic) Hamiltonian describing the electron in the hydrogen atom must be corrected by an extra term $-p^4/8m^3c^2$. The new Hamiltonian is then

$$H = \frac{p^2}{2m} - \frac{1}{4\pi\varepsilon_0}\frac{e^2}{r} - \frac{1}{8}\frac{p^4}{m^3c^2}$$

Use first-order perturbation theory to find the correction to the energy of the ground state.

8. The actual charge distribution of the nucleus of hydrogen (proton) is

$$\rho = \frac{e}{8\pi b^3}\,e^{-r/b}$$

where $b = 0.23 \times 10^{-15}$ m. Use perturbation theory to calculate the change that this charge distribution produces in the energy of the ground state of hydrogen; express your answer in eV.

9. In a one-dimensional anharmonic oscillator, a particle of mass m moves along the x axis in a potential $V(x) = \frac{1}{2}m\omega^2 x^2 + bx^3$. Assume

that the term bx^3 is small. Find the change in the energy of the ground state, in first order and in second order.

10. A particle of mass m bound in a one-dimensional harmonic-oscillator potential $V(x) = \frac{1}{2}m\omega^2 x^2$ is perturbed by a second harmonic–oscillator potential $V'(x) = \frac{1}{2}m\omega'^2(x - b)^2$, centered on the point b. The first excited state has energy $\frac{3}{2}\hbar\omega$ when unperturbed.

    (a) Using second-order perturbation theory, find the change in the energy of this state caused by the perturbation.

    (b) Find the perturbed state vector for the first excited state according to first-order perturbation theory.

11. A hydrogen atom is placed in a uniform magnetic field **B** in the z direction. If we ignore the spin of the electron, the perturbing Hamiltonian is then $(e/2m)B_z L_z$. Find the energies and the coefficients c_{mn} for the excited states of quantum numbers $n = 2$, $l = 1$, $m_l = -1, 0$, $+1$. Draw the energy-level diagram of the hydrogen atom for $n = 1$ and $n = 2$ in the presence of this perturbation. Into how many separate energy levels does the $n = 2$ level split? (This is the *normal* Zeeman effect.)

12. Suppose that the electron in a hydrogen atom is perturbed by a repulsive potential concentrated at the origin. Assume that the potential has the form of a delta function, so the perturbed Hamiltonian is

$$H = \frac{\mathbf{p}^2}{2m} - \frac{1}{4\pi\varepsilon_0}\frac{e^2}{r} + A\,\delta(\mathbf{r})$$

    where A is a constant.

    (a) To first order in A, find the change in the energy of the state with quantum numbers $n \geq 1$, $l = 0$. [Hint: $\psi_{n00}(0) = 2/\sqrt{4\pi}\,(na_0)^{3/2}$.]

    (b) Find the change in the wavefunction.

13. Show that the equation for $c_{mn}^{(2)}$ is, if $m \neq n$,

$$c_{mn}^{(2)} = \frac{1}{E_n^{(0)} - E_m^{(0)}}\left(-\frac{H'_{mn}H'_{nn}}{E_n^{(0)} - E_m^{(0)}} + \sum_{k\neq n}\frac{H'_{mk}H'_{kn}}{E_n^{(0)} - E_k^{(0)}}\right)$$

14. The Hamiltonian for the helium atom is

$$H = \frac{1}{2m}\mathbf{p}_1^2 + \frac{1}{2m}\mathbf{p}_2^2 - \frac{1}{4\pi\varepsilon_0}\frac{2e^2}{r_1} - \frac{1}{4\pi\varepsilon_0}\frac{2e^2}{r_2} + \frac{1}{4\pi\varepsilon_0}\frac{e^2}{|\mathbf{r}_2 - \mathbf{r}_1|}$$

The first two terms in the potential energy represent the interaction of each electron with the nucleus; these terms are hydrogen-like. The last term represents the interaction of the electrons with each other. If the last term were absent, each electron would behave like the electron in a hydrogen atom with nuclear charge $2e$, and the energy of the ground state of the helium atom would be $2 \times 2^2 \times (-13.6\text{ eV})$

$= -108.8$ eV. Treat the interaction between the electrons as a perturbation, and find the first-order correction to the energy. Compare your result for the ground-state energy of the helium atom with the observed value, -78.6 eV. [Hint: To evaluate the double integral

$$\iint \frac{1}{4\pi\varepsilon_0} \frac{e^2}{|\mathbf{r}_2 - \mathbf{r}_1|} |\psi(r_1)|^2 \, |\psi(r_2)|^2 \, dV_1 \, dV_2$$

note that it has the form of the electrostatic interaction energy of two (overlapping) charge distributions, $-e|\psi(r_1)|^2$ and $-e|\psi(r_2)|^2$. Begin by finding the electrostatic potential generated by $-e|\psi(r_1)|^2$; then calculate the energy of the second charge distribution $-e|\psi(r_2)|^2$ immersed in this potential.]

15. An infinite cubical potential well extends from $x = 0$ to $x = L$, $y = 0$ to $y = L$, and $z = 0$ to $z = L$. The first excited state of a particle in this well has an energy

$$E_{211} = \frac{6\hbar^2\pi^2}{2mL^2}$$

This state is threefold degenerate. Suppose that the potential is subjected to a perturbation $H' = bxy$. To lowest order in b, find the perturbed energies that arise from these three degenerate states.

16. Suppose that a particle of mass m is initially in the ground state of a one-dimensional infinite square well, which extends from $x = 0$ to $x = L$. If at time $t = 0$ we suddenly move the right boundary of the well from $x = L$ to $x = 2L$, what is the probability that the particle will later be found in the new ground state? In the first excited state?

17. Suppose that a particle of mass m is initially in the ground state of a one-dimensional harmonic oscillator with a potential $V(x) = \frac{1}{2}m\omega^2x^2$. If at time $t = 0$ we suddenly increase the spring constant by a factor of 4, what is the probability that the particle will later be found in the new ground state? In the new first excited state? In the new second excited state?

18. Consider a cubical infinite potential well of dimensions $L \times L \times L$. The standing waves in this potential well can be characterized by three quantum numbers n_1, n_2, and n_3 which represent the number of half-wavelengths in the x, y, and z directions, respectively. The energy of the particle is

$$E_{n_1n_2n_3} = \frac{\hbar^2\pi^2}{2mL^2} (n_1{}^2 + n_2{}^2 + n_3{}^2)$$

Count the number dN of possible standing-wave modes in an energy interval dE, and show that this leads to the value of dN/dE given in Eq. (79).

19. Suppose that an electron in a one-dimensional harmonic-oscillator potential $\frac{1}{2}m\omega_0^2 x^2$ is subjected to an oscillating electric field $\mathscr{E} = \mathscr{E}(0) \cos \omega t$ in the x direction.

    (a) If the electron is initially in the ground state, what is the probability that the electron will be in the nth excited state at time t?

    (b) If $\omega = \omega_0$, perturbation theory will fail at some time t. What is the critical time?

20. (a) Find the number dN of momentum states in a small momentum range dp for the case of a particle with one-dimensional motion along the x axis. Assume that the states are normalized in a large "box" of length L.

    (b) Find the density of states dN/dE for this case.

21. In the case of an electromagnetic wave of optical frequency ($\omega \simeq 4 \times 10^{15}$/s) incident on a hydrogen atom, what is maximum electric field for which $\mathbf{A}^2$ can be neglected in Eq. (86)?

22. Suppose that a hydrogen atom, initially in the ground state, is placed in an oscillating electric field $\mathscr{E}_0 \cos \omega t$ in the z direction, with $\hbar\omega \gg$ 13.6 eV. Calculate the rate of transitions to the continuum, and compare your result with that obtained in Section 10.6 for an electromagnetic wave.

23. Suppose that a hydrogen atom, initially in the ground state, is placed in an oscillating magnetic field $B_0 \cos \omega t$ in the x direction, with $\hbar\omega \gg$ 13.6 eV. Calculate the rate of transitions to the continuum and compare your result with that obtained in Section 10.6 for an electromagnetic wave.

24. An electron is initially in the ground state of a very deep (but finite) one-dimensional square potential well (see Fig. 10.5) An oscillating electric field $\mathscr{E} = \mathscr{E}_0 \cos \omega t$ parallel to the x axis acts on the electron.

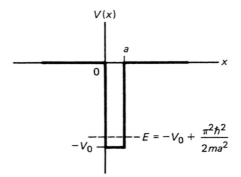

Fig. 10.5 A very deep, but finite potential well.

Assume that the initial state is approximately

$$\psi_1(x) = \begin{cases} \sqrt{\dfrac{2}{a}}\, \sin \dfrac{\pi x}{a} & \text{for } 0 < x < a \\[2ex] 0 & \text{for } x > a \text{ or } x < 0 \end{cases}$$

and that the final state is

$$\psi_p(x) = \frac{e^{ipx/\hbar}}{\sqrt{L}}$$

This final state is normalized in a large one-dimensional box of length L.

(a) What must be the value of p if transitions are to occur?

(b) Find the transition rate for transitions from ψ_1 to ψ_p. The density of states at energy E in the continuum is (see Problem 20) $dN/dE = L\sqrt{m/2E}/2\pi\hbar$.

11

Scattering and Resonances

Most of our knowledge of nuclei and of "elementary" particles and of the forces acting at the nuclear and subnuclear level comes from scattering experiments. In these experiments, a beam of particles is used to bombard a target, and the deflections of the incident particles upon collisions with the target particles are measured. If the energy of the incident particles is sufficiently large, the scattering process can become very complicated, with the creation of new particles in violent reactions.

In this chapter, we will discuss only the simple case of elastic scattering of a particle by a potential. This potential may be thought of as the potential exerted by the target particle on the incident particle; the coordinate x or r that we will use for the incident particle should therefore be thought of as a relative coordinate giving the distance between the particles. The recoil motion of the target particle can be taken into account by introducing the coordinates of the center of mass, as in Section 8.2; but we will not deal with this detail here.

We will first discuss scattering in one dimension. This frees us from the complications imposed by angular momentum, and permits us to bring many of the fundamental concepts into sharper focus.

Throughout this chapter, we will adopt the position representation, since this is the most convenient for scattering problems with potentials defined as functions of position.

11.1 Elastic Scattering in One Dimension

Consider a particle moving in one dimension in a potential that is different from zero only in some finite range. The particle is incident from a large distance, interacts with the potential, and then

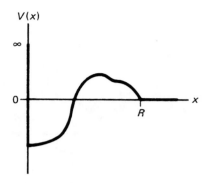

Fig. 11.1   A typical potential in one dimension. The range R of the potential is finite.

again moves out to a large distance. A typical potential is the following (see Fig. 11.1):

$$V = \begin{cases} V(x) & \text{for } 0 < x < R \\ 0 & \text{for } x < R \\ \infty & \text{for } x < 0 \end{cases} \tag{1}$$

The quantity R is called the *range* of the potential.[1]

It might seem strange to use a potential that restricts the motion of the particle to positive values of x. But this case of exceptional importance, because the effective potential for the *radial* motion of a particle moving in three dimensions often has the form given by Eq. (1). If the potential in three dimensions is a function of r only (a central potential), then we can separate the radial motion from the angular motion by assuming that the incident particle is in an eigenstate of angular momentum. The Schrödinger equation for three dimensions then reduces to a Schrödinger equation for one dimension, with r as the one-dimensional coordinate [we saw examples of this reduction of the Schrödinger equation in Eqs. (8.6) and (8.37)]. The radial motion of the particle then reduces to one-dimensional motion with a potential which may be taken as infinite at $r = 0$. Note that although the case of the Coulomb potential fits the pattern given in Eq. (1), the range of

---

[1] For potentials that tend to zero sufficiently quickly as r increases—such as potentials proportional to an exponentially decreasing function—the definition of the range can be relaxed somewhat. But in this section we will concentrate on piecewise constant potentials, for which the range is determined by the point at which the potential disappears.

this potential is infinite, and this introduces some exceptional complications in the scattering problem; we will not deal with the Coulomb potential or other potentials of infinite range.

First we will solve the scattering problem for the trivial case $V(x) = 0$. This means that we are dealing with a free particle which is reflected by the infinitely high barrier at $x = 0$. Suppose that the particle has energy E; then the magnitude of its momentum is $p = \sqrt{2mE}$ and the wave vector is $k = p/\hbar$. The wavefunction that represents the stationary state of positive energy E must have an *incoming* part e^{-ikx} and an *outgoing* part e^{ikx}. These parts describe, respectively, waves traveling to the left and to the right. In the total wavefunction $\phi(x)$, the incoming and outgoing waves must be superposed in such a way that the boundary condition $\phi(0) = 0$ is satisfied. This implies that the only viable superposition is

$$\phi(x) \propto e^{ikx} - e^{-ikx} \qquad (2)$$

In writing this wavefunction, we have omitted the time dependence; it is given by the usual factor $e^{-iEt/\hbar}$.

The wavefunction (2) is not normalizable, but we find it convenient to insert an extra factor $1/2i$, so

$$\phi(x) = \frac{1}{2i} (e^{ikx} - e^{-ikx}) = \sin kx \qquad (3)$$

Next, we must examine the scattering problem with a nonzero potential. We assume that the potential has a finite range R, as in Eq. (1). In the region $x > R$, the wavefunction will again have an incoming and an outgoing part, as in the case of zero potential. For the incoming part, we take $-(1/2i)e^{-ikx}$, exactly the same as in Eq. (3); this will make it easy to see what the effects of the potential are. The outgoing part must be e^{ikx} multiplied by some coefficient. We will use the notation $(1/2i)e^{2i\delta}$ for this coefficient. The wavefunction is then

$$\psi(x) = \frac{1}{2i} (e^{ikx+2i\delta} - e^{-ikx}) \qquad \text{for } x > R \qquad (4)$$

The quantity δ is *real*. Thus, $e^{2i\delta}$ is simply a phase factor, and the intensities of the incoming and outgoing waves are equal. This equality of incoming and outgoing intensities is required by the conservation of probability. The potential cannot destroy or create probability—it cannot destroy or create particles.

The value of δ depends on the potential $V(x)$ and on the energy E; it can be calculated by solving the Schrödinger equation in the interior region, $x < R$. Before we proceed with such a calculation, we briefly consider some general properties of the quantity δ. This quantity is called the *phase shift* for the following reason: Suppose that we compare the probability densities given by Eq. (3) [corresponding to $V(x) = 0$] and by Eq. (4) [corresponding to $V(x) \neq 0$]. These probability densities are

$$|\phi(x)|^2 = \sin^2 kx \tag{5}$$

and

$$|\psi(x)|^2 = \sin^2 (kx + \delta) \tag{6}$$

The comparison of these functions shows that, in the presence of the potential, the maxima and minima of the probability distribution are shifted toward the origin by a distance δ/k relative to their location in the absence[2] of the potential (see Fig. 11.2). As we will see, the value of δ is positive for an attractive potential; the maxima and minima are then shifted toward the origin. The value of δ is negative for a repulsive potential; the maxima are then shifted away from the origin.

We can define a *scattered wave* $\psi_S(x)$ as the difference between $\psi(x)$ and $\phi(x)$:

$$\psi_S(x) = \psi(x) - \phi(x) \qquad \text{for } x > R \tag{7}$$

The scattered wave tells us how much the potential changes the wave from what it is in the absence of the potential. The scattered wave is zero if the potential is absent, which means that there is no scattering.

We can also write $\psi_S(x)$ as follows:

$$\psi_S(x) = \frac{1}{2i} (e^{ikx+2i\delta} - e^{-ikx}) - \frac{1}{2i} (e^{ikx} - e^{-ikx})$$

$$= \frac{e^{ikx}}{2i} (e^{2i\delta} - 1) \tag{8}$$

From this we see that the scattered wave is a purely outgoing wave. The amplitude of this outgoing wave is called the *scatter-*

---

[2] Absence of potential means that the function $V(x)$ in Eq. (1) is zero; but the potential for $x < 0$ remains infinite.

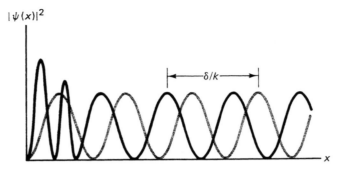

Fig. 11.2 Shift the maxima and minima of the probability distribution. The black curve is $|\psi(x)|^2$; the gray curve is $|\phi(x)|^2$.

ing amplitude:

$$[\text{scattering amplitude}] = \frac{\psi_S}{e^{ikx}} = \frac{1}{2i}\left(e^{2i\delta} - 1\right) \tag{9}$$

The modulus squared of the scattering amplitude is a measure of the probability of scattering, or the strength of scattering:

$$\begin{bmatrix} \text{probability} \\ \text{of scattering} \end{bmatrix} \propto \left|\begin{bmatrix} \text{scattering} \\ \text{amplitude} \end{bmatrix}\right|^2 = |\psi_S|^2 = \left|\frac{e^{2i\delta} - 1}{2i}\right|^2 = \sin^2\delta \tag{10}$$

As we will see in Section 11.4, in three dimensions the cross section is proportional to the square of scattering amplitude. This means that δ or, more precisely, $\sin^2\delta$ is what we can measure by experiment. From the measured value of δ as a function of energy, we can then try to extract information about the interaction responsible for the scattering.

Another interesting property of δ is the following: $2\hbar d\delta/dE$ gives the *time delay* that the incident particle suffers by having to pass through the potential. To understand this, suppose that we send a wave packet toward the origin from some large distance. The packet then returns to us after a certain time, having passed through the potential and having been reflected at the origin. The time delay is defined as the difference between the times taken when there is a potential and when there is no potential. This time "delay" can be either positive or negative. A negative "delay" means that the particle spends less time in the region of the potential than a free particle. This can happen if the potential is attractive, and therefore increases the speed of the particle. But it can also happen for a repulsive potential, since then the particle's path length can be shortened if it is reflected before it reaches the origin.

For the calculation of the time delay, we suppose that our wave packet has an average energy E_0. The incoming wave packet is then

$$\psi_{\text{in}}(x, t) = \int f(E) \, e^{-i\sqrt{2mE}\,x/\hbar} e^{-iEt/\hbar} \, dE \qquad (11)$$

where $f(E)$ is some function peaked around $E = E_0$. Here we have written the wave packet as an integral over energies rather than an integral over momenta (as we did in Chapter 2), because it better suits our present purposes; of course, it is easy to change the variable of integration from p to E, or vice versa. According to Eq. (4), corresponding to an incoming wave $e^{-i\sqrt{2mE}\,x/\hbar}$, there is an outgoing wave $-e^{i\sqrt{2mE}\,x/\hbar + 2i\delta(E)}$. In the latter wave, we have indicated the energy dependence of δ explicitly. Hence

$$\psi_{\text{out}}(x, t) = -\int f(E) \, e^{i\sqrt{2mE}\,x/\hbar + 2i\delta(E)} e^{-iEt/\hbar} \, dE \qquad (12)$$

Provided that the peak of $f(E)$ at $E = E_0$ is sufficiently narrow, we can approximate $\delta(E) \simeq \delta(E_0) + \delta'(E_0)(E - E_0)$ and therefore

$$\psi_{\text{out}}(x, t) = -e^{2i\delta(E_0)-iE_0 t/\hbar} \int f(E) \, e^{i\sqrt{2mE}\,x/\hbar} e^{-i(E-E_0)[t - 2\hbar\delta'(E_0)]/\hbar} \, dE \qquad (13)$$

Let us compare this with the corresponding wave packet in the absence of the potential. The latter is

$$\psi_{\text{in}}(x, t) = -e^{-iE_0 t/\hbar} \int f(E) \, e^{i\sqrt{2mE}\,x/\hbar} e^{-i(E_0-E)t/\hbar} \, dE \qquad (14)$$

Ignoring the overall phase factors that stand in front of these integrals, which have no effect on the motion of the envelope of the packets, we see that $t - 2\hbar\delta'(E_0)$ appears in Eq. (13), where t appears in Eq. (14). Hence the former packet is behind the latter by $2\hbar\delta'(E_0)$, that is,

$$[\text{time delay}] = 2\hbar \, \frac{d\delta}{dE}$$

Furthermore, according to Eqs. (13) and (14), this time delay is the *only* effect the potential has on the outgoing packet. Of course, this depends on the approximation that the packets have a very narrow energy distribution. In general, the packet will suffer both a time delay and a distortion of its shape as it passes through the potential.

11.2 *Scattering by a Square Well*

For a simple example of scattering in one dimension, we take the piecewise constant potential shown in Fig. 11.3:

$$V(x) = \begin{cases} \infty & \text{for } x < 0 \\ -V_0 & \text{for } 0 < x < a \\ 0 & \text{for } x > a \end{cases} \tag{15}$$

The wavefunction in the exterior region, $x > a$, must have the form given in Eq. (4):

$$\psi(x) = e^{i\delta} \sin(kx + \delta) \qquad \text{for } x > a \tag{16}$$

with $k = \sqrt{2mE}/\hbar$. The wavefunction in the interior region, $x < a$, must be such as to vanish at $x = 0$:

$$\psi(x) = A \sin k'x \qquad \text{for } x < a \tag{17}$$

with $k' = \sqrt{2m(E + V_0)}/\hbar$.

The boundary conditions to be imposed at $x = a$ are that the wavefunction and the derivative of the wavefunction are continuous (see Section 3.1). Thus,

$$A \sin k'a = e^{i\delta} \sin(ka + \delta) \tag{18}$$

$$k'A \cos k'a = k e^{i\delta} \cos(ka + \delta) \tag{19}$$

These boundary conditions determine the wave amplitude A and the phase shift δ. The equation for the phase shift is

$$\cot \delta = \frac{\tan ka + \dfrac{k'}{k} \cot k'a}{1 - \dfrac{k'}{k} \cot k'a \tan ka} \tag{20}$$

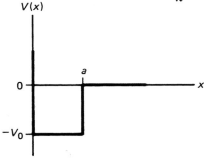

Fig. 11.3 A square potential well.

Exercise 1. Obtain this expression from Eqs. (18) and (19).

From $\cot \delta$ we can readily evaluate factor $e^{2i\delta}$ that appears in the scattered wave [see Eq. (8)]:

$$e^{2i\delta} = \frac{\cot \delta + i}{\cot \delta - i}$$

$$= \frac{\tan ka + \dfrac{k'}{k} \cot k'a + i\left(1 - \dfrac{k'}{k} \cot k'a \tan ka\right)}{\tan ka + \dfrac{k'}{k} \cot k'a - i\left(1 - \dfrac{k'}{k} \cot k'a \tan ka\right)} \tag{21}$$

Figure 11.4a is a plot of δ as a function of ka for the case $V_0 = 3.4\hbar^2/(2ma^2)$. This particular value of V_0 is interesting because the nuclear force between a neutron and a proton with parallel spins (triplet state) can be approximately described by such a potential.[3] Figures 11.4b, c, and d are plots of the square of the scattering amplitude, the wave amplitude $|A|$ in the interior region, and $(1/a)d\delta/dk$. The time delay is related to $d\delta/dk$:

$$[\text{time delay}] = 2\hbar \frac{d\delta}{dE} = \frac{2m}{\hbar k} \frac{d\delta}{dk} \tag{22}$$

which can also be written as

$$\frac{1}{a} \frac{d\delta}{dk} = \frac{\hbar k}{2am} \times [\text{time delay}] \tag{23}$$

Since $\hbar k/m$ is the speed of the particle in the absence of the potential, Eq. (23) indicates that the quantity $(1/a)d\delta/dk$ plotted in Fig. 11.4d is the time delay expressed in units of the "free transit time" $2a/(\hbar k/m)$ that the free particle takes to cover the distance $2a$. Note that for $k \to 0$, the time delay in our example tends to infinity; however, the ratio of the time delay to the free transit time remains finite.

---

[3] The values of V_0 and of a for the neutron–proton interaction in the triplet state are 38.5 MeV and 1.93×10^{-15} m, respectively. However, the interaction can be described by such a potential only at low energies ($E_{cm} \ll 200$ MeV). At higher energies, the hard repulsive core of the nuclear interaction as well as inelastic processes become important. Note that m is the reduced mass of the neutron–proton system.

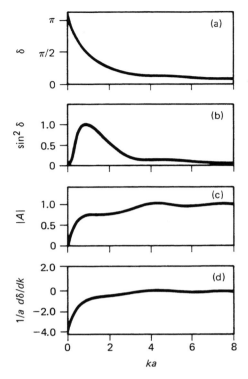

Fig. 11.4 Plots of (a) the phase shift, (b) the square of the scattering amplitude, (c) the wave amplitude $|A|$ in the interior region, and (d) the derivative $(1/a)d\delta/dk$ of the phase shift versus k for $V_0 = 3.4\hbar^2/(2ma^2)$.

In the limit of large energy, Fig. 11.4 suggests that

$$\delta \to 0 \tag{24}$$

$$\sin^2\delta \to 0 \tag{25}$$

$$A \to 1 \tag{26}$$

$$[\text{time delay}] \to 0 \tag{27}$$

It is easy to understand how these limits arise. If the particle has a very large energy ($E \gg V_0$), then it hardly notices the potential, and it behaves pretty much as a free particle, so the scattering tends to zero. In arriving at Eq. (24), we have made use of a convention for the phase shift. Obviously, Eq. (20) determines δ only modulo π, and we can say only that $\delta \to n\pi$ as $E \to \infty$. We adopt the convention that n is to be taken as zero.

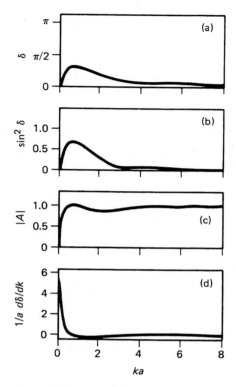

Fig. 11.5 Plots of (a) the phase shift, (b) the square of the scattering amplitude, (c) the wave amplitude $|A|$ in the interior region, and (d) the derivative $(1/a)d\delta/dk$ of the phase shift versus k for $V_0 = 2.1\hbar^2/(2ma^2)$.

As a second example, Fig. 11.5 plots the results of a similar calculation for a somewhat weaker potential, with $V_0 = 2.1\hbar^2/(2ma^2)$. This happens to be the potential that describes the interaction between a neutron and a proton with antiparallel spins (singlet state).[4] Comparing Figs. 11.4a and 11.5a, we see that the behavior of the phase shift at low energy is quite different. For the stronger potential, $\delta(0) = \pi$ and for the weaker potential $\delta(0) = 0$.

This difference between our two examples is a consequence of an interesting theorem that relates the net change of the phase shift between $E = 0$ and $E = \infty$ to the number of bound states in the

---

[4] For the neutron–proton interaction in the singlet state, $V_0 = 14.3$ MeV and $a = 2.50 \times 10^{-15}$ m.

potential. This theorem, known as *Levinson's theorem*, asserts that the phase shift at zero energy is

$$\delta(0) = N\pi \tag{28}$$

where N is the number of possible bound states in the potential. According to the results of Section 3.4 [see Eq. (3.94)], the square well with $V_0 = 3.4\hbar^2/(2ma^2)$ has exactly one bound state; hence $N = 1$ and $\delta(0) = \pi$. The square well with $V_0 = 2.1\hbar^2/(2ma^2)$ is too shallow to have any bound states; hence $N = 0$ and $\delta(0) = 0$.

The proof of the theorem is very simple. Consider a particle moving in some potential, constrained to the region $x > 0$. Pretend that the particle is also constrained so $x < L$, where L is large (ultimately, the limit $L \to \infty$ will be taken). This means the particle is confined in an infinite potential well of width L; all the eigenstates are discrete, which makes it easy to count them. If the potential in the region $0 < x < L$ is zero, then the positive-energy eigenfunctions are $\phi \propto \sin kx$, and since they must vanish at $x = L$,

$$kL = n\pi \tag{29}$$

Hence the number of states in a small momentum interval Δk is

$$\Delta n = \frac{L}{\pi} \Delta k \tag{30}$$

If the potential is not zero, then the positive-energy eigenfunctions *outside the range of the potential* are $\psi \propto e^{i\delta} \sin(kx + \delta)$ [see Eq. (4)], and the condition that they vanish at $x = L$ is

$$kL + \delta = n\pi \tag{31}$$

The number of states in a small momentum interval Δk is therefore

$$\Delta n = \frac{L}{\pi} \Delta k + \frac{1}{\pi} \delta'(k) \, \Delta k \tag{32}$$

where $\delta'(k) = d\delta/dk$. The change, caused by the potential, in the number of positive-energy eigenstates in the interval Δk is then $(1/\pi) \, \delta'(k) \, \Delta k$. The total change in the number of positive-energy states is

$$\Delta N = \frac{1}{\pi} \int_0^\infty \delta'(k) \, dk = \frac{1}{\pi} [\delta(\infty) - \delta(0)] \tag{33}$$

This equation for the change in the number of states is independent of L, and hence is also valid in the limit $L \to \infty$. The change in the number of positive-energy states implies an opposite change in the number of negative-energy states, since states cannot appear or disappear as the strength of the potential is varied. When the potential is zero, there is a certain number of states; turning on the potential cannot change the *number* of states, only their energies (it can be shown quite generally that the energy of a state depends continuously on the strength of the potential). Some of the positive-energy states become negative-energy states (bound states), others are shifted to some different positive energy (see Fig. 11.6). The decrease in the number of positive-energy states must therefore be compensated by an increase in the number of bound states:

$$N = -\Delta N = \frac{1}{\pi} \left[\delta(0) - \delta(\infty) \right] \tag{34}$$

With the convention $\delta(\infty) = 0$, this completes the proof of Eq. (28).

As a corollary of this theorem we obtain another interpretation of the quantity $\delta'(k)$. According to Eq. (33), $(1/\pi)d\delta/dk$ is the *change in the density of states* caused by the potential. For example, Fig. 11.5d can be interpreted as a plot of the change in the density of states and shows that, near zero energy, the continuum states in the presence of the potential are much more densely packed than in the absence of potential.

Before we proceed with our discussion of scattering, let us look at the bound states of the potential given in Eq. (15). There is, of course, a well-known and straightforward procedure for solving the Schrödinger equation for this case (see Section 3.4). But here we will deal with a clever alternative method that makes use of the positive-energy solution we examined above.

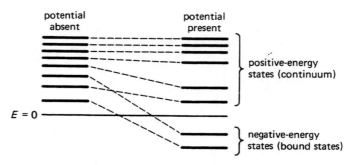

Fig. 11.6   Shifts in the energies of some states produced by the presence of the potential.

If $E < 0$, then

and

$$k' = \frac{\sqrt{2m(E + V_0)}}{\hbar} = \frac{\sqrt{2m(V_0 - |E|)}}{\hbar} \qquad (35)$$

$$k = \frac{\sqrt{2mE}}{\hbar} = i\frac{\sqrt{2m|E|}}{\hbar} \qquad (36)$$

We will use the notation $\kappa = \sqrt{2m|E|}/\hbar$, so

$$k = i\kappa \qquad (37)$$

Now, Eqs. (16) and (17), with δ and A determined by Eqs. (18) and (19), give a solution of the Schrödinger equation with the correct boundary conditions, irrespective of the sign of E. However, if $E < 0$, it is convenient to express this solution in terms of the real variable κ. Equation (16) then becomes

$$\psi(x) \frac{1}{2i} (e^{2i\delta}e^{-\kappa x} - e^{\kappa x}) \qquad (38)$$

where, by Eq. (20),

$$\cot \delta = \frac{\tan i\kappa a + \dfrac{k'}{i\kappa} \cot k'a}{1 - \dfrac{k'}{i\kappa} \cot k'a \tan i\kappa a} \qquad (39)$$

Although Eq. (38) is a solution of the Schrödinger equation, it is not an acceptable solution; it has the wrong behavior at infinity, with $\psi(x) \to \infty$ as $x \to \infty$. We must somehow get rid of the increasing exponential $e^{\kappa x}$. Since the normalization of the bound-state wavefunction must in any case be carried out after we have found an acceptable solution, we may as well ignore an overall factor of proportionality, and replace Eq. (38) by

$$\psi(x) \propto e^{-\kappa x} - e^{-2i\delta}e^{\kappa x} \qquad (40)$$

From this, it is obvious that in order to get rid of the increasing exponential, we need only demand that

$$e^{-2i\delta} = 0 \qquad (41)$$

Since δ as given by Eq. (39) is complex,[5] this equation has solutions. The equation can also be rewritten as a condition on $\cot \delta$:

---

[5] Our previous argument requiring δ to be real applies only when $E > 0$.

$$\cot \delta = i\,\frac{e^{i\delta} + e^{-i\delta}}{e^{i\delta} - e^{-i\delta}} = i\,\frac{1 + e^{-2i\delta}}{1 - e^{-2i\delta}} = i \tag{42}$$

Substituting this into Eq. (39), we obtain

$$\cot k'a = -\frac{\kappa}{k'} \tag{43}$$

or

$$\cot \frac{\sqrt{2m(V_0 - |E|)}\,a}{\hbar} = -\sqrt{\frac{|E|}{V_0 - |E|}} \tag{44}$$

This equation for the energy eigenvalues coincides with Eq. (3.89), obtained by other means.

Exercise 2. Derive Eq. (44) from Eq. (39).

As already mentioned above, the square well with $V_0 = 3.4\hbar^2/(2ma^2)$ has exactly one bound state. In the case of the triplet neutron–proton interaction, this bound state corresponds to the deuteron, with a binding energy $|E| = 2.22$ MeV.

The square well with $V_0 = 2.1\hbar^2/(2ma^2)$ has no bound state. However, it is customary to say that there is a *virtual state,* or an *antibound state,* in singlet neutron–proton scattering. By this is meant the following: We have seen that a bound state is obtained by eliminating the increasing exponential from Eq. (38). A virtual, antibound "state" is obtained by eliminating the *decreasing* exponential, so the wavefunction becomes a purely increasing exponential.[6] This requires

$$e^{2i\delta} = 0 \tag{45}$$

which implies that

$$\cot \frac{\sqrt{2m(V_0 - |E|)}\,a}{\hbar} = +\sqrt{\frac{|E|}{V_0 - |E|}} \tag{46}$$

Exercise 3. Derive Eq. (46).

---

[6] There exists several other, slight different definitions of *virtual state.* The definition given here is the most straightforward. Note that the use of the word *virtual* in relativistic perturbation theory (Feynman diagrams) is unrelated to our use here.

The preceding equation yields a binding energy $|E| = 170$ keV for the virtual singlet state of the deuteron.

It must be emphasized that the presence of a virtual "state" means nothing but that Eq. (45) is satisfied. The virtual "state" is no more than a root of $e^{2i\delta}$; it is not a physical state (hence the name). The wavefunction increases exponentially at large distances and is therefore not acceptable; the system can never be put into this "state." Nevertheless, the presence of the virtual "state" makes itself felt by its influence on the scattering at small (positive) energies. Thus, the behavior of $\sin^2\delta$ near zero energy (see Fig. 11.5b) can be explained in terms of the presence of the nearby virtual "state" at small negative energy.

The quantity $e^{2i\delta}$ by which the outgoing wave is multiplied is usually called the *S-matrix*, or more precisely, an S-matrix element (*S* stands for scattering). In general, each scattering reaction $|\alpha\rangle \rightarrow |\beta\rangle$, where $|\alpha\rangle$ represents some initial state and $|\beta\rangle$ some final state, possibly with a different final set of particles, is assigned an S-matrix element $S_{\alpha\beta}$ which is related to the probability amplitude for the occurrence of the reaction. The S-matrix is the totality of all such elements.

We have seen that $e^{2i\delta} = \infty$ at a bound state. This means that the S-matrix has a pole (singularity) at the bound-state energy:

$$\text{bound state} \Rightarrow \text{pole in } S\text{-matrix} \qquad (47)$$

This rule holds in general and is very important in attempts at constructing the S-matrix ("S-matrix theory"). The program of S-matrix theory is this: Since we often do not know the forces that act between particles, we cannot solve the Schrödinger equation or the relativistic version of this equation. But we can nevertheless find the S-matrix, which contains all the information about scattering, by somehow discovering all the singularities. This would determine the S-matrix completely, because if all the singularities of a function are known, at both real and complex values of its argument, then the function is determined. Since the singularities at both real and complex (unphysical) values of the energy must be discovered, the program of S-matrix theory is ambitious and difficult.

11.3 *Resonances*

A particle incident on the square well of Eq. (15) never suffers any large time delay. Neither does the wave amplitude $|A|$ in the interior region ever become large. We will now examine the case of a more complicated potential, consisting of a square well and a barrier:

$$V(x) = \begin{cases} -V_0 & \text{for } 0 < x < a \\ V_1 & \text{for } a < x < 2a \\ 0 & \text{for } x > a \\ \infty & \text{for } x < 0 \end{cases} \tag{48}$$

This potential is plotted in Fig. 11.7. The barrier in this potential can prevent the escape of a particle from the interior region for a fairly long time, provided the energy is just right, and this leads to the accumulation of a large wave amplitude in the interior region.

The solution of the Schrödinger equation for the potential (48) is of the form

$$\psi(x) = \begin{cases} A \sin k'x & \text{for } 0 < x < a \\ B \sin k''x + C \cos k''x & \text{for } a < x < 2a \\ e^{i\delta} \sin(kx + \delta) & \text{for } x > 2a \end{cases} \tag{49}$$

where $k' = \sqrt{2m(E + V_0)}/\hbar$ and $k'' = \sqrt{2m(E - V_1)}/\hbar$. Note that if $E < V_1$, then k'' will be imaginary, but it is not necessary to treat this case separately.

The usual boundary conditions at $x = a$ and at $x = 2a$ provide four equations that determine the four unknowns A, B, C, and δ. Of these unknowns, A and δ are the most interesting. The formulas for them are rather messy, and it is more instructive to examine plots of these quantities. Figure 11.8 shows plots of δ as a function of ka, and plots of the square of the scattering amplitude, the wave amplitude $|A|$ in the interior region, and $(1/a)\, d\delta/dk$. These plots

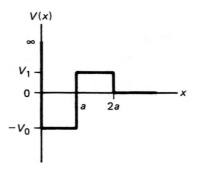

Fig. 11.7   A square well with a barrier.

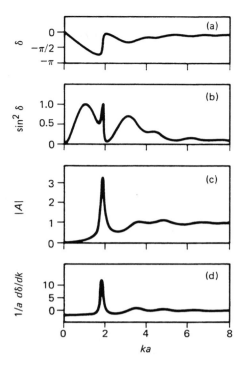

Fig. 11.8 Plots of (a) the phase shift, (b) the square of the scattering amplitude, (c) the wave amplitude $|A|$ in the interior region, and (d) the derivative $(1/a) \, d\delta/dk$ of the phase shift vs. k for the square well with a barrier.

were prepared for $V_0 = 1.0\hbar^2/2ma^2$ and $V_1 = 5.0\hbar^2/2ma^2$. A striking feature of Fig. 11.8 is the *resonance* or *quasi-stationary state* that appears at $ka = 1.8$. At this energy

i. the phase shift suddenly increases by (nearly) π and passes through $\pi/2$.

ii. the square of the scattering amplitude has a sharp maximum and reaches $\sin^2\delta = 1$

iii. the wave amplitude in the interior region has a sharp maximum

iv. the derivative $d\delta/dk$ has a sharp maximum.

The quantity $d\delta/dk$ is related to the time delay by Eq. (23), which we can write as

$$\frac{d\delta}{dk} = R \times \frac{[\text{time delay}]}{[\text{free transit time}]} \tag{50}$$

where R is the range of the potential, and the free transit time is $2R/(\hbar k/m)$. The time delay may be regarded as the lifetime of the resonant state. If this lifetime is very large, the resonance will resemble a stationary state (for the latter the lifetime is infinite). We can also express this in another way: the maximum in $d\delta/dk$ at the resonant energy means that the *density of states is at a maximum*. If this maximum is very large and very narrow, the density of states resembles that of a stationary state (for the latter, the density of states is infinite at the stationary state and zero elsewhere, that is, it is a delta function).

From our knowledge of classical oscillating systems, we expect that at the resonant energy, the amplitude of oscillation of the system should build up to a very large value. In the present case, the quantity $|A|$ may be regarded as the amplitude of oscillation of the system, and it indeed has the expected behavior. Figure 11.9 is a plot of the absolute value squared of the wavefunction near resonance. Clearly, the amplitude of oscillation is large inside the potential well.

Unfortunately, in scattering experiments, measurements are performed on the incident particle only when it is far away from the scattering region; that is, the incident particle is observed only before and after the collision with the target particle. Hence the quantity $|A|$ is not directly accessible in these experiments. What can be measured directly is the scattering probability $\sin^2\delta$. The measurement of $\sin^2\delta$ as a function of energy determines δ and $d\delta/dE$ as a function of energy, except perhaps for an ambiguity in sign.

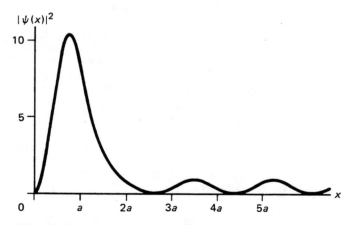

Fig. 11.9   Plot of $|\psi(x)|^2$ vs. x when the energy is near resonance.

The time delay $2\hbar d\delta/dE$ at resonance can be obtained directly by measurement of the time that a particle takes to move though the potential. The wave packet used for such a measurement must have a width Δk that is small compared with the width of the resonance. Since the latter width is approximately $\pi(d\delta/dk)^{-1}$, we need

$$\Delta k \ll \pi \left(\frac{d\delta}{dk}\right)^{-1} \tag{51}$$

and the packet must have a length

$$\Delta x \geq \frac{1}{2\,\Delta k} \gg \frac{d\delta}{dk} \tag{52}$$

This means that the time of departure and of arrival of the particle is uncertain by

$$\Delta t = \frac{\Delta x}{v} \gg \frac{d\delta}{dk}\frac{m}{\hbar k} \tag{53}$$

or

$$\Delta t \gg \hbar\frac{d\delta}{dE} \tag{54}$$

Hence, Δt is much larger than the time delay that we seek to measure. This means that it will be impossible to perform the measurement by observation of a single particle. However, repeated measurements on identically prepared systems will give an *average* time delay which is not affected by the large uncertainties of the individual measurements. The time delay has, therefore, only a statistical significance.

We saw that in the case of the potential given by Eq. (48), the conditions (i)–(iv) all hold at resonance. We will now proceed in general, without any special assumption about the shape of the potential, and examine what relationships we can establish between the conditions (i)–(iv).

First, we show that condition (i) implies all the others. If δ increases sharply by about π in the vicinity of $k = \alpha$, passing through $\delta = \pi/2$ at $k = \alpha$, then we can approximate δ in this vicinity by

$$\delta \simeq \tan^{-1}\frac{\beta}{\alpha - k} \tag{55}$$

where β is a positive constant such that

$$\frac{1}{\beta} = \frac{d\delta}{dk}\bigg|_{k=\alpha} \tag{56}$$

The function (55) has the correct value of δ and the correct value of the derivative $d\delta/dk$ at $k = \alpha$; it is therefore effectively equivalent to a two-term Taylor-series expansion. According to Eq. (55),

$$e^{2i\delta} = \frac{1 + i\tan\delta}{1 - i\tan\delta} \simeq \frac{k - \alpha - i\beta}{k - \alpha + i\beta} \tag{57}$$

and hence the square of the scattering amplitude is

$$|\psi_S|^2 = \sin^2\delta = \frac{1}{4}|e^{2i\delta} - 1|^2 \simeq \frac{\beta^2}{(k - \alpha)^2 + \beta^2} \tag{58}$$

This has a maximum at $k = \alpha$, as required by condition (ii).

If we write Eq. (58) as a function of the energy, with $E_\alpha = \alpha^2\hbar^2/2m$ and with the approximation $(E - E_\alpha) \simeq (k - \alpha)\alpha\hbar^2/m$, then the square of the scattering amplitude becomes

$$|\psi_S|^2 = \sin^2\delta \simeq \frac{\frac{1}{4}\Gamma^2}{(E - E_\alpha)^2 + \frac{1}{4}\Gamma^2} \tag{59}$$

where

$$\Gamma = \frac{2\alpha\beta\hbar^2}{m} \tag{60}$$

and the approximation has been made. Equation (59) is known as the *Breit–Wigner resonance formula*. It gives the behavior of the square of the scattering amplitude near resonance under the assumption that the nonresonant scattering can be neglected. Figure 11.10 shows plots of δ and of $\sin^2\delta$ near resonance, according to the Breit–Wigner formula; note that the shape of these plots is in agreement with Fig. 11.8.

It is obvious that condition (iv) is also satisfied since, with the approximation (55),

$$\frac{d\delta}{dk} = \frac{d}{dk}\tan^{-1}\frac{\beta}{\alpha - k} = \frac{\beta}{(k - \alpha)^2 + \beta^2} \tag{61}$$

which, indeed, has a maximum at $k = \alpha$.

We will now give a simple, but not rigorous, argument that a maximum in $d\delta/dk$ is usually associated with a maximum in the wave amplitude in the interior region. If there is a time delay

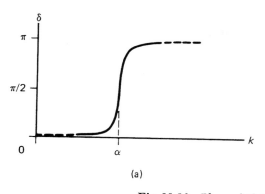

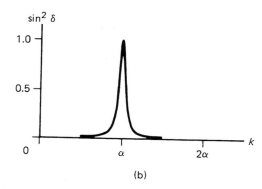

Fig. 11.10 Phase shift δ and square of the scattering amplitude $|\psi_S|^2 = \sin^2\delta$ near resonance.

$2\hbar d\delta/dE$, the emergence of a wave train that enters the potential is delayed, and for a time $2\hbar d\delta/dE$ probability is flowing into the region of the potential, but there is no corresponding flow out of the region of the potential; thus, probability accumulates in this region. The incoming wave has a probability density $|e^{-ikx}/2i|^2 = 1/4;$[7] its probability current is the product of this probability density and the speed $\hbar k/m$:

$$[\text{probability current}] = \frac{1}{4}\frac{\hbar k}{m}$$

The accumulated probability is then

$$[\text{probability current}] \times [\text{time delay}] \simeq \frac{1}{4}\frac{\hbar k}{m} \times 2\hbar \frac{d\delta}{dE} \quad (62)$$

However, we can also express the change of probability that the potential produces in the interior region as

$$\int_0^R (|\psi|^2 - |\phi|^2)\,dx$$

where $\phi = \sin kx$ [see Eq. (3)] and ψ is the solution of the Schrödinger equation for the given potential. Hence[8]

---

[7] The wavefunction is not normalizable, and hence this probability and all the probabilities in the following equations are relative, not absolute.

[8] Equation (63) is not exact since, strictly, the time delay can be defined only for wave packets, and not for the harmonic waves used in the above argument. The exact equation has an extra term $[\sin 2kx - \sin 2(kx + \delta)]/2k$ added to the right side.

$$\frac{1}{2}\frac{\hbar^2 k}{m}\frac{d\delta}{dE} \simeq \int_0^R \left(|\psi|^2 - |\phi|^2\right) dx \tag{63}$$

that is,

$$\frac{d\delta}{dk} \simeq 2\int_0^R \left(|\psi|^2 - |\phi|^2\right) dx \tag{64}$$

Since

$$\int_0^R |\phi|^2 \, dx = \int_0^R \sin^2 kx \, dx < R \tag{65}$$

we see that a large maximum in $d\delta/dk$ must correspond to a large maximum in $\int |\psi|^2 \, dx$, and conversely. This implies that the average wave amplitude in the interior is at maximum, in agreement with condition (iii).

Next we consider the implications of condition (ii). If the square of the scattering amplitude goes through its maximum value ($\sin^2\delta = 1$), then δ must pass through $\pi/2$. However, such a maximum is not a resonance unless δ is increasing with energy. For example, the peak at $ka = 1.0$ in Fig. 11.8b does not correspond to a resonance; the time delay is negative at this energy, and there is no quasi-stationary state. Maxima with negative time delays are not rare. In fact, the number of such maxima is never less than the number of resonances. Levinson's theorem tells us that, on the average, δ does not increase with energy [$\delta(0) - \delta(\infty) \geq 0$]; since at each resonance, δ *increases* through $\pi/2$, it must again *decrease* through $\pi/2$, and therefore give a nonresonant maximum in the square of the scattering amplitude.

Although in general the existence of a maximum in $\sin^2\delta$ is not a sufficient condition for the existence of a resonance, the existence of a *narrow* maximum is a sufficient condition. This is so because a narrow maximum implies that δ is passing through $\pi/2$ very quickly, that is, $|d\delta/dk|$ is large, and this is possible only if $d\delta/dk$ is positive. The reason for a positive value of $d\delta/dk$ is causality. If $d\delta/dk$ were negative, we would have a time *advance*. But, by causality, the largest possible time advance is $2R/v$, which occurs if the incident particle is reflected at the outer edge of the potential (or, equivalently, transverses the potential with very large speed). Therefore, the time delay necessarily satisfies the condition

$$2\hbar\frac{d\delta}{dE} \geq -\frac{2R}{v}$$

or

$$\frac{d\delta}{dk} \geq -R \tag{66}$$

This is known as the *Wigner condition.*[9]

The Wigner condition tells us that if the peak of $\sin^2\delta$ is so narrow that $|d\delta/dk|$ is larger than the range of the potential, then this peak will indeed correspond to a resonance; a peak that is not so narrow may or may not correspond to a resonance. This provides some justification for experimental physicists who, upon observing very narrow peaks in nuclear cross sections, immediately claim that they have found a resonance.

As a next step of our investigation of the relationships among the conditions (i)–(iv), we will show that a sufficiently sharp peak in the quantity $d\delta/dk$—which is proportional to the time delay expressed in units of the free transit time [see Eq. (50)]—implies the existence of a resonance. If $d\delta/dk$ has a maximum at $k = \alpha$, we can approximate this function in the vicinity by

$$\frac{d\delta}{dk} \simeq \frac{\xi\beta}{(k-\alpha)^2 + \beta^2} \tag{67}$$

where ξ and β are positive constants. Equation (67) amounts to no more than the assumption that $(d\delta/dk)^{-1}$ can be expanded in a Taylor series about the point $k = \alpha$:[10]

$$\left(\frac{d\delta}{dk}\right)^{-1} = \frac{\beta}{\xi} + \frac{1}{\beta\xi}(k-\alpha)^2 \tag{68}$$

We assume that the higher-order terms in this expansion can be neglected as long as $|k - \alpha| \leq \beta$. Note that Eq. (61) is a special case of Eq. (67) with $\xi = 1$; it will be interesting to see how this special case arises.

The differential equation (67) can be integrated to give δ as a function of k:

$$\delta(k) = \int_\alpha^k \frac{\xi\beta}{(k-\alpha)^2 + \beta^2}\, dk + \delta(\alpha)$$

$$\simeq \xi \tan^{-1}\frac{\beta}{\alpha - k} - \frac{\xi\pi}{2} + \delta(\alpha) \tag{69}$$

---

[9] Our argument is not rigorous, and neither is the result (66). The exact Wigner condition is $d\delta/dk \geq -(R + 1/k)$.

[10] The Taylor series for $(d\delta/dk)^{-1}$ is more convenient than that for $d\delta/dk$, because $(d\delta/dk)^{-1}$ is small near resonance.

We must now take into account an important restriction on the possible values of ξ. It is obvious from Eq. (4) that δ is defined only modulo π, since an increase of δ by $n\pi$ leaves this equation unchanged. On the other hand, $\xi \tan^{-1} \beta/(\alpha - k)$ is defined modulo $\xi\pi$. For the sake of consistency of these two equations, we must therefore require that ξ is an integer:

$$\xi = 0, 1, 2, 3, \ldots \tag{70}$$

The case $\xi = 0$ is of no interest. The cases $\xi = 1, 2, 3, \ldots$ correspond to resonances of different types. We will for now assume that $\xi = 1$, because this type of resonance seems to be the one that usually occurs in nature. In this special case

$$\delta(k) \simeq \tan^{-1} \frac{\beta}{\alpha - k} - \frac{\pi}{2} + \delta(\alpha) \tag{71}$$

which shows that δ increases by about π in the vicinity of $k = \alpha$. Hence δ must necessarily pass through $\pi/2$ (modulo π) in this vicinity, and we see that all the conditions (i)–(iv) are satisfied.

To simplify matters further, let us assume that the scattering off resonance is small, so the resonant peak in the scattering is well isolated from the nonresonant background. In terms of Eq. (71), this means that the constant term $-\pi/2 + \delta(\alpha)$ is small and that

$$\delta(k) \simeq \tan^{-1} \frac{\beta}{\alpha - k} \tag{72}$$

We conclude that under these circumstances δ passes through $\pi/2$ (modulo π) at $k \simeq \alpha$, so the maxima in $d\delta/dk$ and in $\sin^2\delta$ occur at nearly the same energy. The expression (72) is identical to (55) and leads to the familiar Breit–Wigner formula (59).

11.4 *Elastic Scattering in Three Dimensions*

Many of the concepts we introduced to describe the scattering in one dimension also apply to three dimension. Consider a particle incident on some three-dimensional region in which the potential $V(\mathbf{r})$ is different from zero. We will call this region the interaction region (see Fig. 11.11). The incident particle is represented by a plane wave, and this wave is scattered by the interaction with the potential, that is, the some portion of the wave is spread out over all angles.

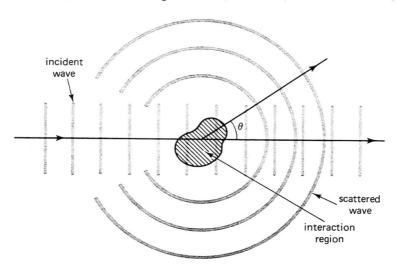

Fig. 11.11 Scattering of an incident plane wave by interaction with a potential. The scattering angle is measured relative to the incident direction.

For convenience, we assume that the incident wave approaches the interaction region from the negative z direction. Thus, the incident wave is

$$\phi(\mathbf{r}) = e^{ikz} \tag{73}$$

As in Section 11.1, we have omitted the time dependence of this wave; it is given by the usual factor $e^{-iEt/\hbar}$ for all our waves.

At large distances from the interaction region, the scattered wave consists of an outgoing spherical wave with a radial dependence of the form e^{ikr}/r. Here, we have included a factor of $1/r$ in this wavefunction because, when the wave travels outward, it spreads over an increasing area, and the probability density in the wave must decrease as $1/r^2$ (this decrease of the intensity of the spherical quantum-mechanical wave is entirely analogous to the decrease of the intensity of a spherical sound wave spreading out from a small source). Since the wave need not have the same intensity in all directions, we also want to include an angular dependence in the scattered wave. We describe this angular dependence by a function $f(\theta)$, where θ is the scattering angle, or the deflection angle, relative to the incident direction (see Fig. 11.11). The scattered wave then takes the form

$$\psi_S(\mathbf{r}) = f(\theta)\,\frac{e^{ikr}}{r} \tag{74}$$

The net wave is the sum of the incident wave and the scattered wave,

$$\psi(\mathbf{r}) = \phi(\mathbf{r}) + \psi_S(\mathbf{r}) = e^{ikz} + f(\theta)\,\frac{e^{ikr}}{r} \qquad (75)$$

In one-dimensional scattering, we defined the scattering amplitude as the amplitude of the scattered wave [see Eq. (9)]. In three-dimensional scattering, the amplitude of the scattered wave is $f(\theta)/r$ [see Eq. (74)]; however, since the factor $1/r$ is always present in three-dimensional scattering and conveys no distinctive information about the scattering, we will omit it from the definition of the scattering amplitude:

$$[\text{scattering amplitude}] = f(\theta) \qquad (76)$$

We will see that the scattering probability and the scattering cross section can be expressed directly in terms of this scattering amplitude $f(\theta)$.

As in Section 10.6, we define the differential cross section $d\sigma$ as the ratio of the number of particles scattered per unit time into a solid angle $d\Omega$ to the flux of incident particles:

$$d\sigma = \frac{\left[\begin{array}{c}\text{number of particles scattered per unit time}\\ \text{into solid angle } d\Omega \text{ at } \theta,\, \phi\end{array}\right]}{[\text{flux of incident particles}]} \qquad (77)$$

The flux of incident particles is the product of the probability density of the incident wave and the speed. Since the incident wave (73) has a probability density $|e^{ikz}|^2 = 1$ and the particles have a speed $\hbar k/m$,

$$[\text{flux of incident particles}] = 1 \times \frac{\hbar k}{m} \qquad (78)$$

The number of scattered particles in a small volume (see Fig. 11.12) between r and $r + dr$ in the solid angle $d\Omega$ is

$$\left| f(\theta)\,\frac{e^{ikr}}{r} \right|^2 r^2\, dr\, d\Omega = |f(\theta)|^2\, dr\, d\Omega \qquad (79)$$

These particles take a time $dt = dr/(\hbar k/m)$ to leave this volume; hence the number emerging per unit time is

$$|f(\theta)|^2\,\frac{\hbar k}{m}\, d\Omega \qquad (80)$$

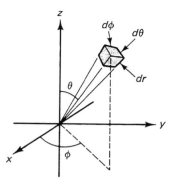

Fig. 11.12   A small volume between r and $r + dr$ in the solid angle $d\Omega = \sin \theta \, d\theta \, d\phi$.

and the differential cross section (77) is

$$d\sigma = \frac{|f(\theta)|^2 \, \hbar k/m}{\hbar k/m} \, d\Omega \tag{81}$$

or

$$\frac{d\sigma}{d\Omega} = |f(\theta)|^2 \tag{82}$$

The total cross section is obtained by integrating the differential cross section over all angles:

$$\sigma = \int d\sigma = \int \frac{d\sigma}{d\Omega} \, d\Omega = \int |f(\theta)|^2 \, d\Omega \tag{83}$$

11.5 *Partial Waves in a Central Potential; The Optical Theorem*

For the calculation of the scattering amplitude $f(\theta)$ and the cross section, we need to specify the potential. Most of the potentials of interest in physics are central potentials, that is, potentials that depend only on the radial coordinate r. For such a central potential, the incident wave and the scattered wave can be expressed as superpositions of eigenfunctions of the orbital angular momentum, and each of these eigenfunctions is scattered independently of the other eigenfunctions. The angular-momentum eigenfunctions contained within the incident and the scattered wave are called *partial waves.* Each partial wave is characterized by the value of its angular-momentum quantum number l. During the scattering, each partial wave acquires a phase shift of its own, and these phase shifts determine the scattering amplitude and the cross section.

To construct the partial waves, we recall from Section 8.3 that the plane wave e^{ikz} can be written as a superposition of angular-momentum eigenfunctions:

$$e^{ikz} = \sqrt{4\pi} \sum_{l=0}^{\infty} \sqrt{2l+1}\; i^l Y_l^0(\theta) j_l(kr) \tag{84}$$

We want to examine the behavior of the wavefunctions at a large distance from the interaction region. Thus, we will assume that r is very large, and that the functions $j_l(kr)$ can be replaced by their asymptotic forms $(1/kr)\sin(kr - l\pi/2)$, according to Eq. (8.114). The plane wave e^{ikz} then becomes

$$e^{ikz} \simeq \sqrt{4\pi} \sum_{l=0}^{\infty} \sqrt{2l+1}\; i^l Y_l^0(\theta) \frac{1}{kr} \sin\left(kr - \frac{l\pi}{2}\right) \tag{85}$$

or

$$e^{ikz} \simeq \frac{\sqrt{4\pi}}{k} \sum_{l=0}^{\infty} \sqrt{2l+1}\; i^l Y_l^0(\theta) \frac{1}{2i}\left(\frac{e^{i(kr-l\pi/2)}}{r} - \frac{e^{-i(kr-l\pi/2)}}{r}\right) \tag{86}$$

This shows that the lth partial wave consists of an incoming spherical wave

$$-\frac{e^{-i(kr-l\pi/2)}}{r} \tag{87}$$

and an outgoing spherical wave

$$\frac{e^{i(kr-l\pi/2)}}{r} \tag{88}$$

If the potential $V(r)$ is zero, then the sum of these incoming and outgoing spherical waves [with the coefficients specified in Eq. (86)] provides us with the solution of the scattering problem.

If the potential is nonzero, then the solution of the scattering problem will still be a sum of incoming and outgoing spherical waves, but with somewhat different coefficients. From our discussion of scattering in one dimension, we know that the incoming wave is unchanged, but the outgoing wave acquires an extra phase factor $e^{2i\delta_l}$. Thus, the solution of the scattering problem is

$$\psi(\mathbf{r}) = \frac{\sqrt{4\pi}}{k} \sum_{l=0}^{\infty} \sqrt{2l+1} \; i^l Y_l^0(\theta) \frac{1}{2i} \left(\frac{e^{i(kr-l\pi/2+2\delta_l)}}{r} - \frac{e^{-i(kr-l\pi/2)}}{r} \right) \tag{89}$$

According to Eq. (75), the scattered wave $f(\theta)e^{ikr}/r$ is the difference between the net wavefunction $\psi(\mathbf{r})$ and the incident wave e^{ikz}:

$$f(\theta) \frac{e^{ikr}}{r} = \psi(\mathbf{r}) - e^{ikz}$$

$$= \frac{\sqrt{4\pi}}{k} \sum_{l=0}^{\infty} \sqrt{2l+1} \; i^l Y_l^0(\theta) \frac{1}{2i} \left(\frac{e^{(ikr-l\pi/2+2\delta_l)}}{r} - \frac{e^{-i(kr-l\pi/2)}}{r} \right)$$

$$- \frac{\sqrt{4\pi}}{k} \sum_{l=0}^{\infty} \sqrt{2l+1} \; i^l Y_l^0(\theta) \frac{1}{2i} \left(\frac{e^{(ikr-l\pi/2)}}{r} - \frac{e^{-i(kr-l\pi/2)}}{r} \right)$$

$$= \frac{\sqrt{4\pi}}{k} \sum_{l=0}^{\infty} \sqrt{2l+1} \; i^l Y_l^0(\theta) \frac{1}{2i}(e^{2i\delta_l} - 1) \frac{e^{i(kr-l\pi/2)}}{r} \tag{90}$$

Hence,

$$f(\theta) = \frac{\sqrt{4\pi}}{k} \sum_{l=0}^{\infty} \sqrt{2l+1} \; Y_l^0(\theta) \frac{1}{2i}(e^{2i\delta_l} - 1)$$

$$= \frac{\sqrt{4\pi}}{k} \sum_{l=0}^{\infty} \sqrt{2l+1} \; Y_l^0(\theta) \; e^{i\delta_l} \sin \delta_l \tag{91}$$

From this expression for $f(\theta)$, we can calculate the differential cross section [Eq. (82)] and the total cross section [Eq. (83)]. The latter takes an especially simple form, in consequence of the orthonormality relation for the spherical harmonics:

$$\sigma = \int |f\theta)|^2 \, d\Omega = \frac{4\pi}{k^2} \sum_{l=0}^{\infty} (2l+1) \sin^2 \delta_l \tag{92}$$

Exercise 4. Derive this formula for the total cross section.

An interesting result emerges from Eq. (91) if we examine the imaginary part of the scattering amplitude at $\theta = 0$, called the *forward scattering amplitude*. With $Y_l^0(0) = \sqrt{(2l+1)/4\pi}$, we obtain

$$\text{Im}\,[f(0)] = \text{Im}\left[\frac{\sqrt{4\pi}}{k}\sum_{l=0}^{\infty}\sqrt{2l+1}\,\sqrt{\frac{2l+1}{4\pi}}\,e^{i\delta_l}\sin\delta_l\right]$$

$$= \frac{1}{k}\sum_{l=0}^{\infty}(2l+1)\sin^2\delta_l \tag{93}$$

Comparison of this with Eq. (92) shows that

$$\sigma = \frac{4\pi}{k}\,\text{Im}\,f(0) \tag{94}$$

This is called the *optical theorem*. (A similar theorem holds for light waves, and it was known long before its rediscovery in quantum mechanics.)

Finally, let us apply some of these results to a simple case of scattering by the potential of a hard elastic sphere of radius a. If the particles have low energy, the dominant contribution to the scattering comes from the partial wave $l = 0$ (s wave scattering); this is so because if the particles have low energy, then they can attain a nonzero angular momentum only if they have a large impact parameter, and then they miss the sphere. Thus, for low-energy particles, the total cross section can be approximated by the first term in the sum (92):

$$\sigma \simeq \frac{4\pi}{k^2}\sin^2\delta_0 \tag{95}$$

The phase shift δ_0 is easy to evaluate. In the presence of the hard-sphere potential, the wavefunction has a node at the surface of the sphere, that is, at $r = a$, whereas in the absence of the potential the wavefunction has a node at $r = 0$. This means that the path of the wave coming in and then moving back out is shortened by $2a$, and the phase of the outgoing wave is therefore shifted by $2ak$ relative to the outgoing wave in the absence of potential.[11] Thus, $2\delta_0 = -2ak$, and Eq. (95) becomes

$$\sigma \simeq \frac{4\pi}{k^2}\sin^2(-ak)$$

or, since k is small,

---

[11] This simple method of calculating the phase shift for the hard sphere works for the $l = 0$ partial wave only. The other partial waves experience an effective centrifugal potential, and their wavelengths are not constant and not equal to $2\pi/k$.

$$\sigma \simeq \frac{4\pi}{k^2}(ak)^2 = 4\pi a^2 \tag{96}$$

This quantum-mechanical scattering cross section is four times as large as the geometrical cross section of the hard sphere.

11.6 *The Born Approximation; Rutherford Scattering*

Equation (92) gives us the cross section if we know the phase shifts. However, the determination of the phase shifts requires the solution of the Schrödinger equation for each partial wave. As we saw in Section 11.1, such a determination of the phase shifts is quite tedious. Perturbation theory sometimes provides an alternative method for the calculation of the cross section. Here we will deal with a simple method of calculation based on Fermi's Golden Rule.

Suppose we regard the potential $V(\mathbf{r})$ that produces the scattering as a perturbation of the free-particle Hamiltonian. According to Fermi's Golden Rule, the rate of transitions from an initial momentum eigenstate $|\psi_{\mathbf{p}}\rangle$ to a final momentum eigenstate $|\psi_{\mathbf{p}'}\rangle$ is then

$$W_{\mathbf{p}\to\mathbf{p}'} = \frac{2\pi}{\hbar}\langle\psi_{\mathbf{p}'}|V(\mathbf{r})|\psi_{\mathbf{p}}\rangle\frac{dN(E)}{dE} \tag{97}$$

With the box normalization for the wavefunctions and with the usual expression (10.79) for the number of states per unit energy interval, this transition rate becomes[12]

$$W_{\mathbf{p}\to\mathbf{p}'} = \frac{2\pi}{\hbar}\left|\int\frac{e^{-i\mathbf{p}'\cdot\mathbf{r}/\hbar}}{\sqrt{V}}V(\mathbf{r})\frac{e^{i\mathbf{p}\cdot\mathbf{r}/\hbar}}{\sqrt{V}}d^3r\right|^2\frac{V}{h^3}m\sqrt{2mE}\,d\Omega \tag{98}$$

From this transition rate into the solid angle $d\Omega$, we immediately obtain the differential cross section by dividing by the incoming flux of particles [compare Eq. (10.101)]. Since the flux of particles associated with the incident wave is $(1/V)\sqrt{2E/m}$, the differential cross section is

$$d\sigma = \left(\frac{m}{2\pi\hbar^2}\right)^2\left|\int V(\mathbf{r})\,e^{i(\mathbf{p}-\mathbf{p}')\cdot\mathbf{r}/\hbar}\,d^3r\right|^2 d\Omega \tag{99}$$

Accordingly, the scattering amplitude is

---

[12] Do not confuse the volume V with the potential $V(\mathbf{r})$.

$$f(\theta) = -\frac{m}{2\pi\hbar^2} \int V(\mathbf{r}) \, e^{i(\mathbf{p}-\mathbf{p}')\cdot\mathbf{r}/\hbar} \, d^3r \tag{100}$$

Here, a minus sign has been inserted to bring Eq. (100) into conformity with the result obtained by an alternative calculation (via a Green's function) of the scattering amplitude. Equation (100) is called the *Born approximation* for the scattering amplitude.

If the potential is central, we can simplify the integral in Eq. (100) by introducing the vector

$$\mathbf{K} = \frac{\mathbf{p} - \mathbf{p}'}{\hbar} = \mathbf{k} - \mathbf{k}' \tag{101}$$

which is the difference between the wave vectors of the incident and the scattered particles. The magnitude of this vector is

$$K = |\mathbf{k} - \mathbf{k}'| = 2k \sin\frac{\theta}{2} \tag{102}$$

where θ is the scattering angle (see Fig. 11.13). The integration in Eq. (100) can then be performed by means of a change to spherical coordinates r, Θ, ϕ with a new z axis along the direction of $\mathbf{K}$. This yields, with $\mu = \cos\Theta$,

$$f(\theta) = -\frac{m}{2\pi\hbar^2} \int_0^{2\pi} d\phi \int_0^\infty \int_0^\pi V(r) e^{iKr\cos\Theta} \, r^2 \sin\Theta \, d\Theta \, dr$$

$$= -\frac{m}{\hbar^2} \int_0^\infty r^2 \, V(r) \int_{-1}^1 e^{iKr\mu} \, d\mu \, dr$$

$$= -\frac{2m}{\hbar^2 K} \int_0^\infty rV(r) \sin Kr \, dr \tag{103}$$

The Born approximation treats the potential as a perturbation of the free-particle Hamiltonian. Hence its validity is restricted to

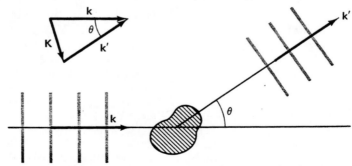

Fig. 11.13 Scattering of an incident plane wave by the interaction region.

kinetic energies that are large compared with the potential energy.[13] This means that the Born approximation to some extent complements the method of partial waves. The Born approximation requires high energies, whereas the method of partial waves is best applied at low energies, where the angular momenta of the particles are small, and the first few terms in the infinite sum (92) suffice for an approximate evaluation of the cross section. Incidentally, the Born approximation fails to satisfy the optical theorem [according to Eq. (103), the scattering amplitude has no imaginary part]. This failure can be traced to the assumption of large kinetic energy and small potential energy implicit in this approximation.

Note that we cannot use the Born approximation for the Coulomb potential, since, with $V(r) \propto 1/r$, the integral (103) fails to converge. However, we can use it for the *screened Coulomb potential*, which has the form of the Coulomb potential with an extra exponential factor $e^{-r/b}$, where b is a positive constant:

$$V(r) = - \frac{Ze^2}{4\pi\varepsilon_0} \frac{e^{-r/b}}{r} \tag{104}$$

This screened Coulomb potential approximately describes the potential experienced by an electron incident on an atom with many electrons, where the charge distribution of the electrons of the atom screens the nucleus from view. The appropriate value of the exponential decay length b for such an atom is approximately $4\pi\varepsilon_0\hbar^2/Z^{1/2}me^2$. A potential of the same mathematical form as (104)—but with a different value of b and a different overall constant of proportionality—also occurs in nuclear physics. This potential, called the *Yukawa potential*, describes the strong force acting between two nucleons.

For the screened Coulomb potential, the Born approximation gives us a scattering amplitude.

$$f(\theta) = \frac{2m}{\hbar^2 K} \frac{Ze^2}{4\pi\varepsilon_0} \int_0^\infty e^{-r/b} \sin Kr\, dr$$

$$= \frac{2m}{\hbar^2} \frac{Ze^2}{4\pi\varepsilon_0} \frac{1}{K^2 + 1/b^2} \tag{105}$$

---

[13] This requirement can be relaxed if the potential is very shallow, roughly, if the potential is so shallow that it admits no bound state. For a simple discussion of the limitations of the Born approximation, see, for example, D. S. Saxon, *Elementary Quantum Mechanics*.

Hence the differential cross section is

$$\frac{d\sigma}{d\Omega} = |f(\theta)|^2 = \left(\frac{2m}{\hbar^2}\frac{Ze^2}{4\pi\varepsilon_0}\right)^2 \frac{1}{(K^2 + 1/b^2)^2} \tag{106}$$

If we suppose that the incident electron has a high energy, so $K^2 \gg 1/b^2$, then we can neglect the term $1/b^2$ in the denominator of Eq. (106), and we obtain

$$\frac{d\sigma}{d\Omega} = \left(\frac{2m}{\hbar^2}\frac{Ze^2}{4\pi\varepsilon_0}\right)^2 \frac{1}{K^4}$$

or, with $K = 2k \sin \frac{1}{2}\theta = 2(\sqrt{2mE}/\hbar) \sin \frac{1}{2}\theta$,

$$\frac{d\sigma}{d\Omega} = \left(\frac{Ze^2}{16\pi\varepsilon_0 E}\right)^2 \frac{1}{\sin^4 \frac{1}{2}\theta} \tag{107}$$

Here, the dependence on the exponential decay length has dropped out, because for a high-energy particle, most of the scattering occurs near the nucleus, where the potential is strongest and almost unscreened.

Equation (107) is the Rutherford cross section for the scattering of a particle of charge $\pm e$ by a nucleus of charge Ze. The cross section for the scattering of an alpha particle of charge $2e$ is four times as large as (107).

PROBLEMS   1. According to Eq. (3.48), the solutions of the time-dependent Schrödinger equation in one dimension satisfy the identity

$$\frac{\partial}{\partial t}(\psi^*\psi) = -\frac{\partial}{\partial x}\left[\frac{\hbar}{2im}\left(\psi^*\frac{\partial\psi}{\partial x} - \psi\frac{\partial\psi^*}{\partial x}\right)\right]$$

(a) Prove that if ψ is the wavefunction for a stationary state, then

$$\frac{\hbar}{2im}\left[\psi^*(x)\frac{\partial\psi(x)}{\partial x} - \psi(x)\frac{\partial\psi^*(x)}{\partial x}\right] = \text{constant}$$

(b) Substitute the wavefunction (4) into this identity and thereby deduce that δ must be real.

2. (a) Prove that if ψ is a solution of the three-dimensional time-dependent Schrödinger equation, then

$$\frac{\partial}{\partial t}(\psi^*\psi) = -\nabla \cdot \left[\frac{\hbar}{2im}(\psi^* \nabla\psi - \psi \nabla\psi^*)\right]$$

(b) According to this equation, the quantity

$$\frac{\hbar}{2im} (\psi^* \nabla\psi - \psi \nabla\psi^*)$$

can be regarded as the probability current. Check that for a plane wave $\psi = e^{i\mathbf{p}\cdot\mathbf{r}/\hbar}$, this probability current coincides with the product of probability density and velocity, that is, $\psi^*\psi \, \mathbf{p}/m$.

3. Figure 11.14 shows a potential barrier of height V_0 and width a placed adjacent to the origin. The potential is

$$V(x) = \begin{cases} \infty & \text{for } x < 0 \\ V_0 & \text{for } x < a \\ 0 & \text{for } x > a \end{cases}$$

Find the phase shift for this potential barrier, and plot as a function of the wave number k. Are there any resonances?

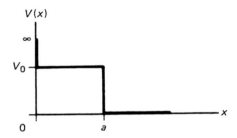

Fig. 11.14

4. Solve Eqs. (18) and (19) for the wave amplitude A in the interior of the square well.

5. Show that for scattering on a square well (see Fig. 11.3), the wave amplitude $|A|$ can never exceed 1. Show that $|A|$ equals 1 if $E = \hbar^2\pi^2(n + \frac{1}{2})^2/2ma^2 - V_0$, and show that this condition corresponds to constructive interference of the de Broglie waves reflected back and forth across the well.

6. Show that the Wigner condition (66) implies that

$$\int_0^R |\psi|^2 \, dx \geq -\frac{1}{4k} \sin 2kR$$

7. Consider particles scattering on a hard sphere of radius a. If the particles have an energy $E = \hbar^2 k^2/2m$, roughly what impact parame-

ter is required to give angular momentum $l\hbar$? Roughly what range of values of l will contribute to the scattering on the hard sphere?

8. If the total cross section (92) is finite, show that δ_l must approach zero at least as fast as $1/l$ as $l \to \infty$.

9. In the scattering of particles of energy $E = \hbar^2 k/2m$ by a nucleus, an experimenter finds a differential cross section

$$\frac{d\sigma}{d\Omega} = \frac{1}{k^2} (0.86 + 3.07 \cos\theta + 2.77 \cos^2\theta)$$

   (a) What partial waves are contributing to the scattering, and what are their phase shifts at the given energy?

   (b) What is the total cross section?

10. A free electron of momentum **p** initially directed along the positive z axis is perturbed by a potential

$$V(x, y, z) = \begin{cases} V_0 & \text{for } |x| < a, \, |y| < a, \, |z| < a \\ 0 & \text{for } |x| > a, \, |y| > a, \, |z| > a \end{cases}$$

   where V_0 is a constant. Use Fermi's Golden Rule to find the scattering amplitude. Find the differential cross section $d\sigma\,(\theta, \phi)$ for transitions into a small solid angle $d\Omega$.

11. (a) In three dimensions, a particle is scattered elastically by a spherical potential well

$$V(r) = \begin{cases} -V_0 & \text{for } r < a \\ 0 & \text{for } r > a \end{cases}$$

   where V_0 is positive. Use the one-dimensional results obtained in Section 11.2 to find the phase shift for the partial wave $l = 0$. If the energy of the particle is low, what is the differential cross section and what is the total cross section?

   (b) Footnote 3 gives the appropriate values of the parameters V_0 and a for the scattering of neutrons by protons in the triplet state. Evaluate the total cross section at an energy $E_{cm} \ll 10$ MeV. (Note that the mass m in our equations must be interpreted as the reduced mass, $m \simeq \frac{1}{2}m_n$.)

   (c) Footnote 4 gives the appropriate values of the parameters V_0 and a for the scattering of neutrons by protons in the singlet state. Evaluate the total cross section at an energy $E_{cm} \ll 10$ MeV.

12. A particle is scattered elastically by a "soft sphere" with a repulsive spherical potential

$$V(r) = \begin{cases} V_0 & \text{for } r < a \\ 0 & \text{for } r > a \end{cases}$$

where V_0 is positive. Evaluate the phase shift for the partial wave $l = 0$. If the energy of the particle is low, what is the differential cross section and what is the total cross section?

13. Integrate the differential cross section (106) over all angles and find the total cross section.

14. Consider the Gaussian potential

$$V(r) = V_0 e^{-r^2/b^2}$$

Use the Born approximation to find the differential cross section and the total cross section.

15. The potential for a "soft sphere" is

$$V(r) = \begin{cases} V_0 & \text{for } r < a \\ 0 & \text{for } r > a \end{cases}$$

where V_0 is positive. Use the Born approximation to find the differential cross section and plot as a function of the scattering angle θ.

12

The Interpretation of Quantum Mechanics

Throughout this book, we have relied on the *Copenhagen interpretation* of quantum mechanics. This is the traditional interpretation of the quantum-mechanical formalism. Its main features were initially sketched by Heisenberg and by Bohr, and its details were later filled in by many collaborators and disciples of Bohr at the Institute for Theoretical Physics at Copenhagen. Because this interpretation provides us with only probabilistic information about the state of a quantum-mechanical system, and because this interpretation has some weird aspects that go counter to our intuition, its adequacy has often been questioned. Several other interpretations of quantum mechanics have been proposed, but none has been judged clearly superior to the Copenhagen interpretation, which still remains the only widely accepted of all the interpretations of quantum mechanics. Of course, scientific issues are not decided by popularity polls, but the wide acceptance of the Copenhagen interpretation means that a physicist who wants to communicate some result or discovery in quantum mechanics will feel compelled to couch the result in the language of the Copenhagen interpretation.

Critics of the Copenhagen interpretation do not challenge the accuracy of the numerical results calculated from quantum mechanics. At a pragmatic level, quantum mechanics works perfectly—the numerical results for, say, the eigenvalues of the angular momentum and the energy of the hydrogen atom are found to be in perfect agreement with experiment. But critics challenge whether the Copenhagen interpretation really gives us the most complete, most exhaustive knowledge of a quantum system we can hope for. For instance, is it really impossible to say anything about the precise instantaneous position of the electron in the hydrogen atom and its motion as a function of time? Or is the inability of quantum mechanics to provide this information an indication of some deficiency of the theory? In the view of some critics, the probabilistic character of the predictions of quantum mechanics is held to reflect our ignorance of the details of the underlying dynamics. Theories that attempt to provide a more detailed knowledge than provided by the Copenhagen interpretation are said to contain *hidden variables*.

The discussion of the interpretation of quantum mechanics and of hidden variables has received a fresh stimulus in recent years, because it has become possible to perform an experiment originally conceived as a *Gedankenexperiment* by Einstein, Podolsky, and Rosen in 1935. A new theoretical analysis of this *Gedankenexperiment* by Bell in 1964 established that it could be used to discriminate between the Copenhagen interpretation and a wide class of theories with hidden variables, and this encouraged experimenters to attempt some actual versions of the experiment. The experimental results fully support the Copenhagen interpretation and contradict theories with hidden variables.

12.1 *The Copenhagen Interpretation*

The main features of the Copenhagen interpretation can be briefly summarized as follows:

1. The state vector $|\psi\rangle$ provides a complete characterization of the state of the system.

2. The state vector tells us the probability distribution for the result of the measurement of any observable quantity. This probability distribution applies to each *individual* quantum particle or quantum system.

3. The uncertainty relations indicate the intrinsic spreads in the values of complementary observables for the *individual* quantum particle or quantum system. These uncertainty relation deny the existence of sharp values of complementary observables.

4. Measurements produce unpredictable, discontinuous changes in the state vector, which do not obey the Schrö-

dinger equation. The outcome of a single measurement of an observable is unpredictable—the outcome can be any of the eigenvalues within the spread of the probability distribution. During the measurement, the state of the system collapses into an eigenstate of the observable.

This list of features overlaps, to some extent, with the axioms of quantum mechanics stated in Chapters 4 and 5. We could include all of these axioms in our list of features of the Copenhagen interpretation, but some of these axioms—for instance, the axiom for the time evolution of the state vector—do not pertain directly to the *interpretation* of quantum mechanics, and this is why we prefer not to include them here. We have used the features of the Copenhagen interpretation in the preceding chapters. Now we will discuss them critically.

The fundamental assumption of the Copenhagen interpretation is that the state vector $|\psi\rangle$ (or, in the position representation, the wavefunction ψ) provides a complete, exhaustive characterization of the state of the system. This means that the state vector encompasses all that can be said about the state of the system. The other assumptions and prescriptions of the Copenhagen interpretation are built upon this fundamental assumption.

In contrast to the classical characterization of the state of a system, where the instantaneous coordinates and momenta give us a detailed picture of the instantaneous configuration of the system, the quantum-mechanical characterization by means of the state vector gives us merely the probabilities for the outcome of measurements that we can perform on the system. For instance, if $|E_n\rangle$ is an energy eigenstate, then $|\langle E_n|\psi\rangle|^2$ gives us the probability that the outcome of an energy measurement is E_n. From the probability distribution for the different energy eigenvalues, we can calculate the expectation value of the energy; alternatively, we can calculate this expectation value, or average value, according to the concise formula

$$\langle E \rangle = \langle \psi | H | \psi \rangle \tag{1}$$

where H is the energy operator. Similar formulas give us the expectation values of all other physical observables. Because the state vector $|\psi\rangle$, or the wavefunction ψ, determines the expectation values of all observables, Schrödinger has called the wavefunction the "expectation-catalog."

We must resist the temptation to regard the wavefunction as some kind of snapshot of the instantaneous configuration of the system, in the way that, say, the classical wavefunction for a standing wave on a string is a snapshot of the instantaneous configuration of the string. The quantum-mechanical wavefunction of, say, an electron in an atom does not give us a picture of the shape of the instantaneous configuration of matter or of electric charge in the atom. It merely gives us the probability distribution of the electric charge; it merely provides us with the means of calculating expectation values. The quantum-mechanical wavefunction makes no assertions about the instantaneous position of the electron or about the instantaneous charge distribution in the atom. One of the advantages of the abstract state vector $|\psi\rangle$ over the wavefunction ψ is that as long as we deal with the abstract state vector we are unlikely to fall into the error of imagining the wavefunction as some kind of actual configuration of electric charge in space.

Quantum mechanics does not supply us with concrete mental pictures of the behavior of atoms and subatomic particles. Quantum mechanics does not tell us what atoms and subatomic particles are like; it merely tells us what happens when we perform measurements. As Heisenberg said: "The conception of objective reality . . . evaporated into the . . . mathematics that represents no longer the behavior of elementary particles but rather our knowledge of this behavior."[1]

The emphasis of the Copenhagen interpretation on measurements and on the procedures for measurements is in accord with the philosophical doctrines of positivism and operationalism. In brief, positivism asserts that the only meaningful statements we can make about a physical system are those that are verifiable by observation and experiment, and thus the only meaningful physical quantities are those that are measurable. And operationalism asserts that the definition of any physical quantity must spell out the experimental, or "operational," procedure for measuring the quantity. According to strict positivist doctrine, the aim of science is to describe and to predict, but not to explain; speculations about unobservable and unmeasurable properties are held to be irrelevant.

This emphasis on measurements is a strength and also a weakness of the Copenhagen interpretation: strength, since its lack of commitment to any detailed model of the atomic and subatomic

---

[1] W. Heisenberg, Daedalus, **87**, 99 (1958).

world makes it nearly impregnable; and weakness, since it fails to satisfy our craving for concrete mental images of atomic and sub-atomic processes. Of course, we can imagine the wavefunction, and this can help us to understand the mathematical properties of ψ; but when you imagine, say, the scattering of an incident proton wave on a nuclear target, you are not seeing the physical behavior of the proton, only the mathematical evolution of the wavefunction.

In the Copenhagen interpretation, the meaning of the quantum-mechanical probability distributions is quite different from that of the probability distributions familiar from classical statistical mechanics. When a classical physicist has recourse to a probability distribution to describe, say, the speed of a molecule in a gas, he does not mean to deny that the molecule has a perfectly well defined speed at each instant of time; but he does not know this speed—he only knows some macroscopic quantities of the gas, such as the average density, temperature, and pressure. Hence, in classical statistical mechanics, the probability distribution for molecular speeds reflects the ignorance of the observer of the precise microscopic conditions in the gas. This kind of probability distribution is called an *ensemble* distribution, since it describes the average conditions for a large number of molecules in a gas. In contrast, the quantum-mechanical probability distribution does not reflect our ignorance of the instantaneous position and momentum, but rather the non-existence of any well-defined position and momentum. The quantum-mechanical system does not consist of particles with well-defined albeit unknown positions and momenta, but of "particles" with intrinsically indeterminate positions and momenta. Thus, the quantum-mechanical probability distributions (and the quantum-mechanical uncertainties Δx and Δp_x; see below) refer to an individual particle, not to an ensemble of particles. An example due to Schrödinger brings this distinction into clear focus: Consider a particle in an energy eigenstate of the isotropic harmonic oscillator, say, a particle in the ground state, with $E = \frac{3}{2}\hbar\omega$. A classical probability distribution for this system with well-defined, but unknown, values of the position and momentum would necessarily require that the distance of the particle from the origin never exceed the distance at which the energy $\frac{3}{2}\hbar\omega$ equals the potential energy (this is the classical turning point of the motion); thus, if we were to assume that the particle had a well-defined instantaneous position and momentum, the probability distribution would have a sharp cutoff at the classical turning point.[2] But the quantum-mechanical probability distribution has no such sharp cutoff—it permits the particle to penetrate into the classically forbidden region beyond the turning point. (As we already discussed in Chapter 3, this penetration into a classically forbidden region leads to no inconsistencies, because, in consequence of the uncertainty relation, any attempt at detecting the presence of the particle in the forbidden region introduces a large uncertainty in the energy and blurs the distinction between the forbidden and the permitted region.) The important lesson we extract from this example is that the consistency of the Copenhagen interpretation demands that the quantum-mechanical probability distribution be associated with an individual particle.

This raises the question of how we can perform an experimental measurement of probabilities when we are dealing with a single particle or a single system. A single trial, say, a single measurement of the position of a particle with some given wavefunction, cannot confirm the quantum-mechanical prediction for the probability distribution of the position. At the most, the single trial could prove quantum mechanics wrong—if the result of the single trial is a position that according to quantum mechanics has zero probability. For a comprehensive examination of the probability distribution, we must repeat the trial again and again, each time starting with the system prepared in the same way, so it has the same initial wavefunction for each trial. In practice, it is more convenient to prepare a large number of identical copies of the system, and measure the distribution of positions across this ensemble of copies. For instance, the measurement of the probability distribution in the diffraction pattern produced by an electron incident on a crystal is routinely performed by means of a beam of many electrons incident on the crystal. The diffracted electrons emerge from the crystal and strike a fluorescent screen, where they generate small flashes of light. Each flash of light amounts to a repeated trial of the experiment. However, under typical experimental conditions, the electrons arrive at the screen in quick succession, and we do not notice the individual flashes of light. What we see on the screen is a more or less steady pattern of

---

[2] Some hidden-variable theories bypass this requirement by modifying the potential energy. Thus, a hidden-variable theory contrived by Bohm adds to the ostensible potential energy $\frac{1}{2}m\omega^2 x^2$ of the harmonic oscillator an extra term depending on the wavefunction $\psi(x)$ (see Problem 3). The classical turning point is then at infinity, and the probability distribution has no sharp cutoff.

bright and dark zones, which give us a direct picture of the probability distribution (see Fig. 1.2). According to the Copenhagen interpretation, the probability distribution for such an ensemble of repeated trials of the diffraction experiment is equal to the probability distribution for each individual electron, and the width of this probability distribution across the screen (or, more precisely, the rms deviation from the mean) is equal to the uncertainty in the position of each individual electron upon arrival at the screen. Of course, after the electron interacts with some atom in the screen and triggers the emission of a flash of light, the uncertainty in its position will be much smaller (this final uncertainty depends on the details of the interaction between the electron and the atom).

The state vector $|\psi\rangle$ presents us with a probability distribution for the possible values for every observable quantity. In general, this probability distribution spans several, or many, values of the observable, and therefore the outcome of a measurement of the observable is afflicted with uncertainties. Only in the exceptional case that $|\psi\rangle$ is an eigenvector of the observable does this uncertainty disappear—the outcome of the measurement is then certain to be the eigenvalue. However, the commutation relations of quantum mechanics place severe restrictions on what observables can simultaneously be free of uncertainties, that is, what observables can have simultaneous eigenvectors. For complementary observables, such as the position x and the momentum p_x, whose commutator has the canonical form $[x, p_x] = i\hbar$, there are no simultaneous eigenvectors, and the certainty in one of these observables implies total uncertainty in the other, in accord with the Heisenberg uncertainty relation

$$\Delta x \, \Delta p_x \geq \frac{\hbar}{2} \tag{2}$$

The uncertainties Δx, Δp_x, and other such quantum-mechanical uncertainties refer to an individual particle, not to an ensemble of particles. These quantum-mechanical uncertainties do not arise from our ignorance of some underlying details of the state of the particle or from some inadequacy of our measuring devices. Instead, the uncertainties reflect the nonexistence of such details; they reflect an intrinsic spread in the position and the momentum of the particle. The position and the momentum are not sharply defined; they are *indeterminate*.

The uncertainty relations are often called *indeterminacy relations*, to distinguish the quantum-mechanical uncertainties from ordinary experimental uncertainties arising from imperfections in the apparatus used in a measurement. To illustrate the distinction between these two kinds of uncertainties, consider the position measurement of an electron by means of a fluorescent screen. In this case, there are two different uncertainties: there is the initial intrinsic uncertainty of the position of the electron, associated with its initial wavefunction; and there is the ordinary experimental uncertainty of the measurement associated with the resolution attainable with the fluorescent screen. We can check the ordinary experimental uncertainty in the measurement of the position of an individual electron by an immediate repetition of this measurement, say, by means of a second fluorescent screen placed adjacent to the first (we will assume that the electron has enough energy to pass through both screens). The electron will then trigger the emission of a flash of light in the second screen, and the difference between the average positions of the flashes triggered in the first and the second screens tells us the ordinary experimental uncertainty. Note that this experimental uncertainty is never smaller than the quantum-mechanical uncertainty associated with the final wavefunction of the electron, after the measurement (in the best conceivable measurement, the experimental uncertainty attains the level of the final quantum-mechanical uncertainties). Whenever we perform a measurement, we must always make a careful distinction between these different kinds of uncertainty.

The Heisenberg uncertainty relation for the position and the momentum of a particle implies that classical determinism fails—the initial values of the position and the momentum (or velocity) of a particle cannot be used to predict the position and momentum at a later time. To make such a prediction for the motion of a particle, the classical physicist would need precise initial values of position and momentum; but Eq. (2) forbids simultaneous precise values of these two observables.

Although quantum mechanics lacks the simple determinism of classical physics, it retains a form of determinism in the state vector $|\psi\rangle$, which evolves in time according to the (general) Schrödinger equation,

$$\frac{\hbar}{i}\frac{d}{dt}|\psi\rangle = H|\psi\rangle \tag{3}$$

This equation expresses determinism and causality, since it permits us to predict the state vector at any later time from a given state vector at the initial time. Thus, in the words of Born: "The motion of particles is subject [only] to probabilistic laws, but the probability itself evolves in accord with causal laws."[3]

12.2 *Measurement and the Collapse of the Wavefunction*

The weirdest feature of the Copenhagen interpretation is that it requires that the wavefunction suffer a discontinuous, unpredictable change during the measurement. Consider, for instance, the impact of an electron on the fluorescent screen in the electron-diffraction experiment. This impact and the flash of light released in it constitute an (approximate) measurement of the position of the electron. Just before this measurement, the wavefunction was spread out all over the screen; immediately after the measurement, the electron position is known to lie within some small spot on the screen, and the wavefunction must therefore have an extent no larger than this spot. Thus, during the measurement, the wavefunction suffers an unpredictable *collapse*, or *reduction*. The collapse is unpredictable, since we have no way of knowing onto what part of the screen the wavefunction will collapse—we know only the probability distribution of the spots on which the wavefunction collapses, that is, the probability distribution of positions for the electron on the screen.

Note that a single measurement tells us very little about the wavefunction *before* the measurement. If a measurement finds an electron at some spot, this tells us only that the wavefunction before the measurement was different from zero at that spot. But the measurement tells us much about the wavefunction just *after* the measurement. In general, a precise measurement of an observable collapses the wavefunction into an eigenstate of that observable. Thus, the wavefunction after such ideal measurement is precisely known. For instance, an ideal measurement of the position of an electron collapses the wavefunction into a delta function (the practical measurement with a fluorescent screen has a limited precision, given by the experimental uncertainty; and the wavefunction after the measurement is not a delta function, but a wave packet of a width of a few Å). A measurement of the energy of an electron collapses the wavefunction into an eigenstate of energy.

---

[3] M. Born, Z. Phys., **38**, 803 (1926).

A measurement of the spin collapses the wavefunction into an eigenstate of spin, and so on. The apparatus plays a crucial role in selecting the kind of eigenstate into which the wavefunction collapses. The apparatus dictates whether the system will collapse into some eigenstate of position, or of momentum, or of spin, and so on. But, of course, the apparatus does not dictate which specific eigenstate of position, or of energy, or of spin, and so on, the system will collapse into; this aspect of the collapse is unpredictable.

Bohr has emphasized that quantum mechanics does not describe quantum systems per se; instead, it describes a whole phenomenon, which includes, in an inextricable way, both the quantum system and the apparatus used to measure it: ". . . an independent reality in the ordinary physical sense can neither be ascribed to the phenomenon nor to the agencies of observation."[4] According to the Copenhagen interpretation, quantum systems in themselves do not have sharply defined attributes, only diffuse potentialities, which are capable of becoming sharply defined when we perform suitable measurements. The attributes of a quantum system depend on the apparatus used to measure them, and they exist only in relation to this apparatus. Thus, the attributes are a joint property of the system and the apparatus. This intimate symbiotic relationship between system and apparatus implies a break with naive realism, according to which the attributes of a physical system belong to the system in itself, and they are supposed to exist independently of the environment surrounding the system. However, in the view of most physicists, the antirealism of the Copenhagen interpretation extends only to the attributes of physical systems, not to the physical systems themselves. The Copenhagen interpretation denies the realism of contingent attributes, but it does not deny the realism of physical systems or of the material world.

In an ideal measurement, the collapse is instantaneous—at one instant the wavefunction has one configuration, at the next instant it has collapsed to a new, drastically different configuration. Such an instantaneous collapse would seem to conflict with the theory of Special Relativity, according to which signals can never exceed the speed of light (if signals can be sent with a speed exceeding the speed of light, then you can send messages into your own past, in blatant violation of causality). But it is easy to

---

[4] N. Bohr, *Quantum Theory and the Description of Nature*, Chapter II.

see that the collapse process cannot be used to transmit messages from one observer to another. For instance, consider an electron wave that has spread out over a very large volume, say several light years, and suppose that an observer at one end of this electron wave detects the electron on her fluorescent screen and brings about the collapse of the wavefunction. This means that it will thereafter be impossible for another, distant observer to detect the electron on his fluorescent screen; but this does not give this other observer a message of any sort, since he has no way of knowing that his attempts at detecting the electron have been condemned to failure. The other observer is now on a fool's errand—he can continue to grope around searching for the electron, but he is unable to conclude that the electron wave has collapsed until he has explored *every* volume element in space, including the vicinity of the first observer, where, of course, he will finally get the message.

The change of the wavefunction during the collapse is not governed by the Schrödinger equation. As we will see in Section 12.3, the Copenhagen interpretation brazenly postulates that this collapse is merely a mathematical procedure, not a physical process. We might be tempted to suppose that the collapse is produced by the dynamics of the interaction of the measured system with the measuring apparatus. But such an interaction, if treated according to the time evolution specified by the Schrödinger equation, is not by itself enough to bring about the collapse of the wavefunction. For instance, when the diffracted electron wave strikes the fluorescent screen, it interacts with *all* the atoms in the screen and scatters off them with some loss of energy (inelastic scattering). This interaction of the electron with the atoms in the screen tells us the probabilities for the emission of flashes of light by the atoms, but it does not tell us which of the many atoms on the screen will actually emit the light, and thereby signal the collapse of the electron wavefunction into its vicinity. Thus, although the interaction of the system and the measuring apparatus is required for the measurement to be possible, the collapse is not produced by this interaction.

It might be argued that single atoms, or groups of atoms, in the fluorescent screen do not constitute a macroscopic measuring device, and that therefore the collapse is to be expected to occur only at the next stage of the measurement process, when the flash of light from the screen triggers a macroscopic measurement device, such as a photomultiplier tube or the human eye. The interaction of an electron with the atoms of a fluorescent screen is complicated, and the ensuing complete chain of events is difficult to analyze in detail. Instead, let us deal with a different example of measurement, in which the interactions are more obvious, and the evolution of the joint system–apparatus state vector can be examined in some detail.

Consider the measurement of the vertical component of the spin of an atom with the Stern–Gerlach apparatus shown in Fig. 12.1. In this measurement, we send the atom through the inhomogeneous magnetic field of a magnet, which displaces the trajectory of the atom vertically upward if the spin is up, and vertically downward if the spin is down. Two detectors serve to discover whether the atom has taken the high road or the low road. In the original Stern–Gerlach experiment, the detector was simply a photographic plate. But such a device absorbs the atom and precludes any further measurements. For our purposes, it will prove more instructive to use some detector that permits the atom to proceed on its trajectory. A suitable detector might consist of a laser beam that intersects the trajectory and a photomultiplier tube aimed at the intersection. If the atom crosses the laser beam, it scatters a

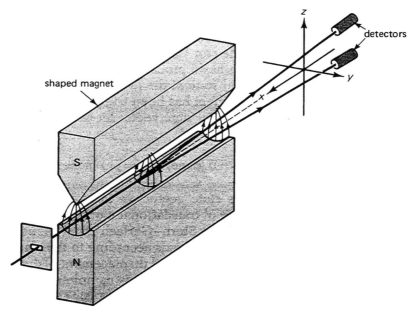

Fig. 12.1 Stern–Gerlach apparatus for the measurement of the vertical component of the spin.

few photons, which reveal the presence of the atom when they trigger the photomultiplier tube. Since the resolution of this simple optical detector is of no concern to us, we can use photons of long wavelength, which hardly disturb the atom at all. Note that this apparatus contains two basic elements: a discriminating device (the inhomogeneous magnetic field) that sends the atom one way or another according to the vertical component of its spin, and an amplifying device (the photomultiplier) that produces a macroscopic pulse of current when triggered by an incident atom. These two basic elements are quite typical of most measuring devices used for measurements on quantum systems.

We can easily see that as long as the atom, the apparatus, and their interactions are governed by the Schrödinger equation, a collapse of the state vector is not possible. We assume that the atom has spin $\frac{1}{2}$ and that the initial state of the atom is some superposition of the spin-up and spin-down states, say, the superposition

$$\frac{1}{\sqrt{2}} \left(|+\rangle + |-\rangle \right) \qquad (4)$$

which corresponds to an initial state of horizontal spin, in the $+x$ direction [see Eq. (9.33)]. According to the usual rules for calculating probabilities for the outcome of a measurement, this initial state of the atom has a probability of $\frac{1}{2}$ for spin up, and $\frac{1}{2}$ for spin down. The initial state of the detectors, before the measurement, is that neither of them has been triggered; we designate this state by $|\text{none}\rangle$. The initial state vector for the joint atom–detector system is then

$$|\psi\rangle = \frac{1}{\sqrt{2}} \left(|+\rangle + |-\rangle \right) |\text{none}\rangle = \frac{1}{\sqrt{2}} |+\rangle |\text{none}\rangle + \frac{1}{\sqrt{2}} |-\rangle |\text{none}\rangle \qquad (5)$$

Note that in this state vector we have not bothered to indicate explicitly the state of translational motion of the atom. Within the approximations of the Stern–Gerlach experiment, the translational motion of the atom proceeds according to the classical mechanics, and the upward or downward displacement of the trajectory for the spin-up or the spin-down state is completely determined by the (classical) parameters of the apparatus. Thus, there is a unique correspondence between spin states and trajectories, and we can pretend that the spin states $|\pm\rangle$ include an implicit specification of translational states.

The atom passes through the apparatus and interacts with the detectors. During this interaction, which is described by some suitable interaction Hamiltonian, the spin-up state triggers the upper detector, but not the lower; we designate the corresponding state vector of the detectors by |upper⟩. The spin-down state triggers the lower detector, but not the upper; we designate the corresponding state vector by |lower⟩. Thus, if the initial state were one or the other of the states of definite spin, the interaction during the measurement would produce the following transition to a definite final state:

$$|+\rangle\,|\text{none}\rangle \rightarrow |+\rangle\,|\text{upper}\rangle \quad \text{or} \quad |-\rangle\,|\text{none}\rangle \rightarrow |-\rangle\,|\text{lower}\rangle \quad (6)$$

Since the Schrödinger equation is linear, the superposition (5) of initial states will therefore produce a corresponding superposition of final states:

$$|\psi\rangle = \frac{1}{\sqrt{2}}\,|+\rangle\,|\text{upper}\rangle + \frac{1}{\sqrt{2}}\,|-\rangle\,|\text{lower}\rangle \quad (7)$$

Thus, the result of the interaction is not a collapse into one or another of the states of definite spin up or down and a definite response from the detector, but a superposition in which the spin-up and spin-down states are correlated with the detector states. But such a superposition cannot be regarded as a completed measurement, since it fails to make a definite choice between the spin-up and spin-down states. In fact, the state vector (7) is merely the initial state vector (5) translated in time. With a slight modification of the apparatus, it is even possible to restore the initial horizontal spin state of the atom by a further translation in time. If we add a second magnet to the apparatus in tandem with the first and with a reversed magnetic field, then the trajectory of the atom will suffer a reversed displacement in the second magnet, and the initial spin state of the atom will be restored when the upper and the lower trajectories again merge into one.[5] Such a restoration of the horizontal spin state demonstrates that our Stern–Gerlach experiment cannot be regarded as a completed measurement of the vertical component of the spin.

---

[5] In order to achieve a complete restoration of the initial state vector, we also have to reset the detectors, so their state vector is restored to |none⟩. Since the detectors are macroscopic devices, a restoration to the exact initial quantum state is difficult, perhaps impossible. Some versions of the Copenhagen interpretation use lack of reversibility as a criterion for what constitutes a completed measurement (see the next section).

Although the state vector (7) does not, in itself, provide an adequate description of the outcome of a measurement, we might hope that we can bring about its collapse into one or another of the two states of this correlated superposition by performing a measurement on the joint atom–apparatus system. For this purpose, we might employ a secondary apparatus that observes the primary apparatus and checks which detector has triggered. The usual rule for calculating probabilities tells us that such a measurement on the state vector (7) has a probability of $\frac{1}{2}$ for the result $|+\rangle$ $|\text{upper}\rangle$ and a probability of $\frac{1}{2}$ for the result $|-\rangle$ $|\text{lower}\rangle$; thus, the probabilities for the outcomes of the secondary measurement are consistent with the probabilities for the primary measurement that we attempted with the Stern–Gerlach apparatus. However, if the secondary apparatus and its interaction with the primary apparatus are, again, governed by the Schrödinger equation, then this attempt at a measurement yields, again, a superposition:

$$|\psi\rangle = \frac{1}{\sqrt{2}}\,|+\rangle\,|\text{upper}\rangle\,|\text{upper confirmed}\rangle$$

$$+ \frac{1}{\sqrt{2}}\,|-\rangle\,|\text{lower}\rangle\,|\text{lower confirmed}\rangle \tag{8}$$

From this example we see that stacking one apparatus on top of another does not bring about the desired collapse of the state vector.

The absence of collapse in any system governed by the Schrödinger equation—and the concomitant impossibility of bringing a measurement to completion, no matter how many apparata are stacked one on top of another—is called *von Neumann's catastrophe of infinite regression*. This absence of collapse was established by von Neumann, who made the first rigorous examination of the mathematical foundations of quantum mechanics. Von Neumann decided that the collapse of the state vector during measurement must be inserted into quantum mechanics as a separate axiom. If we arrange any number of apparata in a sequential stack (with a primary apparatus, a secondary apparatus, a tertiary apparatus, etc.), in which each apparatus checks on the apparatus ranking below it and is, in turn, checked by the apparatus ranking above it, we must postulate that the collapse of the state vector occurs somewhere in this stack. As in the cases of one or two apparata discussed in connection with Eqs. (7) and (8), the probabilities for the different outcomes of measurement are unaffected by whether we postulate that the collapse occurs in the primary apparatus, or the secondary apparatus, or the tertiary apparatus, etc.

Apart from its inadequacy for describing the outcome of a measurement, the state vector (7) has some weird properties. This state vector represents an ambivalent state, in which the detectors are in a schizoid superposition of having triggered and not having triggered. Such superpositions are a familiar feature of the atomic or subatomic world, and since we have no direct experience with that world, our intuition is willing to accept such superpositions in that world. But in Eq. (7) we encounter such a superposition in the macroscopic world, where it directly clashes with our intuition.

The weirdness of such superpositions of macroscopic states is brought to an extreme in a celebrated *Gedankenexperiment* contrived by Schrödinger:[6]

> A cat is locked into a steel chamber, along with the following diabolical device (which one must secure against direct intervention by the cat): In a Geiger counter there is a minuscule amount of radioactive substance, so small, that in the course of one hour *perhaps* one of the atoms decays, but also, with equal probability, perhaps none; if it happens, the counter tube discharges and through a relay releases a hammer which shatters a small flask with hydrocyanic acid. If one has left this entire system to itself for an hour, one would say that the cat still lives *if* meanwhile no atom has decayed. The first atomic decay would have poisoned it. The ψ-function of the entire system would express this by having in it the living and the dead cat (pardon the phrase) mixed or smeared out in equal parts.

Although such a schizoid superposition of a live-cat state vector and a dead-cat state vector does violence to our intuition, we cannot disprove it by any experiment. As soon as we open the chamber, or use any measuring device to detect the life signs of the cat, the state vector collapses into either the live-cat configuration or the dead-cat configuration, with equal probabilities. Thus, we can never "see" the cat in the superposed state. Instead, the act of observation or of measurement does something very drastic to the state of the cat—it flips the cat into either the live state or the dead state.

Wigner has added an extra wrinkle to this *Gedankenexperiment* by proposing that we omit the cyanide capsule and that we

---

[6] E. Schrödinger, Naturwissenschaften, **23**, 807 (1935); a translation of this paper appeared in Proc. Am. Philos. Soc., **124**, 323 (1980).

replace the cat by a human volunteer, Wigner's friend. We let Wigner's friend watch the Geiger counter for a while, and then open the chamber, and ask her to tell us what has happened. If Wigner's friend tells us that the Geiger counter clicked some time ago, we would presumably have to conclude that her presence in the chamber was enough to collapse the state vector, and that our opening of the chamber had no further effect on the state vector.

12.3 *Alternative Interpretations of the Collapse*

Physicists have made a variety of attempts at resolving the conundrum posed by the collapse of the wavefunction during measurement. Most of these attempts accept the main features of the Copenhagen interpretation (listed in Section 12.1), but propose different ways of dealing with the collapse. Here we will briefly discuss four such attempts: the orthodox Copenhagen picture, the popular picture, the subjective picture, and the many-worlds picture.[7]

Orthodox Copenhagen Picture. This is the picture conceived by Bohr and by Heisenberg.[8] An essential feature of this picture is that the results obtained in any experiment are to be described in classical terms. Bohr argued that such a classical view of experimental results is imperative to enable physicists to communicate these results to each other: "However far the phenomena transcend the scope of classical physical explanation, the account of all evidence must be expressed in classical terms. The argument is simply that by the word 'experiment' we refer to a situation where we can tell others what we have done and what we have learned and that, therefore, the account of the experimental arrangement and of the results of the observations must be expressed in unambiguous language with suitable application of the terminology of classical physics."[9] Thus, the apparatus is supposed to indicate the result of a measurement with a well-defined pointer position

---

[7] There is no general agreement on names for these and other different pictures, or even on the number of different pictures.

[8] The following discussion is based mainly on the exposition of the Copenhagen interpretation given by Heisenberg in his *Physics and Philosophy*, Chapter III. Bohr never gave such a systematic exposition of his views; but the isolated statements that he made are consistent with Heisenberg's views.

[9] N. Bohr, *Atomic Physics and Human Knowledge*, p. 39.

on a scale or a well-defined digital readout, without any significant uncertainty.

The orthodox Copenhagen picture does not claim that the laws of quantum physics are inapplicable to the apparatus; on the contrary, Bohr was quite aware that quantum physics is ultimately responsible for all the properties of the materials out of which the apparatus is constructed, and, in his refutation of the *Gedankenexperimente* of Einstein (see Section 12.4), he did not hesitate to apply the uncertainty relations to macroscopic pieces of equipment. But a "good" apparatus is supposed to be designed in such a way that quantum uncertainties in the readout are insignificant. According to Bohr's criterion for a "good" apparatus, if a superposition of different macroscopic apparatus states—such as Eq. (7)—where to occur, it would demonstrate an inadequacy in the design of the apparatus, and the inability of this apparatus to bring the measurement to completion.

In the orthodox Copenhagen picture, the collapse of the wavefunction is not a physical process, merely a mathematical procedure, or a bookkeeping procedure. The wavefunction is not a material entity, merely a mathematical construct. The wavefunction is the expectation catalog that characterizes the quantum system. This catalog lists all the possible outcomes for all the possible experiments we might perform on the system; it tells us that if we perform some experiment, then the outcome will be this or that, with this or that predicted probability. As long as we do not perform any experiment on the system, the expectation catalog evolves continuously in time, according to the Schrödinger equation. But if we perform an experiment on the system and measure some observable, the expectation catalog changes discontinuously. During the measurement, one of the possible outcomes listed in the expectation catalog becomes the actual outcome, and all the other outcomes are rejected. This means that the expectation catalog must suddenly be altered—all the entries in the catalog must be deleted, except, of course, the one entry that became actual. But this sudden alteration, or collapse, of the expectation catalog is merely a reflection of our sudden change of knowledge. The only definite *physical* change during the measurement is the change that occurs in the apparatus, which switches from one well-defined state to another.

Both Bohr and Heisenberg have emphasized that the wavefunction does not tell us what actually happens in the quantum system between one measurement and the next; it does not provide a history of events in the quantum system. The wavefunction, in conjunction with the Schrödinger equation, merely tells us that if we perform a given experiment on the system at one time (say, a measurement of the x component of the spin of an atom), and some other given experiment at a later time (say, a measurement of the z component of the spin), then the outcome of the second experiment is probabilistically related to the outcome of the first. Thus, the mathematical machinery of quantum mechanics provides us with probabilistic connections between one experiment and the next, but it does not provide us with a mental picture of what happens in between (as already mentioned in Section 12.1, we must not confuse a mental picture of the time evolution of the wavefunction with a mental picture of the quantum system itself). Quantum mechanics tells us nothing about the quantum system itself, only about what happens in measurements. In Bohr's words: "The formalism of quantum mechanics is to be considered as a tool for deriving predictions of a . . . statistical character as regards information obtainable under experimental conditions described in classical terms."[10]

The orthodox Copenhagen interpretation insists on a sharp dichotomy between quantum system and apparatus. We must begin any description of an experiment by specifying the system to be measured, the apparatus with which it is to interact, and the dividing line, or the *Heisenberg cut*, between the system and the apparatus. The state of the quantum system is characterized by its wavefunction, but the state of the apparatus (or, at least, the state of the readout end of the apparatus) is characterized by well-defined classical parameters.

Although we must draw a sharp line between the quantum system and the apparatus, we have considerable freedom in just where we draw this sharp line. As is clear from the discussion of Eqs. (7) and (8), the probabilities for outcomes of measurements are not altered when we extend our quantum system so as to include some part of the apparatus with which it interacts during the measurement. For instance, if we attempt to detect a photon with a photomultiplier tube, we can regard the photon as the system and the photomultiplier tube as the apparatus. Alternatively, we can regard the photon and some portion of the photomultiplier tube as the system and the remainder as the apparatus. In the photomultiplier tube, the incident photon ejects a photoelectron from the faceplate, this electron strikes the first dynode and ejects

---

[10] N. Bohr; see Jammer, The *Philosophy of Quantum Mechanics*, p. 204.

several electrons; each of these strikes the second dynode and ejects more electrons, and so on. We can draw the dividing line between system and apparatus at the faceplate, or at the first dynode, or at the second, and so on. Depending on our choice of dividing line, the measured system will consist of a photon or a photon and one or several electrons; accordingly, the wavefunction of the system will have to include the wavefunction of these electrons. However, the Copenhagen picture does not permit us to shift the dividing line all the way to the output end of the photomultiplier, where a *classical* pulse of current emerges. According to Bohr, the classical mode of description becomes compulsory by the occurrence of an irreversible process of amplification; this brings the measurement to completion. This criterion for the completion of a measurement has been enthusiastically advocated by Wheeler who declared: "A phenomenon is not yet a phenomenon until it has been brought to a close by an irreversible act of amplification, such as the blackening of a grain of silver bromide emulsion or the triggering of a photodetector."[11] Wheeler has emphasized that a decisive test for the completion of a measurement is the registration of the information acquired in the measurement, in the form of a permanent, indelible record.

In the *Gedankenexperiment* of Schrödinger's cat, the orthodox Copenhagen interpretation claims that the quantum-mechanical wavefunction collapses when the Geiger counter makes a measurement on the radioactive substance, and therefore the state of the Geiger counter (and the state of the cat) never forms a superposition of two macroscopically different states. At each instant, the Geiger counter either performs an irreversible act of amplification or does not perform such an act, that is, the Geiger counter adopts either a definite state of discharge or a definite state of no discharge. This means that the Geiger counter acquires information about the radioactive decay, and brings about the collapse of the wavefunction of the radioactive substance. The human observer is not required to bring about the collapse. When the observer opens the chamber, he receives the information about the collapsed wavefunction; but since this information was already available in the output of the Geiger counter, he produces no further collapse of the wavefunction.

Although the criterion of irreversible amplification for the

---

[11] J. A. Wheeler in *Quantum Theory and Measurement*, edited by J. A. Wheeler and W. H. Zurek, p. 185.

completion of a measurement and the collapse of the wavefunction seems quite plausible, it is fraught with ambiguities. In a photomultiplier tube the amplification increases by stages. At what stage of this process of successive amplifications with successively increasing irreversibility will we attain sufficient amplification and sufficient irreversibility for the completion of the measurement? Furthermore, the amplification process spans some time, and this raises the question of whether perhaps the collapse of the wavefunction is also spans some time, which would mean that the collapse is not truly discontinuous.

Popular Picture. Physicists have a deep predilection for continuity in nature (*Natura non facit saltus*), and they tend to be uncomfortable with the discontinuous collapse and with the somewhat capricious dichotomy between measured system and apparatus demanded by the orthodox Copenhagen picture. The popular picture is an alternative to the orthodox Copenhagen picture; it is favored by many, perhaps by most, of the physicists of today. In the popular picture, there is no collapse. The state vector evolves continuously at all times, according to the Schrödinger equation. Both the system and the apparatus are treated quantum-mechanically, and they are described by a joint state vector. A measurement is regarded as an interaction between the system and the apparatus, as in our example of the Stern–Gerlach experiment of Section 12.2. During such an interaction, the state vectors of the system and the apparatus become correlated, and the joint state vector forms a superposition of these correlated state vectors. Thus, in our example of the Stern–Gerlach experiment, the result of the measurement is the joint state vector

$$|\psi\rangle = \frac{1}{\sqrt{2}}\,|+\rangle|\text{upper}\rangle + \frac{1}{\sqrt{2}}\,|-\rangle\,|\text{lower}\rangle \tag{9}$$

Consider, now, the expectation value of any operator R that acts on the spin states (but not on the apparatus states). According to the usual prescription for the calculation of an expectation value,

$$\langle\psi|R|\psi\rangle = \tfrac{1}{2}(\langle+|R|+\rangle\,\langle\text{upper}|\text{upper}\rangle + \langle-|R|-\rangle\,\langle\text{lower}|\text{lower}\rangle$$

$$+ \langle+|R|-\rangle\,\langle\text{upper}|\text{lower}\rangle + \langle-|R|+\rangle\,\langle\text{lower}|\text{upper}\rangle) \tag{10}$$

The two apparatus states are, of course, normalized, with $\langle\text{upper}|\text{upper}\rangle = 1$ and $\langle\text{lower}|\text{lower}\rangle = 1$. But the apparatus states are also orthogonal, since the triggering of one detector has no associated probability for triggering of the other detector (if the triggering of one detector tends to produce a triggering of the other detector, there is a defect in the design of the electric connections in the apparatus, and the experimenter must repair the apparatus). Thus, the last two terms in Eq. (10) vanish, and the first two terms reduce to

$$\langle\psi|R|\psi\rangle = \tfrac{1}{2}(\langle+|R|+\rangle + \langle-|R|-\rangle) \tag{11}$$

This result is the same as what we obtain if we take the expectation value of R for the collapsed state vector $|+\rangle|\text{upper}\rangle$ and for the collapsed state vector $|-\rangle|\text{lower}\rangle$, and we average these two possible expectation values. We therefore reach the conclusion that, on the average, for repeated measurements, the expectation value of R is the same as in the Copenhagen picture. Thus, although for each individual measurement, the collapsed state vectors $|+\rangle|\text{upper}\rangle$ and $|-\rangle|\text{lower}\rangle$ differ from the superposed state vector (9), we cannot detect this difference experimentally, because the average expectation values for repeated measurements are indistinguishable.

Exercise 1. The calculation leading to Eq. (11) was based on the simple initial state vector (5), with equal coefficients for the spin-up and spin-down states. Repeat the calculation for a general initial state, consisting of a superposition of spin-up and spin-down states with arbitrary coefficients c_1 and c_2. Show that the expectation values $\langle\psi|R|\psi\rangle$ is, again, an average of the two terms given in Eq. (11), but with weights $|c_1|^2$ and $|c_2|^2$. Thus, the general result is in agreement with the Copenhagen interpretation.

This means that in the popular picture there is collapse without collapse: the state vector does not really collapse, but the results for expectation values are the same as though it had collapsed. When we want to calculate an expectation value, we can therefore use collapse as a shortcut. As Eq. (11) shows, we can omit the state vector of the apparatus in our calculation, and we can *pretend* that the state vector of the system has collapsed to $|+\rangle$ or to $|-\rangle$. This shortcut always yields the same result as the honest calculation with the true, uncollapsed joint state vector given in Eq. (9).

From Eq. (10) we see that what permits the uncollapsed state vector to mimic an average calculated from the collapsed state vectors is the cancellation of the last two terms in this equation, that is, the off-diagonal terms. A somewhat different version of the popular picture attempts to achieve such a cancellation by exploiting unpredictable, random phase differences that are supposedly introduced into the state vector for the system when it interacts with the apparatus during measurement.[12] This version of the popular picture argues that the microscopic quantum state of the apparatus is not known, and is not reproducible from one repetition of the measurement to the next; even if we "reset" the apparatus for each repetition of the measurement, there will be uncontrollable and unpredictable fluctuations in its microscopic quantum state. When the measured system interacts with this apparatus, the different superposed parts of its state vector acquire different, random phase factors, which make the different parts in the superposition incoherent. But an incoherent superposition of several state vectors is equivalent, on the average, to an ensemble of collapsed state vectors. We can understand this equivalence between an incoherent superposition and an ensemble of collapsed state vectors by means of our simple example of measurement of the spin of an atom in a Stern–Gerlach experiment. The initial state vector of the atom [see Eq. (4)] is a coherent superposition of the spin-up and spin-down states. If the interaction with the apparatus inserts extra, random phase factors into this superposition, we obtain a final state vector

$$|\psi\rangle = \frac{e^{i\alpha_1}}{\sqrt{2}}|+\rangle + \frac{e^{i\alpha_2}}{\sqrt{2}}|-\rangle \tag{12}$$

Such a superposition with random phase factors is called a *mixture*. According to Eq. (12), the expectation value of any arbitrary operator R is then

$$\langle\psi|R|\psi\rangle = \frac{1}{2}\langle+|R|+\rangle + \frac{1}{2}\langle-|R|-\rangle$$
$$+ \frac{e^{i(\alpha_1-\alpha_2)}}{2}\langle-|R|+\rangle + \frac{e^{i(\alpha_2-\alpha_1)}}{2}\langle+|R|-\rangle \tag{13}$$

For random phases α_1 and α_2, the factor $e^{i(\alpha_1-\alpha_2)}$ averages to zero when we perform the measurement repeatedly, and therefore the average value of (13) over repeated measurements is simply

$$\langle\psi|R|\psi\rangle = \frac{1}{2}(\langle+|R|+\rangle + \langle-|R|-\rangle) \tag{14}$$

---

[12] D. Bohm, *Quantum Theory*, Chapter 22, Sections 6–12.

Note that here, as in Eq. (10), the off-diagonal terms have canceled. We therefore, again, reach the conclusion that, on the average, for repeated measurements, the expectation value of R is the same as for the Copenhagen picture.

The cancellation of off-diagonal terms by random phases [Eq. (13)] seems simpler and more straightforward than the cancellation by orthogonality of apparatus states [Eq. (10)]. However, the random-phase scheme suffers from a fatal defect. The phases must ultimately arise from the interactions between the system and the apparatus. If we want to calculate these phases, we must begin with an initial joint system–apparatus state vector, such as in Eq. (5), and we must investigate its time evolution during the interaction. The result will then be a correlated joint system–apparatus state vector, such as in Eq. (7), with extra, random phases. But this state vector *cannot* be factored into a product of a system state vector of the form (12) and some apparatus state vector. Thus, interactions cannot lead to a final state vector of the form of Eq. (12) for the system after the measurement.

Although random phases by themselves do not provide a consistent picture of the collapse, random phases could possibly play an ancillary role in suppressing interference effects in the correlated joint system–apparatus state vector given in Eq. (9). One difficulty with this state vector is that, to the extent that the system–apparatus interaction is reversible, the state vector (9) could possibly evolve back into the initial state vector (5). But if the two terms in the state vector (9) acquire extra, random phase factors $e^{i\alpha_1}$ and $e^{i\alpha_2}$, then the measurement becomes irreversible. Once the system has acquired random phases, we have lost essential information about the state vector, and we cannot reverse the evolution of the state vector in time and return to the initial state. Thus, the picture of random phases provides us with an explicit model of how irreversibility might enter the measurement process.

Subjective Picture. Another proposal for the collapse is that it is produced in the mind of the observer, by the intervention of the observer's consciousness. This notion was first proposed by von Neumann. As we saw in the preceding section, no apparatus governed by the Schrödinger equation can bring about the collapse, and neither can a stack of apparata arranged to check on each other. However, experience tells us that if a human observer is looking at one of the apparata in the stack, he always perceives the apparatus in a definite state. This forces us to accept that the collapse occurs, somehow, no later than in this observed apparatus or, at the most, no later than within the human observer. Since any apparatus is built of atomic or subatomic pieces, it presumably obeys the Schrödinger equation, and is free of collapse. As a last resort, von Neumann therefore suggested that the collapse occurs when the signals from the apparatus register in the observer's consciousness. This picture of the collapse process was adopted by London and Bauer[13] and by Wigner,[14] who saw in it the resolution of the conundrum posed by the *Gedankenexperiment* of Wigner's friend (see Section 12.2). Wigner proposed that the collapse is brought about by some (unknown) nonlinear process whenever the quantum system interacts with the consciousness of an observer.

But this proposal raises some awkward questions. Exactly what is meant by "consciousness"? What level of consciousness is sufficient to bring about collapse? Is human consciousness required, or is that of a cat or of a mosquito sufficient? Some of these questions can be bypassed by postulating that there is only one consciousness (my own) in the entire universe. This is the philosophical doctrine of solipsism. It is logically unexceptionable, but it is viewed with distaste by most physicists, whose scientific training tells them to be cautious about accepting claims made by one observer alone.

Many-Worlds Picture. Another, radically different treatment of the collapse problem is the many-worlds picture of Everett.[15] In this picture, as in the popular picture, there is no collapse, and the state vector evolves according to the Schrödinger equation at all times. But the many-worlds picture differs from the popular picture in that it includes the observer as part of the quantum-mechanical system. Thus, the many-worlds picture eliminates the dividing line (Heisenberg cut) between the observer and the apparatus, whereas the popular picture implicitly retains this dividing line. The interaction between measured system, apparatus, and observer produces a joint state vector consisting of a superposition of correlated joint state vectors. For instance, in the example of the Stern–Gerlach experiment in Section 12.2, the state vector for

---

[13] F. London and E. Bauer, *La théorie de l'observation en mécanique quantique* (Hermann, Paris, 1939). Translated in *Quantum Theory and Measurement*, edited by J. A. Wheeler and W. H. Zurek.

[14] E Wigner, *Symmetries and Reflections*, p. 183.

[15] H. Everett, Rev. Mod. Phys., **29**, 454 (1957).

the joint atom–apparatus–observer system after the measurement is a correlated joint state vector of the form given in Eq. (8), where we now regard the states |upper confirmed⟩ and |lower confirmed⟩ as states of the observer. The many-worlds interpretation insists that such a schizoid superposition with two or more terms, or "branches," with different observer states, is the correct description of the outcome of the measurement. The two terms in the superposition (8) are interpreted as one branch in which the apparatus has detected spin up and the observer has seen the apparatus detect spin up, and one branch in which the apparatus has detected spin down and the observer has seen it detect spin down. Thus, in each branch the state of the observer is consistent with the state of the apparatus, and in each branch the observer is unaware that something different has happened in the other branch, or even that there is another branch.

Although all of the branches exist simultaneously, the cloned observers in the individual branches do not interact,[16] and they remain forever unaware of each other. The cloned observers effectively inhabit separate worlds. Whenever there is a measurement, the history of the world splits into two or more branches, corresponding to the different outcomes of this measurement. Note that in the many-worlds picture, any measurement-like interaction occurring anywhere gives rise to new branches; thus the universe is continually splitting into a myriads of branches, and each of us is continually splitting into myriads of clones, even when the measurements are not being performed in our immediate vicinity.

In the many-worlds picture, we cannot directly interpret the coefficient $1/\sqrt{2}$ in Eq. (8) as probability amplitudes, since there is no external (outside-the-universe) observer who can measure the state of the system. When the observer is in the state |upper confirmed⟩, he is not aware of the other state, or of the coefficients $1/\sqrt{2}$. So how can he obtain probabilities in measurements? To answer this question, the many-worlds picture examines what happens if the observer performs a sequence of repeated measurements and records (or remembers) the results. Each measurement generates a new branch of the world, and after all the repeated

---

[16] The matrix element of the Hamiltonian is zero between any two distinct macroscopic states; if this were not so, then the Hamiltonian could produce transitions from one state of the apparatus or observer to the other, that is, it could change what the apparatus has detected or what the observer has seen.

measurements are completed, there are many branches. At the end of each branch sits a clone of the observer with a sequence of results in his memory, for instance a sequence $+ + - + + + - - -$ $+ - + +$ . . . or a sequence $+ - - + - + - + + + - +$ . . . . Everett proved that for almost every one of these many branches, the sequences of $+$'s and $-$'s are random.[17] Thus, almost all the cloned observers will decide that they have verified the prediction of quantum mechanics for repeated measurements of the spin. Everett takes this to mean that in a "typical" branch of the world, the predictions of quantum mechanics will be verified; and he assumes that our branch—that is, the branch at the end of which we sit—is a typical world.

12.4 *The Einstein–Podolsky–Rosen Paradox*

The fundamental assumption of the Copenhagen interpretation is that the physical state of an individual system is completely specified by the wavefunction ψ. This fundamental assumption leads to the uncertainty relations, which tell us that the coordinates and momenta are indeterminate, and that causality, in the sense of classical mechanics, is impossible. Physicists brought up in the traditions of classical mechanics found it hard to accept these features of quantum mechanics. The most illustrious and most severe critic of the Copenhagen interpretation was Einstein, who insisted that even in the realm of the atom there must exist precisely defined dynamical variables and strict deterministic behavior. In view of the practical success of quantum theory, Einstein was willing to accept that $|\psi|^2$ gives a probability distribution, but he insisted that this probability distribution must be interpreted as an ensemble distribution, which does not arise from an intrinsic indeterminacy of the dynamical variables, but only from our ignorance of their values.

Over the years, Einstein challenged the completeness assumption of the Copenhagen interpretation by a variety of clever *Gedankenexperimente*. At first, the thrust of these was directed at the uncertainty relation. Einstein wanted to find a counterexample to these uncertainty relations, by contriving some measure-

---

[17] More generally, if the terms in the superposition have different coefficients c_1 and c_2, the numbers of $+$'s and $-$'s in the sequences are weighted in proportion to $|c_1|^2$ and $|c_2|^2$.

ment procedure that would simultaneously determine the coordinate and the momentum of a particle. One such *Gedankenexperiment*, proposed by Einstein in a discussion with Bohr at the 1928 Solvay Meeting, was based on the momentum exchange between the incident particle and a slotted plate, such as might be used to demonstrate diffraction effects. Figure 12.2 shows the experimental arrangement (in the arrangement actually examined by Einstein and Bohr, another plate with two slots was placed in tandem with the single-slot plate, but this is an unessential complication). The particle is incident on the plate from the left, suffers diffraction while passing through the slot, and lands at some (unpredictable) position on the screen at the right. The passage through the slot amounts to a measurement of the vertical position, with an uncertainty $\Delta y = a$, the width of the slot. In the usual analysis of this *Gedankenexperiment*, the vertical momentum is calculated from the diffraction angle, which is known from the observed impact point on the screen; obviously,

$$p_y = p \sin \theta \tag{15}$$

The uncertainty in the angle θ is roughly given by the width of the central diffraction peak, $\Delta(\sin \theta) = \lambda/a = h/ap$, which leads to an uncertainty Δp_y:

$$\Delta p_y = p\,\Delta(\sin \theta) \simeq p\,\frac{h}{ap} = \frac{h}{a} \tag{16}$$

The product of the uncertainties in y and p_y is therefore

$$\Delta y\,\Delta p_y \simeq a\,\frac{h}{a} = h \tag{17}$$

which is consistent with the uncertainty principle.

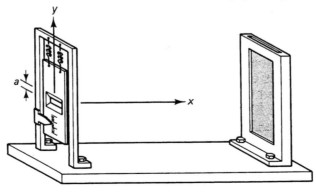

Fig. 12.2 A slotted plate used for a diffraction experiment.

However, Einstein proposed to modify this *Gedankenexperiment;* he proposed to measure the momentum p_y, not by the impact point on the screen, but by the recoil suffered by the plate. For this purpose, he suggested that the plate be loosely suspended (by the springs in Fig. 12.2), so it can move and its recoil motion can be determined. Since the recoil momentum of the plate, which is a large macroscopic body for which the laws of classical mechanics ought to hold, can presumably be measured with arbitrary precision, it should be possible to violate the uncertainty relation. But Bohr was quick to notice that the plate is itself subject to the uncertainty principle, and if its momentum is measured with an uncertainty of Δp_y smaller than h/a, then its position will become uncertain by an amount in excess of $h/\Delta p_y = a$, and this means that the momentum and the position of the particle deduced from the position and the momentum of the plate will, again, obey the uncertainty relation.

In fact, it is easy to see that the uncertainty relations are self-consistent: if all bodies obey the uncertainty relations, then a measurement of one body by another can never lead to a result that violates the uncertainty relation. But it is crucial for this consistency that *all* bodies in the universe obey the uncertainty relation; if there were some purely classical body somewhere, with perfectly well-defined position and momentum, then by examining the collision of this body with another body, we could violate the uncertainty relations for the position and momentum of this other body.

Blocked in his direct attacks on the uncertainty relations, Einstein, in a joint venture with Podolsky and Rosen,[18] launched a more subtle attack on the completeness assumption on which the uncertainty relations are based. Their argument, which became known as the Einstein–Podolsky–Rosen (EPR) paradox, begins with the hypothesis that the quantum-mechanical predictions for the results of measurements are correct and tries to show, by means of a *Gedankenexperiment,* that the quantum-mechanical description of the state of the system is incomplete, that is, the system is endowed with physical properties that go beyond those permitted by quantum mechanics.

The EPR paradox hinges on the examination of the joint quantum-mechanical state of two particles that are initially correlated in such perfect way that a measurement performed on one of the

---

[18] A. Einstein, B. Podolsky, and N. Rosen, Phys. Rev., **47**, 777 (1935).

particles immediately tells us the state of the other particle, without any need to measure or disturb this other particle. Einstein, Podolsky, and Rosen contemplated two particles of opposite momenta, released initially at one point. But, as remarked by Bohm, the EPR paradox can equally well be stated in terms of two particles of spin $\frac{1}{2}$ in a state of net spin zero, that is, in a state in which their spins are opposite. We will discuss the EPR paradox in terms of such a spin state, because this eases the mathematics and because actual laboratory trials of the experiment have made use of spin states.

Consider two particles of spin $\frac{1}{2}$, such as two protons or two neutrons, in a state of net spin zero. Suppose that the particles are initially close together, but then they move apart to a large distance, while they remain in the original state of net spin zero. Once they are widely separated, we measure the spin of one of these particles. Since the net spin is zero, the measurement of the spin of the first particle immediately allows us to infer the spin of the other particle—it must always be opposite to the spin of the first particle. For instance, if we measure the z component of the spin of the first particle and find $\hbar/2$, then we immediately know that the z component of the spin of the second particle must be $-\hbar/2$. The crucial step in the argument of Einstein, Podolsky, and Rosen is now this: Since our measurement did not touch this second particle, its state before the measurement ought to be the same as after, and therefore this particle ought to have had a well-defined z component of spin even *before* we performed the measurement. But we can now apply the same argument to a measurement of the x component of the spin; if we repeat the experiment and measure the x component instead of the z component, then our argument leads us the conclusion that the second particle ought to have had a well-defined x component of the spin before the measurement. And we can apply the same argument to the y component, and conclude that the second particle ought to have had a well-defined y component of the spin before the measurement. Thus, *all* of the components of the spin of the second particle ought to be well defined, in contradiction to quantum mechanics, which asserts that if one component is well defined, then the others are indeterminate. Accordingly, Einstein, Podolsky, and Rosen claimed that the quantum-mechanical description provided by the state vector cannot be complete. In their view, the state vector must be supplemented or replaced by some extra "hidden variables," and the spin components must be expressed as functions of these hidden variables, so all the spin components are simultaneously well defined.

Note that the crucial step in the EPR argument hinges on the reality of the attributes of the particles and on the locality of the measurement procedure. The spin of the second particle is supposed to exist, in itself, even if we do not measure it; and the measurement performed on the first particle is supposed to produce no effect on the second, distant, particle. Quantum mechanics refutes this paradox by denying both of these suppositions. The Copenhagen interpretation tells us that the particles do not have attributes in themselves, but only in relation to a measurement procedure. Furthermore, it tells us that a measurement performed on one portion of a wavefunction, at one place, affects the entire wavefunction, even its very distant portions.

According to quantum mechanics, the state vectors of the two particles are so intimately intertwined that it makes no sense to speak of the state vector of each individual particle, and it makes no sense to speak of a real value of spin of each. We can see this from the expression for the eigenstate of net spin zero ($s = 0$, $m_s = 0$) formed from the two states of spin one-half:

$$|0,0\rangle = \frac{1}{\sqrt{2}} (|+\rangle|-\rangle - |-\rangle|+\rangle) \tag{18}$$

Here, the first bra in each term indicates the spin state of the first particle, and the second bra that of the second [see Eq. (9.95)]. For each individual particle, this state $|0,0\rangle$ is neither an eigenstate of the individual z component of spin, nor even a simple superposition of the eigenstates $|+\rangle$ and $|-\rangle$. There is no definite state vector for the individual particles—only a joint state vector for the system. Thus, it is not surprising that a measurement of the spin of one particle affects the other particle. The measurement of the spin of one particle changes the *whole* state.

We cannot measure one portion of the quantum-mechanical wavefunction and leave the rest undisturbed. When we measure any portion of the wavefunction, the *whole* wavefunction collapses. The strange simultaneous collapse of the spin states of both particles in the EPR *Gedankenexperiment* is no more remarkable than the simultaneous collapse of all parts of the wavefunction of a single particle. When we place a detector in one part of such a single-particle wavefunction, and we find (or do not find) the particle in the volume of the detector, this affects the *entire* wavefunction throughout space. The spin state vectors of the two particles are just as intertwined as the portions of the wavefunction of a single particle. The system of the two spinning particles has a single wavefunction, which happens to depend on two variables. The wavefunction cannot be regarded as consisting of separate, disjoint pieces. Our intuitive expectation that we can measure one portion of a wavefunction without disturbing its distant portions is brought about by an overemphasis on the position representation. Excessive reliance on this representation misleads us into expecting analogies between the behavior of the quantum-mechanical wavefunctions and classical wavefunctions. If we use the abstract state vector, we find it easier to resist this temptation.

Note that the simultaneous collapse of the spin states of the two particles cannot be used to transmit signals from one location to the other. If the measurement of the spin of one of the particles reveals it to be in the state $|+\rangle$, then the other particle collapses into the state $|-\rangle$; but an observer who then measures the spin of this second particle and finds it to be $|-\rangle$ has no way of knowing that this result was enforced by the previous measurement at the other location—he will only know this after he has received a telegram or a letter from the other location informing him of the previous result.

Although quantum mechanics gives a perfectly logical answer to the EPR paradox, it does not give an answer that satisfies our intuition. The EPR paradox shows that the weirdness of quantum mechanics can be found even in systems involving macroscopic distances. The two spinning particles are separated by a large distance, and they do not interact—nevertheless, they form a single, mathematically inseparable system. Quantum mechanics asks us to ignore our intuition and to accept the weird intertwined, nonlocal behavior of the particles in this *Gedankenexperiment*.

12.5 *Bell's Theorem*

In hidden-variable theories, the unpredictable results for a sequence of repeated measurements are attributed to our lack of knowledge of the values of the hidden variables. The average value obtained in a sequence of measurements is taken to equal an average over the unknown (and unknowable?) values of the hidden variables; thus, the average value obtained in a measurement is taken to equal an ensemble average. Of course, the hidden variables and the ensemble used in the averaging are chosen so as to obtain agreement with the expectation values calculated from quantum mechanics. Einstein and other physicists who favored hidden variables took it for granted that the predictions of quantum mechanics can always be duplicated by adopting some sufficiently large set of hidden variables with a sufficiently complicated ensemble distribution. However, in 1964, Bell[19] demonstrated that not all of the subtleties of the probabilistic predictions of quantum mechanics can be duplicated by hidden variables. He demonstrated that the correlations among spin measurements on two particles of spin one-half in a state of zero net spin cannot be duplicated by local hidden variables.

Consider a sequence of measurements of the components of the spins of the two particles along two different directions. The component of the spin of particle 1 is measured along the direction of the unit vector **a**, and the component of the spin of particle 2 is measured along the direction of the unit vector **b**. The results of these measurements are S_{a1} and S_{b2}, respectively, where the spin components S_{a1} and S_{b2} take the usual values $\pm\frac{1}{2}\hbar$. If the directions **a** and **b** are the same, measurements on the quantum-mechanical spin state exhibit a perfect correlation, or rather, a perfect anticorrelation: whenever the measurement of the spin of one of the particles yields the value $S_{a1} = \frac{1}{2}\hbar$, the measurement of the spin of the other particle yields the opposite value $S_{a2} = -\frac{1}{2}\hbar$.

However, if the directions **a** and **b** are not the same, then the anticorrelation of the paired spin measurements will not be perfect. In general, we can characterize the amount of correlation observed in a sequence of a large number of repeated measurements by a *correlation coefficient,* defined as the average value of the product $(4/\hbar^2)S_{a1}S_{b2}$:

$$C(\mathbf{a}, \mathbf{b}) = \left[\frac{4}{\hbar^2} S_{a1}S_{b2}\right]_{av} \tag{19}$$

Note that for each paired spin measurement, the value of $(4/\hbar^2)S_{a1}S_{b2}$ is either $+1$ or -1; hence, the correlation coefficient is the average of a sequence of $+1$'s and -1's, and necessarily falls within the range $-1 \leq C(\mathbf{a}, \mathbf{b}) \leq +1$. If for each paired spin measurement in our sequence, the observed values of S_{a1} and S_{b2} are exactly opposite, then the correlation coefficient will be $C(\mathbf{a}, \mathbf{b}) = -1$; this characterizes a perfect anticorrelation. If for

---

[19] J. S. Bell, Physics, **1**, 195 (1964).

each paired spin measurement, the observed values of S_{a1} and S_{b2} are equal, then the correlation coefficient will be $C(\mathbf{a}, \mathbf{b}) = +1$, a perfect correlation. If some pairs of measurements yield opposite spins and some pairs equal spins, then the correlation coefficient will fall between the extreme values $+1$ and -1.

The average in Eq. (19) has been indicated with a square bracket to emphasize that it is calculated directly from the experimental data. Thus, this definition of the correlation coefficient does not hinge on any particular theory of the spin. But, of course, any theory of spin will make a prediction for the value of the correlation coefficient. In quantum mechanics, the average $[(4/\hbar^2)S_{a1}S_{b2}]_{av}$ over the experimental data for a long sequence of repeated measurements is predicted to equal the expectation value $\langle (4/\hbar^2)S_{a1}S_{b2} \rangle$. By evaluating this expectation value in the quantum-mechanical state of net spin zero, we can show that the correlation coefficient is

$$C(\mathbf{a}, \mathbf{b}) = \left\langle \left(\frac{4}{\hbar^2} \right) S_{a1}S_{b2} \right\rangle = -\cos\theta \qquad (20)$$

where θ is the angle between the directions of $\mathbf{a}$ and $\mathbf{b}$. Note that for $\theta = 0$, this yields $C(\mathbf{a}, \mathbf{b}) = -1$, as expected. And for $\theta = 90°$ it yields $C(\mathbf{a}, \mathbf{b}) = 0$. This is also expected, since the second spin is always opposite the first, and therefore has equal probabilities for the two possible eigenstates ($S_{b2} = \pm\frac{1}{2}\hbar$) of spin at right angles; consequently, there is no correlation between S_{b2} and S_{a1}.

For a derivation of the formula (20), let us assume that $\mathbf{a}$ is along the $+z$ direction and that $\mathbf{b}$ is in the z-x plane, at an angle θ with the z axis. The spinor for the zero-spin state is

$$\frac{1}{\sqrt{2}} (|+\rangle |-\rangle - |-\rangle |+\rangle) \qquad (21)$$

The correlation coefficient is the expectation value of $(4/\hbar^2)S_{z1}S_{b2}$ in this state:

$$C(\mathbf{a}, \mathbf{b}) = \frac{1}{2} (\langle+|\langle-| - \langle-|\langle+|) \left(\frac{4}{\hbar^2} S_{z1}S_{b2} \right) (|+\rangle |-\rangle - |-\rangle |+\rangle)$$

$$= \frac{2}{\hbar^2} (\langle+| S_{z1} |+\rangle \langle-| S_{b2} |-\rangle - \langle+| S_{z1} |-\rangle \langle-| S_{b2} |+\rangle$$

$$- \langle-| S_{z1} |+\rangle \langle+| S_{b2} |-\rangle + \langle-| S_{z1} |-\rangle \langle+| S_{b2} |+\rangle) \qquad (22)$$

Here, the second and the third terms are zero, since $\langle+| S_{z1} |-\rangle = 0$. In the first and fourth terms, the expectation values of S_{z1} are easy to evaluate: $\langle+| S_{z1} |+\rangle = \hbar/2$ and $\langle-| S_{z1} |-\rangle = -\hbar/2$. However, the expectation values of S_{b2} are more difficult, since $|+\rangle$ and $|-\rangle$ are not eigenstates of S_{b2}, but of S_{z2}. To get around this difficulty, we use the vector properties of the spin operator $\mathbf{S}$ and express S_{b2} as a superposition of S_{z2} and S_{x2}:

$$S_{b2} = S_{z2} \cos \theta + S_{x2} \sin \theta \qquad (23)$$

This equation is merely the standard formula for the transformation of the z component of a vector when the z axis is rotated by an angle θ toward the x axis. Thus, the expectation values of S_{b2} in the first and fourth terms of Eq. (22) are

$$\langle+| S_{b2} |+\rangle = \langle+| S_{z2} |+\rangle \cos \theta + \langle+| S_{x2} |+\rangle \sin \theta = \frac{\hbar}{2} \cos \theta + 0 \quad (24)$$

and

$$\langle-| S_{b2} |-\rangle = \langle-| S_{z2} |-\rangle \cos \theta + \langle-| S_{x2} |-\rangle \sin \theta = -\frac{\hbar}{2} \cos \theta + 0$$

$$(25)$$

Combining these results, we find

$$C(\mathbf{a}, \mathbf{b}) = \frac{2}{\hbar^2} \left(-\frac{\hbar^2}{4} \cos \theta - \frac{\hbar^2}{4} \cos \theta \right) = -\cos \theta$$

This establishes the validity of the formula (20).

Bell examined the correlation coefficients for measurements of the spin components along three (or more) different directions. He proved that in any local hidden-variable theory the correlation coefficients are restricted by an inequality, and that this inequality is *not* satisfied by the correlation coefficient predicted by quantum mechanics.

Consider three different directions specified by the unit vectors **a**, **b**, and **c**, and suppose that we perform paired measurements of the spin components of the two particles along these directions, taking two directions at a time. We begin with a sequence of paired measurements along the directions **a** (for particle 1) and **b** (for particle 2); then a sequence of measurements along the directions **a** and **c**; and finally, a sequence of measurements along the directions **b** and **c**. The correlation coefficients for these sequences of measurements are $C(\mathbf{a}, \mathbf{b})$, $C(\mathbf{a}, \mathbf{c})$, and $C(\mathbf{b}, \mathbf{c})$, respectively. According to hidden-variable theory, the predicted values of these correlation coefficients are ensemble averages over the hidden variables, with some distribution function (weight function). The number and the kind of hidden variables and the shape of the distribution function depend on the details of the hidden-variable theory. But Bell proved that in any local hidden-variable theory, the correlation coefficients necessarily obey the inequality

$$|C(\mathbf{a}, \mathbf{b}) - C(\mathbf{a}, \mathbf{c})| - C(\mathbf{b}, \mathbf{c}) \le 1 \qquad (26)$$

This result, known as *Bell's theorem,* is independent of the details of hidden-variable theory; it makes no difference how many hidden variables the theory contains, and how their probability distributions are contrived.

For a proof of the theorem, it is convenient to start with an examination of the quantity

$$g = -\frac{4}{\hbar^2} S_{a1} S_{b1} \left(1 - \frac{4}{\hbar^2} S_{b1} S_{c1}\right) \qquad (27)$$

which, as we will see, is closely related to the correlation coefficients. In a local hidden-variable theory, a measurement at one place does not affect what happens at another, distant, place; and it follows from this, by the EPR argument, that each particle has well-defined simultaneous spin components. Thus, in such a theory (but not in quantum mechanics) the components of the spin along all three directions $\mathbf{a}$, $\mathbf{b}$, and $\mathbf{c}$ are well defined, although they are not necessarily known to us, and they are likely to be different for each repetition of the measurement, because the values of the hidden variables are likely to be different. But for the purposes of Bell's theorem, it suffices that, for each particle 1 in our sequence of repeated measurements, the quantity g has some well-defined value.

Since $(S_{b1})^2 = \hbar^2/4$, we can rewrite g as

$$g = -\frac{4}{\hbar^2} (S_{a1} S_{b1} - S_{a1} S_{c1}) \qquad (28)$$

We can relate this expression to the correlation coefficients by taking into account that, in the configuration of net spin zero for the two particles, their spins must be opposite, and hence their spin components along any direction must also be opposite:

$$S_{b1} = -S_{b2}$$
$$S_{c1} = -S_{c2} \qquad (29)$$

Substituting these equations into Eq. (28), we obtain

$$g = \frac{4}{\hbar^2} (S_{a1} S_{b2} - S_{a1} S_{c2}) \tag{30}$$

From this we see that the ensemble-average value of g is

$$[g]_{av} = C(\mathbf{a, b}) - C(\mathbf{a, c}) \tag{31}$$

Next, we examine the absolute value of $[g]_{av}$. Since the absolute value of the average of g is less than or equal to the average of the absolute value of g, and since $|S_{a1} S_{b2}| = \hbar^2/4$, Eq. (27) leads to

$$|[g]_{av}| \le [|g|]_{av} = \left[\left| \frac{4}{\hbar^2} S_{a1} S_{b1} \left(1 - \frac{4}{\hbar^2} S_{b1} S_{c1} \right) \right| \right]_{av}$$

$$= \left[\left(1 - \frac{4}{\hbar^2} S_{b1} S_{c1} \right) \right]_{av}$$

$$= 1 + \left[\frac{4}{\hbar^2} S_{b1} S_{c2} \right]_{av} \tag{32}$$

But the second term on the right side of the last equation is $C(\mathbf{b, c})$, and thus

$$|[g]_{av}| \le 1 + C(\mathbf{b, c})$$

Combining this with Eq. (31), we immediately obtain the inequality (26) for the correlation coefficients.

The inequality (26), called *Bell's inequality*, is obeyed by every local hidden-variable theory. But this inequality is *not* obeyed by quantum mechanics. For the sake of simplicity, let us consider the special case with $\mathbf{a, b}$, and $\mathbf{c}$ in the same plane, say the z-x plane, and with $\mathbf{a}$ along the $+z$ axis, $\mathbf{b}$ at an angle of θ with respect to the $+z$ axis, and $\mathbf{c}$ at an angle of 2θ with respect to the $+z$ axis. The quantum-mechanical correlation coefficients are then [see Eq. (20)]

$$C(\mathbf{a, b}) = -\cos \theta$$

$$C(\mathbf{a, c}) = -\cos 2\theta$$

$$C(\mathbf{b, c}) = -\cos \theta$$

Thus, the quantum-mechanical expression for the left side of Bell's inequality is

$$|C(\mathbf{a, b}) - C(\mathbf{a, c})| - C(\mathbf{b, c}) = |-\cos \theta + \cos 2\theta| + \cos \theta$$

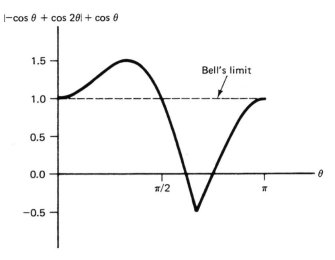

$|-\cos\theta + \cos 2\theta| + \cos\theta$

Fig. 12.3 Plot of $|-\cos\theta + \cos 2\theta| + \cos\theta$ vs. θ. The dashed line is the upper limit set by Bell's inequality.

Figure 12.3 shows a plot of this expression as a function of θ. We see that in the interval $0 < \theta < \pi/2$, the quantum-mechanical result exceeds the limit set by Bell's inequality. Thus, the quantum-mechanical result is inconsistent with all local hidden-variable theories.

Bell's inequality provides us with a way to discriminate experimentally between the predictions of quantum mechanics and those of local hidden-variable theories. Before Bell's theorem, such a discrimination was thought to be nearly impossible, since hidden-variable theories are designed to mimic the results of quantum mechanics as best they can. After Bell's theorem, several experiments were performed to test the quantum-mechanical prediction for correlation coefficients vs. the hidden-variable prediction. Most of these experiments studied the correlations of the polarizations of paired photons of net spin zero emitted by an atom. Inequalities similar to the Bell inequality (26) can be derived for the correlations of the polarizations of such photons. The inequalities tested in these experiments actually were generalized Bell inequalities, that make allowances for the less-than-perfect counting efficiencies of the photon polarizers and detectors, and for the experimental errors introduced by the apparatus. One weakness of the original Bell inequality is that, like the EPR argument, it relies on an ideal apparatus capable of measuring a spin component with absolute precision—we can use the relation $S_{a2} = -S_{a1}$ to find the spin of the second particle from a measurement performed on the first particle only if the first measurement is *absolutely precise*. The generalized Bell inequalities do not make such extreme demands of the apparatus.

The pairs photons used in experimental tests of Bell's inequality are emitted in a cascade process, in which an atom quickly makes two successive transitions from an upper state of angular momentum $j = 0$, to an intermediate state of $j = 1$, and finally to a lower state of $j = 0$. Since the initial and the final states have angular momenta zero, the net angular momentum carried away by the two photons emitted in these two transitions must be zero, and their polarizations are therefore perfectly correlated, like the spins of the two particles we considered above.[20] Experiments have also been performed with pairs of protons obtained by proton–proton scattering at low energies. From our discussion of partial waves in scattering (see Section 11.5), we know that when a low-energy proton is incident on a target proton, most of the scattering is contributed by the partial wave of zero orbital angular momentum. This means the protons are in a symmetric orbital state, and the Pauli exclusion principle demands that they must then be in an antisymmetric spin state, that is, a state of net spin zero.

With one exception, attributed to systematic experimental errors, all these experiments found correlations that agreed with the predictions of quantum mechanics and that exceeded the upper limit demanded by Bell's inequality. In the best of these experiments, by A. Aspect and his associates at the Institut d'Optique d'Orsay,[21] the experimental results exceeded the Bell inequality by more than 40 standard deviations. These experimental results conclusively rule out local hidden-variable theories. Thus, nature tells us that the weird nonlocal character of quantum mechanics brought out in the EPR paradox must be accepted. This does not necessarily mean that we have to accept quantum mechanics. We could still think of contriving a *nonlocal* hidden-variable theory,

---

[20] The polarization of a photon is in direct correspondence to its spin state. The two states of circular polarization correspond to spin parallel and spin antiparallel to the momentum of the photon. The two states of plane polarization correspond to superpositions of these spin states.

[21] For a review of recent experimental results, see the article by A. Aspect and P. Grangier in *Quantum Concepts in Space and Time*, edited by R. Penrose and C. J. Isham. For a comprehensive review of earlier experiments, see J. F. Clauser and A. Shimony, Rep. Prog. Phys., **41**, 1881 (1978).

even though such a theory would be somewhat pointless, since, according to the EPR argument, the main rationale for a hidden-variable theory is the attainment of locality.

In a nonlocal theory, a measurement at one place can affect a measurement at another place. In such a theory, some hidden variables might generate a (nonlocal) influence between the detectors, so the orientation of the axis of polarization of detector 1 alters the behavior of detector 2, in such a way that the measured correlations match those predicted by quantum mechanics. However, a modification of the two-photon correlation experiment by Aspect established that if this influence exists, it must propagate from one detector to the other at a speed exceeding the speed of light. In the modified experiment, the polarizations of the detectors were independently switched from one direction to another in a time shorter than the light travel time between detectors. The experimental results were, again, in agreement with quantum mechanics, and in disagreement with Bell's inequality. This establishes that any nonlocal influence exerted by one detector on the other would have to proceed via superluminal action-at-a-distance. Furthermore, this action-at-a-distance would have to be contrived in such a way that an experimenter can never use it to send deliberate signals, which would violate causality. These features of a nonlocal hidden-variable theory would be even more weird than the features of quantum mechanics.

PROBLEMS
1. In one attempt at finding a counterexample to the energy–time uncertainty relation, Einstein proposed that a closed box full of radiation be equipped with a shutter operated by a clock (see Fig. 12.4). The box is first weighed precisely with a spring balance, then the clock opens the shutter for an interval Δt and releases a photon, and then the box is again weighed precisely. This would seem to permit a precise determination of the energy of the photon ($\Delta E = 0$), in contradiction with the uncertainty relation $\Delta E \, \Delta t \geq \hbar/2$. Bohr refuted this counterexample by noting that the vertical uncertainty in the position of the box during the weighing leads to an uncertainty in the rate of the clock, via the gravitational time-dilation effect of Einstein's theory of general relativity. A simpler refutation can be constructed by taking into account that the *clock and the shutter* are a quantum system, which is subject to the energy–time uncertainty relation.

(a) Prove that the clock and shutter cannot be in an energy eigenstate. (Hint: Consider that the hands of the clock and the shutter have time-dependent positions.)

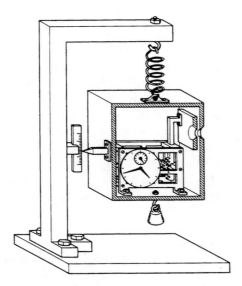

Fig. 12.4 A box with a clock-operated shutter. The box is suspended from a spring balance.

(b) Show that the energy of the clock-and-shutter system must be un-certain by $\Delta E \geq \hbar/2\Delta t$. Deduce from this that a precise measure-ment of the net energy of the box does *not* determine the energy of the radiation precisely, and that the photon energy satisfies the usual energy–time uncertainty relation.

2. Consider a classical body, with perfectly well-defined position and momentum at each instant. Show that by measuring the position and the momentum of this body before and after a collision with a quan-tum-mechanical particle, you can determine the position of this parti-cle *exactly*, and violate the uncertainty principle. (Hint: The collision does not increase the uncertainty in the momentum.)

3. In a hidden-variable theory first formulated by de Broglie and later revised by Bohm, electrons and other particles have a well-defined position and momentum at each instant and they move according to classical mechanics, but their motion is affected by the wavefunction, which acts as a "guiding wave," or "pilot wave," for the particle. The pilot wave exerts a force on the particle, with a potential

$$U = -\frac{\hbar^2}{2m}\frac{1}{|\psi|}\nabla^2|\psi|$$

The net potential is then the sum of the ostensible potential (such as the harmonic-oscillator potential, or the Coulomb potential) and the potential of the pilot wave. The probability distribution arises because we do not know where in the wave the particle is placed initially. The particle has a high probability of being found at places where its speed is low (or its potential high), because it spends more time there; and it has a low probability of being found at places where its speed is high (or its potential low), because it spends little time there.

(a) Roughly sketch the net potential $\frac{1}{2}m\omega^2x^2 - \hbar^2/2m \; 1/|\psi| \; \nabla^2|\psi|$ for the ground state of the harmonic oscillator.

(b) Roughly sketch the potential for an electron at a screen placed in the diffraction pattern produced by two narrow slits. The wavefunction of the electron at the screen is approximately $\psi = \cos(2\pi ax/\lambda L)$, where x is the distance measured along the screen, a is the separation of the slits, and L is the distance from the slits to the screen.

4. In some experimental arrangements, it is possible to extract information about the state of a system from the *absence of a detection event*. Such measurements without measurement were first contrived by Renninger. For example, suppose that the observer removes the lower detector in the Stern–Gerlach experiment shown in Fig. 12.1. The observer can then still use this apparatus to measure the spin, because if the atom does *not* trigger the upper detector, he can unambiguously conclude, after waiting some sufficiently long time, that the atom must have taken the lower road, and that the spin is therefore down. But if this is the case, then the apparatus has made no response whatsoever, and there has been no process of any sort in the apparatus! Consider each of the different interpretations of the collapse we examined in Section 12.3, and discuss how each of these interpretations copes with this kind of measurement.

5. Suppose that we measure the spins of the two particles of spin $\frac{1}{2}$ along four different directions, specified by the unit vectors **a**, **a'**, **b**, and **b'**.

(a) In a local hidden-variable theory, consider the quantity

$$g = \frac{4}{\hbar^2} (S_{a1}S_{b2} + S_{a1}S_{b'2} + S_{a'1}S_{b2} - S_{a'1}S_{b'2})$$

Show that this equals

$$g = \frac{4}{\hbar^2} S_{a1}(S_{b2} + S_{b'2}) + \frac{4}{\hbar^2} S_{a'1}(S_{b2} - S_{b'2})$$

Substitute the values $\pm\hbar/2$ for all the spin components, and show that always $g = \pm 2$, and hence $|g| = 2$.

(b) Take the ensemble average of g, and show that this can be expressed as follows in terms of correlation coefficients:

$$[g]_{av} = C(\mathbf{a}, \mathbf{b}) + C(\mathbf{a}, \mathbf{b'}) + C(\mathbf{a'}, \mathbf{b}) - C(\mathbf{a'}, \mathbf{b'})$$

(c) From (a) and (b) deduce the Clauser–Holt–Horne–Shimony inequality

$$|C(\mathbf{a}, \mathbf{b}) + C(\mathbf{a}, \mathbf{b}') + C(\mathbf{a}', \mathbf{b}) - C(\mathbf{a}', \mathbf{b}')| \leq 2$$

(d) Consider the special case with $\mathbf{a} = \mathbf{b}$ and with $\mathbf{a}'$ and $\mathbf{b}'$ in the same plane as $\mathbf{a}$, at angles of $+45°$ and $-45°$, respectively, relative to $\mathbf{a}$. Show that then the quantum-mechanical correlation coefficient for a pair of particles of net spin zero violates the inequality.

6. Show that the Bell inequality (26) can be deduced from the Clauser–Holt–Horne–Shimony inequality stated in Problem 5. (Hint: The minus sign in the CHHS inequality can be shifted to any one of the four terms on the left side.)

7. Suppose that $\mathbf{a}$ is along the z axis and $\mathbf{b}$ is in the z-x plane at an angle θ with respect to the z axis. Suppose that the two particles of spin $\frac{1}{2}$ are in a state of net spin 1, and zero z component ($s = 1$, $m_s = 0$). What is the quantum-mechanical correlation coefficient $C(\mathbf{a}, \mathbf{b}) = \langle (4/\hbar^2) S_{a1} S_{b2} \rangle$ for this state?

APPENDIX 1: *Fundamental Constants*

The values in the following table were taken from the report of the CODATA Task Group on fundamental constants by E. R. Cohen and B. N. Taylor, *The 1986 Adjustment of the Fundamental Physical Constants*, CODATA Bulletin No. 63, November 1986. The digits in parentheses are the one-standard deviation uncertainty in the last digits of the given value.

Quantity	Symbol	Value	Units	Relative uncertainty (parts per million)
UNIVERSAL CONSTANTS				
speed of light in vacuum	c	299792458	ms^{-1}	(exact)
permeability of vacuum	μ_0	$4\pi \times 10^{-7}$	NA^{-2}	(exact)
permittivity of vacuum	ε_0	$8.854187817\ldots$	$10^{-12}\ Fm^{-1}$	(exact)
Newtonian constant of gravitation	G	6.67259(85)	$10^{-11}\ m^3\ kg^{-1}s^{-2}$	128
Planck constant	h	6.6260755(40)	$10^{-34}\ Js$	0.60
in electron volts		4.1356692(12)	$10^{-15}\ eVs$	0.30
	$\hbar = h/2\pi$	1.05457266(63)	$10^{-34}\ Js$	0.60
in electron volts		6.5821220(20)	$10^{-16}\ eVs$	0.30
ELECTROMAGNETIC CONSTANTS				
elementary charge	e	1.60217733(49)	$10^{-19}\ C$	0.30
magnetic flux quantum, $h/2e$	Φ_0	2.06783461(61)	$10^{-15}\ Wb$	0.30
Josephson frequency–voltage ratio	$2e/h$	4.8359767(14)	$10^{14}\ HzV^{-1}$	0.30
quantized Hall conductance	e^2/h	3.87404614(17)	$10^{-5}\ \Omega^{-1}$	0.045
Bohr magneton, $e\hbar/2m_e$	μ_B	9.2740154(31)	$10^{-24}\ JT^{-1}$	0.34
nuclear magneton, $e\hbar/2m_p$	μ_N	5.0507866(17)	$10^{-27}\ JT^{-1}$	0.34
ATOMIC CONSTANTS				
fine-structure constant, $e^2/4\pi\varepsilon_0\hbar c$	α	7.29735308(33)	10^{-3}	0.045
inverse fine-structure constant	α^{-1}	137.0359895(61)		0.045
Rydberg constant, $\frac{1}{2}m_e c\alpha^2/h$	R_∞	10973731.534(13)	m^{-1}	0.0012
Bohr radius, $4\pi\varepsilon_0\hbar^2/m_e e^2$	a_0	0.529177249(24)	$10^{-10}\ m$	0.045
Electron				
electron mass	m_e	9.1093897(54)	$10^{-31}\ kg$	0.59
		5.48579903(13)	$10^{-4}\ u$	0.023
$m_e c^2$ in electron volts		0.51099906(15)	MeV	0.30
Compton wavelength, $h/m_e c$	λ_c	2.42631058(22)	$10^{-12}\ m$	0.089
classical electron radius, $e^2/4\pi\varepsilon_0 m_e c^2$	r_e	2.81794092(38)	$10^{-15}\ m$	0.13
Thomson cross section, $(8\pi/3)r_e^2$	σ_e	0.66524616(18)	$10^{-28}\ m^2$	0.27
electron magnetic moment	μ_e	928.47701(31)	$10^{-26}\ JT^{-1}$	0.34
in Bohr magnetons	μ_e/μ_B	1.001159652193(10)		1×10^{-5}
electron g-factor, $2\mu_e/\mu_B$	g_e	2.002319304386(20)		1×10^{-5}

APPENDIX 1: *Fundamental Constants (concluded)*

Quantity	Symbol	Value	Units	Relative uncertainty (parts per million)
Proton				
proton mass	m_p	1.6726231(10)	10^{-27} kg	0.59
		1.007276470(12)	u	0.012
$\quad m_p c^2$ in electron volts		938.27231(28)	MeV	0.30
proton–electron mass ratio	m_p/m_e	1836.152701(37)		0.020
proton magnetic moment		1.41060761(47)	10^{-26} JT^{-1}	0.34
Neutron				
neutron mass	m_n	1.6749286(10)	10^{-27} kg	0.59
		1.008664904(14)	u	0.014
$\quad m_n c^2$ in electron volts		939.56563(28)	MeV	0.30
neutron–electron mass ratio	m_n/m_e	1838.683662(40)		0.022
neutron magnetic moment	μ_n	0.96623707(40)	10^{-26} JT^{-1}	0.41
PHYSICO-CHEMICAL CONSTANTS				
Avogadro constant	N_A	6.0221367(36)	10^{23} mol^{-1}	0.59
atomic mass constant, $m_u = \frac{1}{12}m(^{12}\text{C})$	m_u	1.6605402(10)	10^{-27} kg	0.59
$\quad m_u c^2$ in electron volts		931.49432(28)	MeV	0.30
Faraday constant	F	96485.309(29)	C mol^{-1}	0.30
molar gas constant	R	8.314510(70)	J mol^{-1}K^{-1}	8.4
Boltzmann constant, R/N_A	k	1.380658(12)	10^{-23} JK^{-1}	8.5
$\quad$ in electron volts		8.617385(73)	10^{-5} eVK^{-1}	8.4
molar volume (ideal gas), RT/p $T = 273.15$ K, $p = 101325$ Pa	V_m	22.41410(19)	liter/mol	8.4
Loschmidt constant, N_A/V_m	n_0	2.686763(23)	10^{25} m^{-3}	8.5
Stefan–Boltzmann constant, $(\pi^2/60)k^4/\hbar^3 c^2$	σ	5.67051(19)	10^{-8} Wm^{-2}K^{-4}	34
Wien displacement law constant	b	2.897756(24)	10^{-3} mK	8.4

APPENDIX 2: *Clebsch–Gordan Coefficients*

NOTE: A $\sqrt{}$ is to be understood over *every* coefficient: e.g., $-8/15$ read $= -\sqrt{8/15}$.

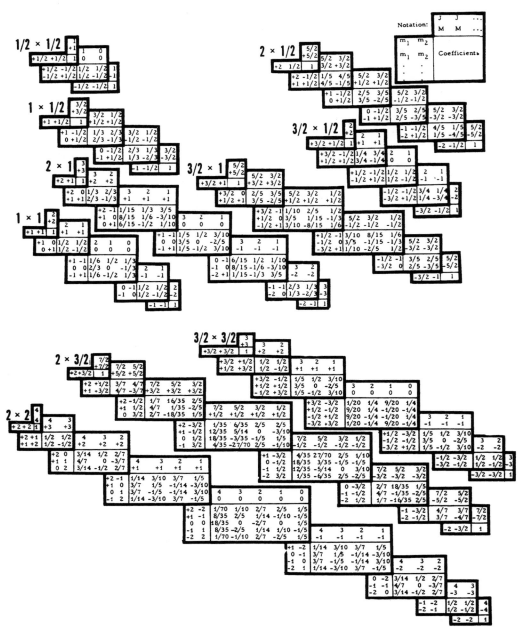

Bibliography

---

1. Experimental Foundations of Quantum Physics

Many introductory textbooks on atomic or modern physics give a good account of the experimental evidence that led to the discovery of quantization. Some of these textbooks are:

Bitter, F., and Medicus, H. A., *Fields and Particles* (Elsevier, New York, 1973)

Enge, H. A., Wehr, M. R., and Richards, J. A., *Introduction to Atomic Physics* (Addison-Wesley, Reading, Massachusetts, 1972)

Eisberg, R. M., and Resnick, R., *Quantum Physics of Atoms, Molecules, Solids, and Nuclei* (Wiley, New York, 1985)

French, A. P., and Taylor, E., *An Introduction to Quantum Physics* (Norton, New York, 1978)

Ohanian, H., *Modern Physics* (Prentice Hall, Englewood Cliffs, New Jersey, 1987)

Tipler, P. A., *Modern Physics* (Worth, New York, 1978)

Weidner, R. T., and Sells, R. I., *Elementary Modern Physics* (Allyn and Bacon, Boston, 1980)

Two older books, first published in the 30's, have become classics and still remain valuable:

Born, M., *Atomic Physics*, translated from the German by J. Dougall (Hafner, New York, 1962)

Herzberg, G., *Atomic Spectra and Atomic Structure*, translated from the German by J. W. T. Spinks (Prentice Hall, Englewood Cliffs, New Jersey, 1937; reprinted by Dover, New York, 1944)

2. History

Boorse, H. A., and Motz, L., *The World of the Atom* (Basic Books, New York, 1966) is an extensive anthology of reprints of the seminal papers on atomic physics, with commentaries and with pertinent biographical sketches.

French, A. P., and Kennedy, P. J., eds., *Niels Bohr, a Centenary Volume* (Harvard University Press, Cambridge, Massachusetts, 1985) is a captivating collection of critical articles, affectionate reminiscences, and anecdotes about Bohr, his scientific work, his philosophy, his politics, and his Institute of Theoretical Physics (now the Niels Bohr Institute) at Copenhagen.

Gamow, G., *Thirty Years that Shook Physics* (Doubleday, Garden City, New York, 1966) tells the story of the revolution in atomic physics in the early part of this century. Gamow was a great popularizer, and he wrote many delightful books on science for the general reader. This short book displays his usual charming style, and it is enlivened with his personal recollections of the physicists that wrought the revolution.

Jammer, M., *The Conceptual Development of Quantum Mechanics* (McGraw-Hill, New York, 1966) gives a detailed survey of the historical development of quantum mechanics. It supplies abundant references to all the original papers, and quotes or paraphrases many relevant passages. Unfortunately, Jammer fails to provide adequate critical guidance, and he often leaves the reader wondering which of the speculative theoretical developments detailed in the book were ultimately consigned to the trash heap.

Segré, E., *From X-Rays to Quarks* (Freeman, San Francisco, 1980) is an informative history of the circumstances surrounding the main discoveries of modern physicists. It is based on public lectures given by the author, and it is written at a more or less popular level.

van der Waerden, B. L., ed. *Sources of Quantum Mechanics* (Dover, New York, 1968) is a collection of reprints of the original papers that led to the discovery of matrix mechanics. It includes an excellent critical analysis of the role of each paper.

Weart, S. R., and Phillips, M., *History of Physics* (American Institute of Physics, 1985) consists of a selection of articles extracted from *Physics Today* dealing with diverse aspects of the history of physics in this century.

At a somewhat more personal level, the following books give us reminiscences and letters by some of the physicists who participated in the development of quantum mechanics:

Born, M., *The Born-Einstein Letters* (Walker, New York, 1971)

Heisenberg, W., *Physics and Beyond: Encounters and Conversations* (Harper and Row, New York, 1971)

Einstein, A., "Autobiographical Notes," in P. A. Schilpp, ed., *Albert Einstein: Philosopher-Scientist* (Harper & Brothers, New York, 1959)

Gamow, G., *My World Line: An Informal Autobiography* (Viking, New York, 1970)

3. Textbooks on Quantum Mechanics

Schrödinger's factorization method is carefully expounded in

Green, H. S., *Matrix Methods in Quantum Mechanics* (Barnes and Noble, New York, 1968)

Heisenberg invented a somewhat different matrix method for the calculation of the energy eigenvalues of the hydrogen atom. His method, based on the Runge-Lenz vector, is nicely described in

Jordan, T. F., *Quantum Mechanics in Simple Matrix Form* (Wiley, New York, 1986)

The traditional method for handling Schrödinger's equation is to concentrate on the position representation, where the Schrödinger equation is a partial differential equation, which can be solved by separation of variables and the standard power-series technique. This traditional method is used in all of the following textbooks:

Capri, A. Z., *Nonrelativistic Quantum Mechanics* (Benjamin/Cummins, Menlo Park, California, 1985)

Bohm, D., *Quantum Theory* (Prentice Hall, Englewood Cliffs, New Jersey, 1951)

Dicke, R. H., and Wittke, J. P., *Introduction to Quantum Mechanics* (Addison-Wesley, Reading, Massachusetts, 1960)

Liboff, R. L., *Introductory Quantum Mechanics* (Addison-Wesley, 1980)

Matthews, P. T., *Introduction to Quantum Mechanics* (McGraw-Hill, New York, 1968)

Powell, J. L., and Craseman, B., *Quantum Mechanics* (Addison-Wesley, Reading, Massachusetts, 1961)

Rae, A. I. M., *Quantum Mechanics* (Hilger, Philadelphia, 1986)

Saxon, D. S., *Elementary Quantum Mechanics* (Holden-Day, San Francisco, 1968)

A lavish collection of interesting plots and contour pictures illustrating the behavior of wave packets and eigenfunctions will be found in

Brandt, S., and Dahmen, H. D., *The Picture Book of Quantum Mechanics* (Wiley, New York, 1985)

4.   Advanced Textbooks on Quantum Mechanics

Chester, M., *Primer of Quantum Mechanics* (Wiley, New York, 1987) gives a concise review of quantum mechanics in the language of bras and kets. Written in an enjoyably casual style, it attempts to summarize all of quantum mechanics in a collection of (mildly) amusing aphorisms, such as "Microscopic is not miniaturized macroscopic. . . . The hamiltonian with hats is the energy operator. . . . At home the hat comes off". . . , etc. Although intended as an introductory text, it is better suited as a review for advanced students, who can better appreciate the wit.

Cohen-Tannoudji, C., Diu, B., and Laloë, F., *Quantum Mechanics*, translated from the French by S. Reid Hemley, N. Ostrowsky, and D. Ostrowsky (Wiley, New York, 1977) is the longest, most encyclopedic of the available textbooks, which contains just about everything you might ever want to know about quantum mechanics. It is written in a rather telegraphic style, with a maximum of formulas and derivations, and a minimum of explanations. The chapters are augmented by "complements" that treat dozens of topics that the authors failed to fit in the bodies of the chapters; these "complements" are preceded by "reader's guides," to rescue those readers who get lost.

Davydov, A. S., *Quantum Mechanics*, translated from the Russian by D. ter Haar (Pergamon, Elmsford, New York, 1976) is a thorough and competent, but somewhat dry and plodding, text.

Dirac, P. A. M., *The Principles of Quantum Mechanics* (Oxford University Press, Oxford, 1958) is one of the great classics in the literature of physics. First published in 1930, it quickly came to be regarded as "the standard work in the fundamental principles of quantum mechanics, indispensable both to the advanced student and to the mature worker, who will always find it a fresh source of knowledge and stimulation." (NATURE). Even today, the book gives a modern impression which belies its age.

Feynman, R. P., and Hibbs, A. R., *Quantum Mechanics and Path Integrals* (McGraw-Hill, New York, 1965) presents an unusual technique of quantization pioneered by Feynman, and much used in current attempts at a superstring theory. Feynman's technique uses neither the Schrödinger equation nor operator algebra, but instead directly evaluates the probability for a particle to move from one position to another by a summation over all the possible classical paths connecting these points.

Frauenfelder, H., and Henley, E. M., *Subatomic Physics* (Prentice Hall, Englewood Cliffs, New Jersey, 1974) covers applications to nuclear and particle physics.

Merzbacher, E., *Quantum Mechanics* (Wiley, New York, 1970) emphasizes scattering problems and the S matrix.

Messiah, A., *Quantum Mechanics*, translated from the French by G. M. Temmer (Interscience, New York, 1961) is almost as encyclopedic as Cohen-Tannoudji, op. cit., but much more readable.

Schiff, L. I., *Quantum Mechanics* (McGraw-Hill, New York, 1968) is distinguished by good qualitative explanations.

Sudbery, A., *Quantum Mechanics and the Particles of Nature* (University Press, Cambridge, 1986) emphasizes mathematical and axiomatic aspects and includes a chapter with a meticulous analysis of all the many interpretations of quantum mechanics.

5. Mathematical Foundations

Jordan, T. F., *Linear Operators for Quantum Mechanics* (Wiley, New York, 1969) is a concise introduction to the mathematics of linear operators used in quantum theory.

Erdelyi, A., ed., *Bateman Manuscript Project, Higher Transcendental Functions* (McGraw-Hill, New York, 1953) is a monumental compendium of tables of transcendental functions and the differential equations and other identities satisfied by them.

Friedman, B., *Principles and Techniques of Applied Mathematics* (Wiley, New York, 1956) contains a rigorous yet simple introduction to the delta function and Schwartz's theory of distributions.

Jahnke, E., and Emde, F., *Tables of Functions* (Dover, New York, 1945) gives tables, plots, and identities of the most important transcendental functions. This paperback book is a good bargain.

von Neumann, J., *Mathematical Foundations of Quantum Mechanics*, translated from the German by R. T. Beyer (Princeton University Press, Princeton, New Jersey, 1955) provides rigorous mathematical proofs of many of the results that physics textbooks take for granted or "derive" in a somewhat cavalier fashion. It also contains a profound analysis of the measurement process in quantum mechanics.

Whittaker, E. T., and Watson, G. N., *A Course of Modern Analysis* (University Press, Cambridge, 1958) is a standard reference on transcendental functions and the derivation of their properties.

6. Interpretation of Quantum Mechanics

Belinfante, F. J., *Measurements and Time Reversal in Objective Quantum Theory* (Pergamon, Oxford, 1975) argues that the state vector does not characterize an individual particle, but merely describes the probability distribution for an ensemble of such particles.

Bell, J. S., *Speakable and Unspeakable in Quantum Mechanics* (University Press, Cambridge, 1987) is a collection of all of Bell's published papers on conceptual problems in quantum mechanics. It includes the original version of Bell's theorem (paper #2), and it also includes a delightful discussion that exploits a simple example to show how inequalities arise among the correlations in the EPR experiment (paper #16).

Bohm, op. cit. includes several discursive chapters that deal with the interpretation of quantum mechanics and with the role of the measurement process.

Bohr, N., *Atomic Theory and the Description of Nature* (University Press, Cambridge, 1961) contains reprints of four early papers by Bohr dealing with conceptual and philosophical issues in quantum theory.

Bohr, N., *Atomic Physics and Human Knowledge* (Science Editions, New York, 1961) is a collection of papers and lectures by Bohr on the implications of quantum theory for biology and for philosophy. It includes an account of the dramatic discussions between Bohr and Einstein on the uncertainty principle.

Clauser, J. F., and Shimony, A., "Bell's theorem: experimental tests and implications," Rep. Prog. Phys., **41**, 1881 (1978) gives a clear summary of experimental results. Later experiments are described in Penrose, op. cit.

Colodny, R. G., *Paradigms and Paradoxes* (University of Pittsburgh Press, 1972) contains essays on conceptual problems in quantum mechanics, among which the longest is an essay by C. A. Hooker covering the EPR paradox and the replies to it by Bohr and by others. Unfortunately, Bohr's writing is often ponderous and foggy, and the debates over the meaning of his reply have led to little enlightenment.

d'Espagnat, B., ed., *Foundations of Quantum Mechanics* (Academic Press, 1971) is a collection of lectures on the interpretation of quantum mechanics given at the 1971 Enrico Fermi Summer School. The first lecture is an excellent survey by E. Wigner of the different pictures of collapse.

d'Espagnat, B., *Conceptual Foundations of Quantum Mechanics*, 2nd ed. (Benjamin, Reading, Massachusetts, 1976) is the only available textbook on the foundations of quantum mechanics. It presents a critical examination of various quantum measurement theories and of the implicit philosophical postulates on which they rely.

DeWitt, B. S., and Graham, N., *The Many-World Interpretation of Quantum Mechanics* (Princeton University Press, Princeton, 1973) contains reprints of the papers by Everett who originated this interpretation of quantum mechanics, and also a reprint of a paper by DeWitt from PHYSICS TODAY giving a clear and simple introduction to this interpretation.

Gibbins, P., *Particles and Paradoxes* (University Press, Cambridge, 1987) is a short and mostly nonmathematical book that covers the logical foundations and the interpretations of quantum mechanics. It includes a critical examination of attempts at a new "quantum logic," which eliminates the paradoxes afflicting quantum measurements by redesigning the rules of logic.

Gribbin, J., *In Search of Schrödinger's Cat* (Bantam Books, New York, 1984) deals with some of the puzzling aspects of quantum mechanics at a popular, somewhat superficial level.

Heisenberg, W., *The Physical Principles of Quantum Theory* (University of Chicago Press, Chicago, 1930), translated from the German by C. Eckhardt and F. C. Hoyt, gives some detailed examples of applications of the uncertainty principle and the collapse of the wavefunction in more or less realistic measurements.

Heisenberg, W., *Physics and Philosophy* (Harper & Row, New York, 1962) provides a simple, clear, and coherent exposition of the Copenhagen interpretation of quantum mechanics.

Herbert, N., *Quantum Reality* (Doubleday, Garden City, New York, 1985) is a simple and fairly clear discussion of the interpretations of quantum mechanics and the EPR paradox written at a popular level. However, the book is blighted by the author's penchant for attaching the word *reality* to just about anything; for instance, the different interpretations, or pictures, of the measurement process are called different *realities*.

Hiley, B. J., and Peat, F. D., eds., *Quantum Implications* (Routledge & Kegan Paul, New York, 1987) is a collection of essays dealing with diverse conceptual, mathematical, philosophical, and interdisciplinary aspects of quantum mechanics. Especially useful among these is an essay by A. J. Leggett, which provides a concise critical examination of the different interpretations of quantum mechanics.

Hughes, R. I. G., *The Structure and Interpretation of Quantum Mechanics* (Harvard University Press, Cambridge, Massachusetts, 1989) covers the logical structure of quantum mechanics and the roles of probabilities and measurements. It includes a detailed discussion of quantum logic.

Jammer, M., *The Philosophy of Quantum Mechanics* (Wiley, New York, 1974) is a comprehensive historical account of the diverse interpretations of quantum mechanics with emphasis on philosophical questions. As in his earlier book on the history of quantum mechanics (op. cit), Jammer has labored hard to collect references to all the papers, and he quotes or paraphrases many relevant passages; but he supplies little guidance to steer the reader through the morass of claims and counterclaims. The book includes a simple, elegant derivation of the Bell inequality taken from a paper by E. Wigner.

Penrose, R., and Isham, C. J., *Quantum Concepts in Space and Time* (Clarendon Press, Oxford, 1986) contains two interesting essays that deal with the possible role of gravity in the collapse of the wavefunction. It also includes a review by A. Aspect and P. Grangier of recent correlation experiments with pairs of photons.

Rae, op. cit., includes a chapter with a clear and concise overview of the different interpretations of quantum mechanics and a good discussion of Bell's theorem.

Rae, A. I. M., *Quantum Physics, Illusion or Reality?* (University Press, Cambridge, 1986) describes the quantum measurement problem at a more or less popular level. It is clearly written and includes a beautifully simple proof of the Bell inequality.

Redhead, M., *Incompleteness, Nonlocality, and Realism* (Oxford University Press, Oxford, 1987) is a concise and clear examination of the logical structure of quantum mechanics and of the general difficulties faced by hidden-variable theories in light of Bell's and other theorems.

Sudbery, op. cit. includes a chapter on quantum metaphysics which lists no less than *nine* different interpretations of quantum mechanics. Sudbery pays attention to differences among interpretations of the state vector, besides differences among the pictures of the collapse process, and this accounts for some of the fine distinctions in his list.

Tarozzi, G., and van der Merwe, A., *Open Questions in Quantum Mechanics* (Reidel, Dordrecht, 1985) contains some papers concerned with diverse interpretations of quantum mechanics.

Wheeler, J. A., and Zurek, W. H., eds., *Quantum Theory and Measurement* (Princeton University Press, 1983) is an extensive collection of reprints dealing with the interpretations of quantum mechanics, hidden variables, irreversibility, and quantum limitations in measurements. It also contains a complete translation of *La theorie de l'observation en mechanique quantique* by F. London and E. Bauer (Hermann, Paris, 1939), a translation of Schrödinger's paper with the "cat paradox," and papers by Bell and by E. Wigner dealing with the Bell theorem.

Wigner, E., *Symmetries and Reflections* (Indiana University Press, Bloomington, Indiana, 1967) contains some of Wigner's less technical papers, among them several incisive essays on philosophical questions and the interpretation of quantum mechanics.

Answers to
Odd-Numbered Problems

Chapter 1

1. 3.1 eV; 1.6 eV
7. 2.5×10^{15} Hz; 6.6×10^{15} Hz
9. 6.2×10^{-11} m; 8.7×10^{-12} m
11. 1.1×10^{-10} m
13. 2.4×10^{-4} radian
15. (b) $h/\sqrt{m_e kT} \simeq 1.1 \times 10^{-8}$ at $T = 300$ K, $n^{-1/3} \simeq 10^{-10}$; no
17. 2×10^{-3} radian
19. $\hbar/\sqrt{k/m}$

Chapter 2

3. $\phi(k) = \dfrac{1}{\sqrt{2\pi}} \dfrac{2b}{b^2 - k^2} \cos \dfrac{k\pi}{2b}$

5. (b) $\phi(k) = \dfrac{e^{-ika}}{\sqrt{2\pi}} \dfrac{2}{kb} \sin \dfrac{kb}{2}$

7. $\dfrac{ik}{\sqrt{2\pi}}$

11. (b) $v_g = \hbar k c^2/E =$ [particle velocity]
15. (a) $p_0^2/2m + b^2/2m$;
    (b) no, expectation value is $b^2/2m$; wave packet has components with nonzero momentum
17. (a) $\delta(x^2 - a^2)$, no;

    (b) $\phi(p) = \dfrac{1}{\sqrt{2\pi\hbar}\, a} \cos \dfrac{pa}{\hbar}$;

    (c) 0

19. $\sqrt{\dfrac{m}{2\pi i\hbar(t_2 - t_1)}}\ \exp\left[\dfrac{im(x - x_1)^2}{2\hbar(t_2 - t_1)}\right]$

21. (a) $\dfrac{1}{8}$;

    (b) $\phi(p) = \dfrac{1}{2\pi}\ \sqrt{\dfrac{b}{\hbar}}\ \dfrac{2ib}{b^2 - p^2/\hbar^2}\ \sin\dfrac{2\pi p}{b\hbar}$;

    (c) $p = \pm\hbar b$;

    (d) $(1/\hbar b)\ dp$

Chapter 3

1. $\psi(x, t) = e^{-iEt/\hbar}\ \sin kx$; $E = \hbar^2 k^2/2m$, not quantized

3. $|\phi(p)|^2\ dp = \dfrac{4\pi}{\hbar L^3}\ \dfrac{\cos^2 pL/2\hbar}{(\pi^2/L^2 - p^2/\hbar^2)^2}\ dp$

5. (a) No;

    (b) $\psi(x, t) = \dfrac{1}{\sqrt{3}}\ \sqrt{\dfrac{2}{L}}\ e^{-iE_2 t/\hbar}\ \sin\dfrac{2\pi x}{L} + \sqrt{\dfrac{2}{3}}\ \sqrt{\dfrac{2}{L}}\ e^{-iE_3 t/\hbar}\ \sin\dfrac{3\pi x}{L}$;

    (c) $0, 1/3, 2/3, 0$;

    (d) $\langle x \rangle = \dfrac{L}{2} - \dfrac{32\sqrt{2}}{25\pi^2}\ L \cos\dfrac{(E_3 - E_2)t}{\hbar}$;

    (e) $\langle p \rangle = \dfrac{16\sqrt{2}}{5}\ \dfrac{\hbar}{L}\ \sin\dfrac{(E_3 - E_2)t}{\hbar}$

7. (a) $\langle x \rangle = L/2,\ \langle p^2/2m \rangle = 5\hbar^2/mL^2$;

    (b) $\Delta x = L/2\sqrt{7}$

9. $\psi(x) = \sqrt{\dfrac{2}{L}}\ \sin\dfrac{2\pi x}{L}$, $\phi(p) = \sqrt{\dfrac{\pi}{\hbar}}\ \dfrac{2}{L^{3/2}}\ \dfrac{1 + e^{-ipL/2\hbar}}{4\pi^2/L^2 - p^2/\hbar^2}$, $(0, 1, 0, 0, \ldots)$

13. (a) $\psi(x, t) = \displaystyle\sum_{n=1}^{\infty} \dfrac{\sqrt{2}}{n\pi}\ [1 - (-1)^n]\ e^{-iE_n t/\hbar}\ \psi_n(x)$;

    (b) $\dfrac{8}{9\pi^2}$

15. $\omega = 3E_1/\hbar,\ 8E_1/\hbar$, and $5E_1/\hbar$; $T = 2\pi\hbar/E_1$

21. $\psi(x) = \dfrac{2k'}{k' + k}\ e^{-ikx}$ for $x < 0$; $\psi(x) = e^{-ik'x} + \dfrac{k' - k}{k' + k}\ e^{ik'x}$ for $x > 0$,

    where $k' = \sqrt{2m(E - V_0)}/\hbar$;   transmission probability is $\dfrac{4kk'}{(k' + k)^2}$,

    reflection probability is $\left(\dfrac{k' - k}{k' + k}\right)^2$

23. (a) $\psi(x) = e^{ikx} + Re^{-ikx}$ for $x < -L$;
       $\psi(x) = Ce^{ik'x} + De^{-ik'x}$ for $-L < x < L$;
       $\psi(x) = Te^{ikx}$ for $x > L$

    (c) $|T|^2 = \dfrac{(2k'/k)^2}{(1 + k'^2/k^2)^2 \sin^2 2k'L + (2k'/k)^2 \cos^2 2k'L}$

25. 1.1×10^{-31}

27. $P = \exp\left[-\dfrac{2}{\hbar}\sqrt{\dfrac{2m}{B}}\,(A - E)\right]$

Chapter 4

11. $1 + a + 2a^2$

13. (a) $\dfrac{1}{2}\sqrt{\dfrac{2}{L}}\sin\dfrac{\pi x}{L} - \dfrac{\sqrt{3}}{2}\sqrt{\dfrac{2}{L}}\sin\dfrac{2\pi x}{L}$;

  (b) $\phi(p) = \sqrt{\dfrac{\pi}{\hbar}}\,\dfrac{1}{2L^{3/2}}\left[\dfrac{1 + e^{-ipL/\hbar}}{\pi^2/L^2 - p^2/\hbar^2} - 3\sqrt{3}\,\dfrac{1 + e^{-ipL/3\hbar}}{9\pi^2/L^2 - p^2/\hbar^2}\right]$;

  (c) $\left(\dfrac{1}{2}, 0, -\dfrac{\sqrt{3}}{2}, 0, 0, \ldots\right)$;

  (d) $H = -\dfrac{\hbar^2}{2m}\dfrac{d^2}{dx^2}$, $H = \dfrac{p^2}{2m}$, $H = \begin{pmatrix} E_1 & 0 & 0 & \cdot \\ 0 & E_2 & 0 & \cdot \\ 0 & 0 & E_3 & \cdot \\ \vdots & \vdots & \vdots & \vdots \end{pmatrix}$;

  (e) $\langle H \rangle = \dfrac{7\hbar^2\pi^2}{2mL^2}$;

  (f) $\Delta E = \sqrt{\langle H^2 \rangle - \langle H \rangle^2} = \dfrac{\sqrt{3}\,\hbar^2\pi^2}{mL^2}$

15. $|\beta'\rangle = |\beta\rangle - 0.3\,|\alpha\rangle$; $|\gamma'\rangle = |\gamma\rangle - 0.813\,|\beta\rangle + 0.244\,|\alpha\rangle$

19. $\Delta E\,\Delta x = \dfrac{\hbar}{2m}\sqrt{p_0^2 + \dfrac{3}{4}b^2}$ for Gaussian packet

Chapter 5

3. $\Delta E = \sqrt{2}\,b^2/m$; $\Delta E\,\Delta t = \sqrt{6}\,\hbar/2$;   yes

7. (a) $\langle x \rangle = L/2$, $\langle p \rangle = 0$;   0, 0;   yes;

  (b) $\langle x \rangle = \dfrac{L}{2} - \dfrac{16}{9}\dfrac{L}{\pi^2}\cos\dfrac{3\hbar\pi^2 t}{2mL^2}$, $\langle p \rangle = -\dfrac{4}{3}\dfrac{\hbar}{L}\sin\dfrac{3\hbar\pi^2 t}{2mL^2}$;

  $\dfrac{d\langle x \rangle}{dt} = -\dfrac{4}{3}\dfrac{\hbar}{mL}\sin\dfrac{3\hbar\pi^2 t}{2mL^2}$, $\dfrac{d\langle p \rangle}{dt} = -2\dfrac{\hbar^2\pi^2}{mL^3}\cos\dfrac{3\hbar\pi^2 t}{2mL^2}$;   yes

9. (a) $\Delta E = \dfrac{\sqrt{27}}{8}\dfrac{\pi^2\hbar^2}{mL^2}$;

  (b) $(\Delta x)^2 = L^2\left(\dfrac{1}{12} - \dfrac{7}{32\pi^2} + \dfrac{4\sqrt{3}}{9\pi^2}\cos\dfrac{3\hbar\pi^2 t}{2mL^2}\right)$;

  $\dfrac{d\langle x \rangle}{dt} = \dfrac{4\sqrt{3}\,\hbar}{3mL}\sin\dfrac{3\hbar\pi^2 t}{2mL^2}$; $\dfrac{4\sqrt{3}\,\hbar}{3mL}$

(c) $\Delta E \, \Delta t = 0.69\hbar$;   yes

(d) $\dfrac{\Delta p}{d/dt \, \langle p \rangle}$

11. $\dfrac{\hbar}{i} x \dfrac{d}{dx}, \dfrac{\hbar}{i} \left(x \dfrac{d}{dx} + \dfrac{1}{2} \right);$

$i\hbar \left(1 + p \dfrac{d}{dp} \right), i\hbar \left(p \dfrac{d}{dp} + \dfrac{1}{2} \right); \quad \dfrac{1}{2} \left(x_{op} p_{op} + p_{op} x_{op} \right)$

Chapter 6

1. $\delta E_j = 2\varepsilon > 0$

3. (a) $\Delta x = \sqrt{\hbar/2m\omega} = 8.9 \times 10^{-12}$ m;
        $\Delta x = \sqrt{7\hbar/2m\omega} = 2.4 \times 10^{-11}$ m

5. (a) $\dfrac{23}{10}\hbar\omega$;

    (b) $|\psi(t)\rangle = \dfrac{1}{\sqrt{5}} e^{-3i\omega t/2} |E_1\rangle + \dfrac{2}{\sqrt{5}} e^{-5i\omega t/2} |E_2\rangle$;   no

    (c) $\langle x \rangle = \dfrac{4}{5} x_0 \cos \omega t$; ω

7. $\langle x^2 \rangle = \dfrac{3}{2} x_0{}^2, \langle x^3 \rangle = 0, \langle x^4 \rangle = \dfrac{15}{4} x_0{}^4$

13. $(\Delta x)^2 = \dfrac{\hbar}{m\omega} \left(\dfrac{7}{6} - \dfrac{4}{9} \cos^2 \omega t \right), (\Delta p)^2 = \hbar m\omega \left(\dfrac{7}{6} - \dfrac{4}{9} \sin^2 \omega t \right)$;   yes

17. (a) Same as at $t = 0$, but with opposite sign

    (b) $(\Delta x)^2 = a^2/2$;   same;   larger

19. η_1 is singular at $x = 0$

Chapter 7

1. $E = \dfrac{\hbar^2 \pi^2}{2m} \left(\dfrac{n_1{}^2}{a^2} + \dfrac{n_2{}^2}{b^2} + \dfrac{n_3{}^2}{c^2} \right),$

$\psi(x, y, z) = \sqrt{\dfrac{8}{abc}} \sin \dfrac{n_1 \pi x}{a} \sin \dfrac{n_2 \pi y}{b} \sin \dfrac{n_3 \pi z}{c};$

$\dfrac{19}{2} \dfrac{\hbar^2 \pi^2}{2ma^2}, \dfrac{25}{2} \dfrac{\hbar^2 \pi^2}{2ma^2}, \dfrac{65}{4} \dfrac{\hbar^2 \pi^2}{2ma^2}$

11. $|1, 1\rangle = \tfrac{1}{2}(|1, 1\rangle' + |1, -1\rangle' + \sqrt{2}\,|1, 0\rangle')$, where the primes denote the eigenstates of the x component of angular momentum

17. (a) $\dfrac{1}{\sqrt{6}} (Y_2{}^2 + Y_2{}^{-2}) + \dfrac{1}{\sqrt{3}} (Y_3{}^2 + Y_3{}^{-2})$;

        1/3 for $l = 2$, 2/3 for $l = 3$, 1/2 for $m_l = 2$, 1/2 for $m_l = -2$;

    (b) $\langle L^2 \rangle = 10\hbar^2, \langle L_z \rangle = 0$;

    (c) $\Delta L^2 = 2\sqrt{2}\,\hbar^2, \Delta L_z = 2\hbar$

Chapter 8

1. $\frac{1}{2}(n + 1)(n + 2)$

3. $\psi_{111}(r, \theta, \phi) = \frac{1}{\sqrt{2}} \psi_{100}(x, y, z) + \frac{i}{\sqrt{2}} \psi_{010}(x, y, z)$

5. $\langle p_r \rangle = 0$

9. $\langle H \rangle = -\frac{1}{4} mc^2\alpha^2, \quad \langle L^2 \rangle = \frac{4}{3} \hbar^2, \quad \langle L_z \rangle = \frac{2}{3} \hbar, \langle P \rangle = -\frac{1}{3}$

11. $\langle p_r \rangle = 0, \left\langle \frac{1}{r} \right\rangle = \frac{1}{4\pi\varepsilon_0} \frac{me^2}{\hbar^2} \frac{1}{n^2}$

13. $\langle \cos^2 \theta \rangle = \dfrac{(l + m_l + 1)(l - m_l + 1)}{(2l + 1)(2l + 3)} + \dfrac{(l + m_l)(l - m_l)}{(2l - 1)(2l + 1)}$

17. Maximum at $\theta = 90^0$ and at $r = 9a_0$;   nodal line at $\theta = 0^0$ and 180^0; nodal surface at $r = 6a_0$

19. $E_1 = -\dfrac{\hbar^2\kappa^2}{2m} \left(\dfrac{2mV_0}{\hbar^2\kappa^2} - \dfrac{1}{2} \right)^2$

Chapter 9

1. (a) 1/9, 8/9;
    (b) $-7\hbar/18$;
    (c) $2\sqrt{2}\,\hbar/9$

3. (a) $\dfrac{1}{\sqrt{2}} \begin{pmatrix} 1 \\ i \end{pmatrix}, \quad \dfrac{1}{\sqrt{2}} \begin{pmatrix} 1 \\ -i \end{pmatrix}$;

    (b) $\frac{1}{2}$

7. (c) Note that $v_x \gg \Delta v_y$ is needed to keep electron in magnet.

9. $j_1 + j_2 + j_3$; $j_1 - (j_2 + j_3)$ or 0, whichever is the largest

11. 1/3

13. (a) $j = 5/2, m_j = -3/2$;

    (b) $\sqrt{\dfrac{3}{10}} \left| \frac{3}{2}, \frac{1}{2} \right\rangle |1, -1\rangle + \sqrt{\dfrac{3}{5}} \left| \frac{3}{2}, -\frac{1}{2} \right\rangle |1, 0\rangle + \sqrt{\dfrac{1}{10}} \left| \frac{3}{2}, -\frac{3}{2} \right\rangle |1, 1\rangle$

17. zero

Chapter 10

1. $\dfrac{kL^2}{2} \left(\dfrac{1}{3} - \dfrac{1}{2\pi^2} \right); \quad k \ll \dfrac{\hbar^2\pi^2}{mL^4}$

3. (a) Unchanged;

(b) $c_{n-1,n} = -\dfrac{e\mathcal{E}}{\hbar\omega}\sqrt{\dfrac{n\hbar}{2m\omega}}$, $c_{n+1,n} = \dfrac{e\mathcal{E}}{\hbar\omega}\sqrt{\dfrac{(n+1)\hbar}{2m\omega}}$;

(c) $E = \left(n + \dfrac{1}{2}\right)\hbar\omega - \dfrac{1}{2}m\omega^2\left(\dfrac{e\mathcal{E}}{m\omega^2}\right)^2$

5. $b(\hbar/2m\omega)^2$

7. $-\frac{5}{8}mc^2\alpha^4$

9. $0; \ -\dfrac{21}{8}\dfrac{b^2\hbar^2}{m^3\omega^4}$

11. $E = -\dfrac{1}{8}mc^2\alpha^2 + \dfrac{e}{2m}B_z m_l\hbar, \ c_{mn} = \delta_{mn};$ three

15. $E = E_{211}, E = E_{211} \pm \left(\dfrac{16}{9\pi^2}\right)^2 bL^2$

17. $2\sqrt{2}/3; \ \ 0; \ \ \sqrt{2}/27$

19. (a) $\dfrac{\mathcal{E}^2 e^2\hbar}{8m\omega_0}\left|\dfrac{e^{i(\omega_0+\omega)t}-1}{\hbar(\omega_0+\omega)} + \dfrac{e^{-i(\omega-\omega_0)t}-1}{\hbar(\omega_0-\omega)}\right|^2$ for $n=1$;   zero for $n>1$

(b) $t = \dfrac{1}{e\mathcal{E}}\sqrt{2m\hbar\omega_0}$

21. 5×10^{10} V/m

23. Zero

Chapter 11

3. Same as Eq. (21), but with $k' = \sqrt{\dfrac{2m(E-V_0)}{\hbar}}$; no resonances

7. $b = l/k; \ \ l = 0$ to $l = ka$

9. (a) $l = 0, l = 1; \ \ \delta_0 = 68°, \delta_1 = 74°;$
    (b) $3.63/k^2$

11. (a) $\dfrac{d\sigma}{d\Omega} = \left(\dfrac{1 - k'a\cot k'a}{k'\cot k'a}\right)^2, \sigma = 4\pi\left(\dfrac{1 - k'a\cot k'a}{k'\cot k'a}\right)^2,$

     where $k' \simeq \sqrt{\dfrac{2mV_0}{\hbar}}$ and $k \to 0$

     (b) $\sigma = 3.7 \times 10^{-28}$ m^2;
     (c) $\sigma = 2.5 \times 10^{-27}$ m^2

15. $\dfrac{d\sigma}{d\Omega} = \left(\dfrac{2mV_0}{\hbar^2 K^3}\right)^2 (\sin Ka - Ka\cos Ka)^2$, where $K = 2k\sin\dfrac{\theta}{2}$

Chapter 12

5. (d) $|\langle g\rangle| = 2.41$

7. $-\cos\theta$

Index